Stochastic Discrete Event Systems

T0181454

Armin Zimmermann

Stochastic Discrete Event Systems

Modeling, Evaluation, Applications

With 86 Figures and 25 Tables

 Springer

Author

Armin Zimmermann
Technische Universität Berlin
Real-Time Systems and Robotics
Einsteinufer 17
10587 Berlin, Germany
zimmermann@cs.tu-berlin.de

ACM Classification: G.3, C.4, I.6

ISBN 978-3-642-09350-0   e-ISBN 978-3-540-74173-2

This work is subject to copyright. All rights are reserved, whether the whole or part of the material is concerned, specifically the rights of translation, reprinting, reuse of illustrations, recitation, broadcasting, reproduction on microfilm or in any other way, and storage in data banks. Duplication of this publication or parts thereof is permitted only under the provisions of the German Copyright Law of September 9, 1965, in its current version, and permission for use must always be obtained from Springer. Violations are liable for prosecution under the German Copyright Law.

Springer is a part of Springer Science+Business Media
springer.com
© Springer-Verlag Berlin Heidelberg 2010

The use of general descriptive names, registered names, trademarks, etc. in this publication does not imply, even in the absence of a specific statement, that such names are exempt from the relevant protective laws and regulations and therefore free for general use.

Cover design: KünkelLopka, Heidelberg

To Anja and Emma

Foreword

Stochastic discrete event systems (SDES) arise frequently in the design, analysis, and optimization of man-made engineered systems. The dominant dynamics of such systems tend to be modeled not through the application of physical laws but through the specification of protocols, algorithms, and managerial rules. The key role of such protocols, algorithms, and rules as a determining factor in the performance of engineered systems has become increasingly evident as technology has advanced over the last half century.

Opportunities for utilizing SDES as a design tool clearly arise in virtually all settings in which software finds application, because software often is generated in the course of formally specifying the algorithms and protocols to be followed by a man-made system. But SDES also have become a standard means of evaluating managerial policies and rules arising in higher-level systems in which human beings interact with technology, as in the setting of call centers, large manufacturing systems, and corporate supply chains. Given the importance of SDES, the development of a systematic and comprehensive mathematical and computational toolset for the specification and design of such systems has become a key issue that has attracted a great deal of attention within the engineering community.

The last twenty years has seen significant progress in developing and building good computational tools capable of creating rich model-based representations of such SDES, with sophisticated associated mathematical methods capable of analyzing both performance and optimizing across designs. A particular challenge in the SDES context is that because the dynamics are algorithmically defined (as opposed to being governed, for example, by physical laws), the spectrum of mathematical problem structures that can be generated in the SDES setting is enormously varied. As a consequence, there are significant challenges that arise in building computational solvers that can fully leverage the problem structure associated with a particular application. Nevertheless, as indicated above, much progress has been made in developing model-based representations that can form the basis for effective and efficient computation. This computational toolset is capable of analyzing both the performance of a given design and optimizing across designs. Because so many such systems are subject to random variation, such a toolset should be capable of systematically incorporating and analyzing the associated stochastic effects. It comes as no surprise that the solution methodology for SDES includes both methods that reduce the necessary equations to be solved to systems that can be computed via conventional numerical analysis methods (as in the setting of Markov jump process representations of a SDES) and methods based on the use of stochastic simulation (in which trajectories of the system are simulated by drawing random numbers).

This monograph by Armin Zimmermann skillfully describes the key ideas and methods that have developed within this discipline, starting first with various model-based representational tools and showing that they share a common modeling worldview. Both simulation-based and non simulation-based solution methodologies for SDES are then explored, including a discussion of the key ideas that arise in the use of iterative approximations in this setting. The book also includes an ambitious discussion of some of the methods available for numerical optimization of SDES, and offers some carefully worked-out examples that illustrate the power of these ideas.

To those of us who work within the SDES discipline and to the larger community that bases their interest in SDES on the potential application of these ideas to domain-specific modeling problems, this monograph serves as a welcome addition to the literature.

Stanford University, USA   Peter W. Glynn
                                           Thomas Ford Professor of Engineering

Preface

The behavior of many technical systems that are increasingly important in our every-day life can be described by discrete states and state-changing events. *Discrete event systems* follow this view of a system, and have gained a lot of interest in the past due to their wide range of applicability. Numerous model classes, analysis algorithms as well as software tools have been developed and successfully put into practice. Activity delays and probabilities of decisions are added to evaluate quantitative issues like performance and dependability. *Stochastic discrete event systems* (SDES) capture the consequential randomness in choices and over time.

There are numerous challenges for an engineer throughout the life span of a technical system. Planning, documentation, functional and quantitative design as well as control are examples. Complexity and size of the considered systems requires supporting tools for these tasks, which have been developed over the last decades with the enormous increase in available computing power.

The starting point is a formal description, requiring to describe the system of interest in a model. Numerous well-known model classes are available today, including queuing networks, Petri nets and automata. Their descriptional power allows to capture stochastic delays as well as random choices. Every one of them represents a different way of describing a SDES with individual abstraction level. The dominant theme of this work is that they share a common underlying view of a system, and can thus be treated alike.

This text is about modeling with and quantitative evaluation of stochastic discrete event systems. An abstract model class for SDES is presented as a pivotal point. Several important model classes are presented together with their formal translation into the abstract model class. Standard and recently developed algorithms for the performance evaluation, optimization and control of SDES are presented in the context of the abstract model class afterwards. The final part comprises some nontrivial examples from different application areas, and demonstrates the application of the presented techniques.

This text evolved from the author's habilitation thesis. Its development would not have been possible without the support of many colleagues that I had the privilege to work with over the last years.

First and foremost, I thank my advisor Günter Hommel, who has continuously supported my work already since the Ph.D. thesis. The open atmosphere in his group allowed the kind of independent work that is demanded for post-doctoral research groups these days. Several of my colleagues at TU Berlin influenced this text. Christoph Lindemann originally introduced me to performance modeling and evaluation, and his as well as the ideas of Reinhard German and Christian Kelling are still the basis for many analysis methods and evaluation projects of my daily work. Dietmar Tutsch and Jörn Freiheit shared my office room in Berlin, and were always a source of discussions and joy in work. The progress of TimeNET over the last years is in large part due to Michael Knoke, Alexander Huck and Jan Trowitzsch. It was a pleasure to supervise their PhD theses.

I gratefully acknowledge the financial support of the German Research Council (DFG), from which I benefited in various projects and especially our graduate school "Stochastic Modelling and Quantitative Analysis of Complex Systems in Engineering". The lively discussions among the fellows opened my mind for many new topics. I also thank Shang-Tae Yee for the insights in supply-chain operation and GM for funding of our joint projects. I also had the privilege to meet and work with Peter Glynn in these projects; our discussions throughout this time are very much appreciated.

I especially owe to the external reviewers of the habilitation thesis. Manuel Silva is a constant source of ideas, which I had the privilege to be shared with during my research visits in Zaragoza. Your group still feels like a second scientific home to me, for which I also want to thank Diego Rodriguez and the other spanish colleagues. Eckehard Schnieder has promoted my work as a reviewer and through discussions during several intellectually stimulating conference meetings in Braunschweig. Furthermore I would like to acknowledge time and effort spent by the further committee members, namely Günther Seliger and Thomas Sikora.

An important part of work over the last years has gone into implementing new ideas in our software tool TimeNET, which would not have been possible without the work and programming skills of numerous master students.

I thank my parents for laying the seed of scientific curiosity in me by answering all the "why?" questions of a child. Finally I want to thank my wife Anja and daughter Emma, with whom I could not spend the time necessary to prepare this text. I dedicate it to you.

Technische Universität Berlin,   *Armin Zimmermann*
June 2007

Contents

1

Introduction

Engineers and designers of technical systems have to deal with increasingly complex projects. The design objects may contain diverse mechatronic elements and distributed components. Embedded control computers add to this complexity, making the overall system design a very hard task. Unplanned influences like failures and concurrency as well as synchronization issues increase the challenge. Whatever application field we look at – telecommunication, logistics, computers, transportation, work flows, information systems, or production – the number of involved parts and their connections make it impossible to oversee the global effects of local decisions.

On the other hand, verification of a functionally correct design as well as the prediction of system performance and dependability is becoming increasingly important due to the tremendous resources that are spent during the design and life time of a technical system. Model-based evaluation of technical systems has turned out to be a powerful and inexpensive way of predicting properties before the actual implementation. Costly design changes can thus be minimized, and the planning time is reduced. A model is a simplified representation of a real-life system, independent of whether it already exists or is only envisioned. In contrast to a fixed model like for a building of an architect, we are mainly interested in the *behavior* of a technical system design and models that describe system dynamics. The overall behavior is not known in advance as opposed to measuring a prototype system. We can only describe the planned structure, architecture, connections, local behavior of a system, etc. The global behavior of such a system must then be predicted with an evaluation method. Models are thus required to be interpretable such that their dynamics are clearly defined.

Modeling is a key feature to understand complex systems in all application areas, just like a mathematical formula describing a natural phenomenon. We accept it as being correct as long as it "explains" the effects that we observe, either until a counter example has been found or there is a more elegant way of description. Modeling is very much like programming on an abstract level: the modeler needs to know the basic descriptional elements, there are state variables as well as methods that change them, as well as structural modules and behavioral aspects. A model as well as a program is based on an abstraction of a real system. The decision of the right abstraction level as well as the correct boundary of considerations are important and require human skills and experience. A deep understanding of the real system is necessary to create a correct model. *Programming is understanding* (Kristen Nygaard) applies to modeling alike. Because of its importance and the requirement of background knowledge of the application area, modeling is a core discipline of many engineering and natural sciences.

A system is always in some kind of state and independently of what we are interested in during the system design, it will depend on the state and its trajectory over time. A general way of describing system behavior thus defines the structure of a state and when and how state-changing activities influence it. Passage of time as well as state values may be continuous, assume only discrete values or be a hybrid mixture of both. The state information is continuous in natural environments that follow the laws of nature as in physics, chemistry, or biology. State changes can then be described by differential equations, which is the subject of systems and control engineering.

1.1 Stochastic Discrete Event Systems

The nature of many man-made technical systems differs from that. Structure and control rules designed artificially, and their dynamic behavior is triggered by *discrete events*. This text focuses on models where system states and events are discrete. This is useful in many application areas mentioned earlier, where individual entities (customers, vehicles, communication packets) and their discrete states and locations are the subject of analysis. Other applications allow the discretization of continuous values.

Systems and their models are dynamic by changing their state over time, which passes independently and out of our control. Activities proceed if their preconditions are met, until their finishing events happen after some delay. This leads to subsequent states, new activities, and corresponding events following the causal dependencies. Time is nature's way of setting the speed of causality. In our understanding, delays in a system are related to time spans that need to pass during an ongoing activity. Time is therefore associated with activities.

The description of a discrete event system on the causal level is already sufficient to answer *qualitative* questions, such as if a certain (dangerous) state can ever be reached or if a control system may deadlock. Corresponding analysis techniques are often based on the model structure and do not require to visit every state of the system individually. This makes them efficient and leads to statements about every theoretically possible system behavior. A good part of the literature on discrete event models and analysis techniques deals with these kinds of problems. It is, however, out of the scope of this text, which concentrates on *quantitative* properties. A model needs to describe additional properties of a system in this case, which are described later.

Environment and inner parts of a model are subject to uncertainty. This is independent of the question whether there is true randomness in nature or not. The reason for this lies in the process of modeling already: whatever abstraction level is chosen, there is always some level of detail hidden from our view, which results in events that are unpredictable from within the model. In many cases it is possible to create a more detailed model, but this might lead to a worse tradeoff between model complexity and analytical tractability.

One source of unpredictable behavior are conflicts between activities. They occur if concurrently running activities change the system state such that others are disabled. The outcome of its solution may not be known inside a model. Probabilistic choices are an appropriate way of describing these situations, in which it is not known what activity will happen. Another level of uncertainty is introduced because the exact time of an event is unknown. Delays need to be described by probability distribution functions to describe their stochastic nature. Spontaneous failures and human interaction are examples.

Stochastic delays and probabilities of decisions are a prerequisite for quantitative modeling and evaluation of systems. Together with the causal relationships between states and events, the dynamic behavior of a stochastic discrete event system (SDES) model can be described by a *stochastic process*. It would, however, be a tedious task for complex applications to specify a model at the level of abstraction of this process. The traditional way of finding an exact mathematical description is restricted to simple models, and often requires a new model for every design change.

Many model classes with a higher level of abstraction have been developed inside the SDES family, for instance, stochastic automata, queuing models, and process algebras. They do restrict the modeling power in different ways, but have the significant advantage of easier use and understanding. There are advantages and disadvantages for the different levels of abstraction just like for high-level programming languages vs. assembler programming. The choice of a good model class for a system design depends on the necessary complexity and available analysis techniques. One of the reasons for this is the existence of more powerful evaluation methods for less complex model classes.

The first part of this text describes a selection of prominent model classes with increasing complexity. Modeling a discrete event system with a stochastic automaton works well as long as the number of states remains manageable. Typical modular configurations of technical systems can be handled with networks of automata. However, more complex systems lead to problems. A system with K customers can for instance not be parameterized on the model level, because the number of automata states depends on the parameter K. Queuing models are able to describe systems with resource allocation and sequences of operations on a much higher level. If synchronizations need to be captured in addition to that, Petri nets can be the right choice. Colored

Petri nets should be chosen if different physical objects with attribute values need to be treated individually in the model.

Research in the field of stochastic discrete event systems goes into many directions: model classes and their theoretical foundation, systematic ways of model construction, input data derivation, graphical representation of models, their states and dynamic behavior, analysis tools for the prediction and optimization of a variety of qualitative and quantitative properties, interfaces between models, and real world, to name a few. Several scientific disciplines contribute to this area, and most problems can only be solved interdisciplinary. Mathematics, statistics, and operations research obviously plays a major role in the formal foundations and analysis techniques. Computer science adds formal and numerical methods as well as the hard- and software tools for their implementation. Numerous application areas from engineering and others are necessary to understand and properly describe systems and design problems.

1.2 Applications and Goals of Modeling

A SDES model of a planned system requires additional work for system understanding, parameter extraction, model construction, and debugging as well as an experienced modeler. This investment pays off only because the model can be used in the design process as follows.

A formal model is unambiguous and can thus be used for documentation together with its graphical representation (provided that it has one). Functional requirements as well as rough system architecture and behavior are modeled and evaluated during the early design stages. Visualization of model behavior and verification of qualitative properties help to understand the system better and to find errors. The model is enhanced by further details to describe the resources. Quantitative evaluation predicts the performance and dependability, and an optimization of design parameters is possible. Control rules may be automatically derived and checked using the model. A control interpretation of the model allows to directly control a real system and to check the start-up phase. The current behavior and state of a running system may be visualized for a supervision.

Strategic and operational questions can be answered easier and less costly based on the model. Figure 1.1 sketches an iterative design process based on model and evaluation techniques. The system is described with a model class, and quantitative measures are defined to express design goals. An evaluation method computes the values of the measures for the current system model, to check if the designed system behaves as required. If it does not, the model is changed and evaluated again. This process is iteratively continued until the model fulfills the goals. The main advantage of using a model instead of a real system in this context is the time and money that would have to be spent on prototypes otherwise.

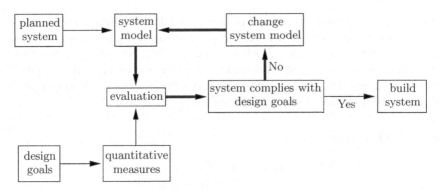

Fig. 1.1. Model-based iterative design

The underlying idea of model evaluation during design has been nicely captured in the following quotation. Although meant for an architect's model of a building, it touches all elements of a model-based design.

> When we mean to build,
> We first survey the plot, then draw the model,
> And when we see the figure of the house,
> Then must we rate the cost of the erection,
> Which if we find outweighs ability,
> What do we then but draw anew the model
> In fewer offices, or at least desist
> To build at all?
>
> Shakespeare, King Henry IV, Part II

1.3 Overview of Topics

The text in hand aims at a presentation of stochastic discrete event systems along the steps of their use. Its organization thus follows the stages of model-based design and operation of technical systems: Part I covers modeling, Part II describes use of models, mostly containing evaluation methods. The necessary software tool support is covered as well. A set of actual application examples, their models, and usage of evaluation techniques is contained in Part III. An abstract model class SDES is introduced as an embracing framework for individual modeling classes. Evaluation techniques based on this description are therefore applicable to any member of the SDES family of stochastic discrete event systems.

There are numerous modeling techniques that could not be considered in this text to limit its space. Among them are message sequence charts (MSC), event-driven process chains (EPC), specification languages, hybrid models, and process algebras to name some of them. Model classes were selected only if they are based on discrete states and events, possess a graphical representation and a clear formal semantics. Stochastic colored Petri nets probably represent the most complex and powerful example of this model class today. Other classes were selected because they paved the ground in the development of more complex classes on the way towards colored Petri nets.

Part I: Modeling

Chapters inside the modeling and evaluation parts are ordered following an increasing complexity of the covered material, if possible. The modeling part describes different modeling classes as well as their integration into one unified modeling framework. It starts with Chap. 2 covering an abstract model class for SDES, into which all later example classes are embedded. The model class SDES is not a real one; it serves as an abstraction of classes like Petri nets or automata instead and does not have a graphical representation. Static model elements of an SDES are explained and defined. The dynamic behavior that follows from such a model is described, resulting in a definition of the underlying stochastic process by construction. The specification of quantitative results by reward measures and their relation to the stochastic process is covered subsequently.

The individual model classes covered are automata (Chap. 3 with a digression on UML Statecharts and their transformation into Petri nets), queuing models (Chap. 4), simple Petri nets (with uncolored tokens; Chap. 5), and colored Petri nets (Chap. 6), covering variants with and without arc variables. For each modeling method an informal introduction of model elements, their use, and graphical appearance is given with toy examples. The model classes are formally defined and their relation to the general SDES model is specified subsequently. Specific details on the dynamic behavior of model classes are given if necessary, although the definition of the dynamic behavior of an SDES is sufficient in the general case after the interpretation and their relation to SDES. The specification of quantitative measures is explained. Final notes in each chapter point at related text parts as well as a selection of significant references.

Part II: Evaluation

Part II is devoted to methods that make use of the models, most importantly for the derivation of quantitative measures. Formulas and algorithms are given to avoid a purely theoretical coverage. Wherever possible, the presented methods take as input an SDES model, which makes them applicable to models from a wide variety of descriptions. Some of the presented methods are restricted to specific model classes, because they require structural information from the model class, which might not be available in a different representation.

The evaluation part begins by recalling some of the most widely used standard methods for the quantitative evaluation of stochastic discrete event models in Chap. 7, and presents their implementation for the abstract SDES model class. Different simulation and numerical analysis approaches are covered. The subsequent chapters of Part II are based on previous results of the author and coworkers, which are presented within the SDES environment whenever possible.

Chapter 8 describes an approximative numerical evaluation method for model classes of stochastic Petri nets. It aims at relaxing the state space explosion problem, which disables the numerical analysis of models with a huge state space. Models are cut into parts and structurally simplified. An iterative technique computes an approximation of quantitative measures based on the simplified model parts. The tradeoff between result accuracy, remaining state space size, and computation speed makes this technique an alternative to simulation and standard numerical analysis techniques.

The subsequent Chap. 9 covers two additional simulation methods, which are much more efficient than standard techniques for certain problem settings. One possibility to get results with sufficient statistical confidence faster is parallel simulation with a distribution of model parts over individual nodes. The complex semantics of general SDES models, however, forbid the application of the usual optimistic time-warp algorithms. To capture the causal relationships of model elements correctly, an appropriate model time management is proposed. Fine-grain partitioning and automatic heuristic load balancing are possible with this technique. Another well-known problem of simulation algorithms appears when rare events have a significant impact on the quantitative results. The RESTART method efficiently speeds up the estimation of such performance measures. It is applied to the SDES model environment in the second part of Chap. 9.

Quantitative evaluation of a model is often only one step of a what-if analysis, which aims at a selection of a good parameter setting of a planned system. Automatic optimization methods can take this burden from the designer. Because of the inherent complexity of the considered models, only indirect optimization based on an evaluation technique can be used. This does, however, lead to an unacceptable computational complexity in many cases. Chapter 10 presents a technique that speeds up such an optimization by two orders of magnitude for typical examples. A heuristic optimization scheme is executed in two phases, where the first one uses a fast approximation to identify a promising region for the later thorough optimization phase. The method uses results for the bounds of performance measures of Petri nets, which can be efficiently obtained from the model structure.

Structure and behavior of discrete event systems make it easy to use them for the direct control of a modeled system. It is moreover a natural step to reuse a model from the design and performance optimization stages for this operational step as well. Errors might otherwise be introduced if different models or software tools are used. Chapter 11 shows how SDES models can be used to control a technical system directly. A corresponding interpretation of the SDES model elements is given after an intuitive Petri net example.

Some notes on software tool support for the creation and use of SDES models are given in Chap. 12. It describes the tool TimeNET, which has been used for the application examples and is being developed by the group of the author.

Part III: Applications

Four case studies are presented as application examples in Part III, each using different modeling methods and evaluation techniques. They are ordered in the same way as the corresponding models in Part I, and cover a broad range of the previous theoretical and methodological parts of the text. Despite some individual differences in the structure, each application chapter presents and explains the application example and design goals or steps. A model of the system is created using a model class out of Part I, which is analyzed or otherwise used subsequently with the methods from Part II. Results for the application examples underline applicability and efficiency of the methods. Most examples are adapted versions of real-life industrial problems.

Optimization of a manufacturing system example is covered in Chap. 13. A generalized stochastic Petri net is used for the modeling, and the technique presented in Chap. 10 is applied to obtain near-optimal parameter settings. One of the subsections explains in detail how typical optimization functions for manufacturing systems can be specified with Petri net performance measures. An overview of types of manufacturing systems and design problems is given in that section as well. Numerical results for the fast approximation technique and the overall optimization approach are presented.

A performability evaluation of the future European train control system ETCS is the topic of Chap. 14. Train operation is briefly explained both for the traditional fixed block system vs. the planned moving block scheme. A model of train operation and communication link is stepwise built using extended deterministic and stochastic Petri nets. An alternative specification of the communication link behavior is presented with a UML Statechart model, which is transformed into an equivalent Petri net with the algorithm of Sect. 3.5. Numerical analysis leads to a condensed link failure model, which is added to the train operation model for a complete description of the problem. The resulting model can not be evaluated in practice using standard numerical analysis or simulation methods. Rare-event simulation allows to efficiently obtain performance results. The tradeoff between communication link dependability and minimum train distance is formally captured and evaluated.

Chapter 15 shows how a supply chain is designed based on a model to achieve certain performance goals. Colored Petri nets allow the natural specification of its complex operations. The model is presented in a modular and hierarchical way for the example, covering the behavior of customers, dealership, plant, and transport logistics in the car maker example setting. The time between order and delivery of a vehicle is analyzed. Several supply chain design changes are implemented in the model and evaluated. As a result, proposals for a more efficient supply chain organization are derived and their impact on the delivery time is quantified.

A physical model of a multimachine production cell is considered in Chap. 16. The individual machines and other resources are modeled with a hierarchical variable-free colored Petri net. It is briefly shown how workplan-specific information can be integrated into the structural model using results that were developed earlier [339, 354, 361]. The throughput of the production cell is analyzed with the iterative approximation technique of Chap. 8. Accuracy of results and computational efforts are compared with results obtained using standard numerical analysis and simulation techniques. Finally, the model is enhanced by a control interpretation following Chap. 11, and is used with the software tool TimeNET to directly control the production cell.

Concluding Parts and Related Work

Concluding remarks as well as an outlook to future research directions are given in the final summary.

The notes at the end of most chapters contain pointers to related work. For the modeling part the most significant references include [9, 42, 168] for automata, [42, 149, 207] for queuing models, [4, 130] for stochastic Petri nets, and [188] for Colored Petri nets. The idea of an abstract framework for different discrete event model classes has been implemented in the Möbius tool and described in [81, 82].

The definition of the stochastic process and of quantitative measures of an SDES uses work presented in [67, 130, 143, 284]. Numerical performance evaluation techniques are described in [4, 59] for Markovian models and more general ones in [130]. The estimation of quantitative measures is also possible by standard simulation techniques [105, 150, 219], iterative approximation [108, 109, 112, 260], fine-grain parallel simulation [210, 211], or acceleration techniques [321] among others. Near-optimal parameter sets for selected SDES models can be efficiently found with the techniques described in [351–353], and the control interpretation is based on [336, 342, 344].

1.4 Notation and Selected Background

Throughout the book, **important terms** are set in bold when they are defined or appear for the first time. Issues of lesser importance for the presentation are *emphasized*, while model elements are set in `typewriter style`. The used symbols are listed and explained starting in p. 345. Algorithms are set using a format that is introduced together with the first algorithm in p. 135.

The text assumes that the reader is familiar with probability theory and stochastic processes. Numerous textbooks cover this field, including [31, 42,

$71, 150, 245, 305]$. Without trying to be exhaustive, some issues significant to the text are briefly recalled.

Random Experiments

Uncertainty in stochastic discrete event systems is expressed in the selection of enabled or executed activities in a certain state (if there has to be a selection), and in the selection of a possibly random activity delay. Any of these decisions is called a **random experiment** in probability theory, because they lead to a result that is previously unknown. The set of all possible outcomes is called the **sample space** S of an experiment. Random experiments are commonly characterized by either having a finite or countably infinite sample space, in which case they are called **discrete** or otherwise **continuous**.

An **event** E (in the sense of probability theory) is a subset of the sample space S ($E \subseteq S$). We may form corresponding subsets of interest out of S to obtain an **event space** $\mathbb{E}$, that must satisfy two conditions. The **complement** $\overline{E}$ of an event E is defined as $\overline{E} = S \setminus E$ and must be an event as well: $\overline{E} \in \mathbb{E}$. Second, for any set of events $E_1, E_2, \ldots$, the **union** must be an event as well: $\bigcup_i E_i \in \mathbb{E}$.

A **probability** $P\{E\}$ is associated with every event E to measures its relative likelihood. One interpretation is the following. If we would conduct the underlying experiment infinitely often, the ratio between the number of results in E and the overall number of experiments will converge to $P\{E\}$. The axiomatic definition of a probability function $P\{\cdot\}$ requires

$$\forall E \in \mathbb{E}: \quad P\{E\} \geq 0$$
$$P\{S\} = 1$$
$$\forall E_1, E_2 \in \mathbb{E}, E_1 \cap E_2 = \emptyset: \quad P\{E_1 \cup E_2\} = P\{E_1\} + P\{E_2\}$$

A **probability space** is then a tuple $(S, \mathbb{E}, P)$ of sample space, event space, and probability function.[1]

The **conditional probability** $P\{E_1 \mid E_2\}$ measures the probability of an event E_1 under the precondition that another event E_2 has happened. This restricts the sample space of the experiment to E_2 and leads to the definition $P\{E_1 \mid E_2\} = P\{E_1 \cap E_2\} / P\{E_2\}$. If the previous occurrence of E_2 does not change $P\{E_1\}$, obviously $P\{E_1 \mid E_2\} = P\{E_1\}$, and we say that the two events are **independent**.

An important special case of a random experiment has two outcomes, which are often interpreted as success and failure. Then $S = \{\text{success}, \text{failure}\}$, and $\mathbb{E} = \{\emptyset, \{\text{success}\}, \{\text{failure}\}, \{\text{success}, \text{failure}\}\}$. We consider a probability space in which $P\{\text{success}\} = p$ and $P\{\text{failure}\} = q$ (obviously $p+q = 1$). Such a random experiment is called a **Bernoulli trial**, and a sequence of independent

---

[1] More rigid treatment requires $\mathbb{E}$ to be a **measurable subset** of the whole set of events.

trials is an adequate model for many processes of stochastic discrete event systems. Of special interest are the probability of k successes in a sequence of n trials as well as the probability that k is the first successful trial.

$$\mathrm{P}\{k \text{ successes in } n \text{ trials}\} = \binom{n}{k} p^k q^{n-k}$$

$$\mathrm{P}\{k \text{ is the first success}\} = q^{k-1} p$$

Random Variables and Probability Distribution Functions

Instead of taking the elements of the sample space as results of a random experiment, it is more convenient for a mathematical treatment to assign real-valued numbers to them. Such a mapping $X : S \to \mathbb{R}$ is called a **random variable** on a sample space S. Consider the relation between elements of a discrete sample space for which the random variable results in the same value. This is an equivalence relation, which defines a partition of the sample space resulting in an event space. The set of sample space elements with equal random variable is required to be an event for a discrete random variable.

$$\forall x \in \mathbb{R} : \{E \in S \mid X(E) = x\} \in \mathbb{E}$$

The probability of the discrete random variable to assume a certain value x is written as $\mathrm{P}\{X = x\}$.

Things are a bit different for non-denumerable sample spaces S, i.e., when the random variable is **continuous**. For such a random variable there are subsets of the sample space for which we cannot define a reasonable probability value. It is thus only required that the set of sample space elements for which the random variable returns a value less or equal than a certain x is an event, which then has a well-defined probability. A (continuous) random variable is thus a function X for which

$$\forall x \in \mathbb{R} : \{E \in S \mid X(E) \leq x\} \in \mathbb{E}$$

The **(cumulative) distribution function** F_X of a random variable X returns the probability that the random variable assumes a value less than or equal to a real x.

$$\forall x \in \mathbb{R} : F_X(x) = \mathrm{P}\{X \leq x\}$$

The **probability density function** f_X of a random variable X is given by the derivative of F_X if it exists.

$$\forall x \in \mathbb{R} : f_X(x) = \frac{\mathrm{d}}{\mathrm{d}x} F_X(x) \quad \text{and} \quad F_X(x) = \int_{-\infty}^{x} f_X(y)\, \mathrm{d}y$$

For a proper F_X holds $\lim_{x \to \infty} F_X = 1$. The distribution function is monotonic, which is similar to a nonnegative density function. Continuous random variables are used in this text to describe random delays of activities. Delays are obviously greater than or equal to zero, thus the support of the distribution function is a subset of $[0, \infty)$. We define the set of nonnegative probability distribution functions $\mathcal{F}^+$ as

$$\mathcal{F}^+ = \{F_X \mid \forall x \in \mathbb{R} : x < 0 \longrightarrow F_X(x) = 0\}$$

Discrete and continuous delays as well as mixed ones can be uniformly described with generalized distributions and density functions. A discrete probability mass at a point x leads to a step in the distribution function and a Dirac impulse in the density function. The step function $s(x)$ is defined as

$$\forall x \in \mathbb{R} : s(x) = \begin{cases} 0 & \text{if } x \leq 0 \\ 1 & \text{otherwise} \end{cases}$$

The Dirac impulse Δ can be formally defined by a rectangular function with constant area of one, for which the length of the basis is taken to zero. $\Delta(x)$ denotes a function with an area of size one at 0, and represents a generalized derivative of the step function. Step function and Dirac impulse can both be shifted to any $b \in \mathbb{R}$ and multiplied by a $a \in \mathbb{R}^+$ to be combined with other parts of a probability distribution or density function.

$$\frac{\mathrm{d}}{\mathrm{d}x}\left(as(x-b)\right) = a\Delta(x-b) \quad \text{and} \quad \int_{-\infty}^{x} a\Delta(y-b)\,\mathrm{d}y = as(x-b)$$

Distribution functions of discrete random variables can therefore be captured by a weighted sum of step functions. Their distribution is otherwise described by a **probability mass function** instead of a density function.

For the later specification of delays in a stochastic discrete event system we define subsets of the set of all allowed delay distributions $\mathcal{F}^+$.

Zero or immediate delays are allowed as a special case. The set of **immediate probability "distribution" functions** $\mathcal{F}^{im}$ is defined accordingly; all other are called **timed**, and required to have no probability mass at zero $(F_X(0) = 0)$.

$$\mathcal{F}^{im} = \{F_X \in \mathcal{F}^+ \mid F_X(0) = 1\}$$

Important cases of timed delay distributions include the following. Exponential distributions have the form

$$\mathcal{F}^{exp} = \{F_X \in \mathcal{F}^+ \mid \exists \lambda \in \mathbb{R}, \forall x \in \mathbb{R}^+ : F_X(x) = 1 - \mathrm{e}^{-\lambda x}\}$$

and the density is $f_{\exp}(x) = \lambda \mathrm{e}^{-\lambda x}$.

The deterministic "distribution" always results in a fixed value τ.

$$\mathcal{F}^{det} = \{F_X \in \mathcal{F}^+ \mid \exists \tau \in \mathbb{R}^+ : F_X(x) = s(x - \tau)\}$$

Its density is given by $\Delta(x - \tau)$.

A class of **expolynomial** distributions is allowed for the numerical analysis of stochastic discrete event systems following [130] and denoted by $\mathcal{F}^{gen}$. Such a function is a weighted sum of n expressions in the form $x^m e^{-\lambda x}$. The weighting factor as well as a truncated support of each expression is obtained with a rectangular distribution $R(a, b] = s(x - b) - s(x - a)$. Thus each $F_X \in \mathcal{F}^{gen}$ has the form

$$\forall F_X \in \mathcal{F}^{gen} : F_X(x) = \sum_{i=1}^{n} x^{m_i} e^{-\lambda_i x} R(a_i, b_i]$$

with $m_i \in \mathbb{N}$, $\lambda_i \in \mathbb{R}^{0+}$, and $a_i, b_i \in \mathbb{R}^+$.

The geometric distribution is discrete and counts the number of Bernoulli trials until the first success. If the time between two trials is denoted by Δt and the individual success probability by p,

$$\forall x \in \mathbb{R}^+ : F_X^{\text{geo}}(x) = 1 - (1 - p)^{\lfloor x/\Delta t \rfloor}$$

An important aspect of some distributions is the **memoryless property**. Consider a random delay X after a time t, which has not yet elapsed (i.e., it is known that $X > t$). How is the remaining delay $X' = X - t$ distributed? If the distribution of the conditional probability that $X' \leq y$ is independent of t and thus identical to the original distribution, we say it is memoryless. It can be shown that the only memoryless distributions are the exponential distribution in continuous time and the geometric one in discrete time. They are of special importance because random effects like failures with a constant rate can be modeled using them and due to their analytical simplicity.

Stochastic Processes

When we observe the behavior of a stochastic discrete event system over time, there will be a (usually random) sequence of states and state changes. The mathematical abstraction for this is a **stochastic process**, which is formally a collection of random variables $\{X(t) \mid t \in T\}$ that are indexed by the time t. The parameter t may have a different interpretation in other environments. Index set T denotes the set of time instants of observation. The set of possible results of $X(t)$ is called **state space** of the process (a subset of $\mathbb{R}$), and each of its values corresponds to a **state**.

Stochastic processes are characterized by the state space and the index set T. If the state space is discrete (countable), the states can be enumerated with natural numbers and the process is a **discrete-state process** or simply a **chain**. T is then usually taken as the set of natural numbers $\mathbb{N}$. Otherwise, it is called a **continuous-state process**. Depending on the index set T the process is considered to be **discrete-time** or **continuous-time**. Four combinations are obviously possible. In our setting of stochastic discrete event systems, we are interested in systems where the flow of time is continuous ($T = \mathbb{R}^{0+}$) and the state space is discrete. The stochastic process is thus a continuous-time chain and each $X(t)$ is a discrete random variable.

A discrete-state stochastic process $\{X(t) \mid t \in T\}$ is called **Markov chain** if its future behavior at time t depends only on the state at t.

$$P\{X(t_{k+1}) = n_{k+1} \mid X(t_k) = n_k, \ldots, X(t_0) = n_0\} =$$
$$P\{X(t_{k+1}) = n_{k+1} \mid X(t_k) = n_k\}$$

It is easier to analyze than more general processes because information about the past does not need to be considered for the future behavior.

Markov chains can be considered in discrete or continuous time, and are then called **discrete-time Markov chain** (DTMC) or **continuous-time Markov chain** (CTMC).

From the memoryless property of a Markov process it immediately follows that all inter-event times must be exponentially (CTMC) or geometrically (DTMC) distributed. Different relaxations allow more general times. Examples are **semi-Markov processes** with arbitrary distributions but solely state-dependent state transitions and **renewal processes** that count events with arbitrary but independent and identically distributed interevent times.

A **generalized semi-Markov process** (GSMP) allows arbitrary interevent times like a semi-Markov process. The Markov property of state transitions depending only on the current state is achieved by encoding the remaining delays of activities with nonmemoryless delay distributions in the state, which then has a discrete part (the system states) and a continuous part that accounts for the times of running activities.

For further details of stochastic processes the reader is referred to the literature mentioned earlier in this section.

Part I

Modeling

2

A Unified Description for Stochastic Discrete Event Systems

Various discrete event model classes with stochastic extensions have been proposed, which all share some common characteristics. Many of the algorithms that have been developed for one or the other class are in principal applicable to all or most of them. This requires some kind of abstract description for stochastic discrete event systems, which this section aims at. We will use the term SDES to refer to this unified description in the following. Different model classes are defined in terms of SDES later on in this part. Evaluation algorithms that use the SDES description as their input are explained in Part II. The definition of SDES thus marks some kind of interface between model description and model evaluation. The level of detail of the SDES description was hence set such that algorithmic parts should only be required in the evaluation algorithms, while the static model descriptions are captured in the SDES definition. From an implementation-oriented point of view one can also think of the SDES definition as a blueprint for an abstract data type with virtual elements, which are instantiated for a certain model class by substituting the attributes with net-class dependent values and functions.

The goal of the SDES definition is also to underline the similarity of the different model classes in stochastic discrete event systems. It should be useful for gaining insights into this area, requiring it to be understandable. The definition therefore refrains from being able to capture all details of every known model class, the restrictions are pointed out later. However, this is only done for simplicity of description, because those additional elements could be put into the SDES description technically simple. Popular model classes like automata, queuing networks, and Petri nets of different kinds (with stochastic extensions) are subclasses of stochastic discrete event systems and can be translated into the introduced SDES description. The future will see more model classes of stochastic discrete event systems coming up, requiring a unified description to be flexible enough to capture the properties of these model classes as well.

After an informal description of the common properties of stochastic discrete-event systems in Sect. 2.1, the SDES model definition and its underlying behavior are formally introduced in Sects. 2.2 and 2.3. A simplified process is defined as an abstraction of the full stochastic process. The type of the processes depending on the model elements are mentioned and restrictions are discussed. Section 2.4 describes how quantitative measures can be specified for a model and their formal relationship to the stochastic process. The chapter ends with some notes on related work.

2.1 Informal Description

A **discrete event system** is a system which is in a **state** during some time interval, after which an atomic **event** might happen that changes the state of the system immediately. The state does not change between two subsequent events, which is the main difference to continuous dynamic models where the change in time can for instance be described by differential equations. A system is usually composed of parts that contribute to the set of possible states with their local states. The states need to be captured, which is done by **state variables**. A **system state** is then characterized by the association of a certain **value** to each of the state variables. State variables usually correspond to passive elements in the model classes, like places in Petri nets or queues in queuing systems. The set of theoretically possible values of a state variable is reflected by its associated **sort**. This sort might be quite complex, think, e.g., of multisets of tokens with sophisticated attributes in a colored Petri net.

Discrete event systems are studied because they capture both static and dynamic information, which makes them useful for evaluation of the modeled system's behavior. The behavior is characterized by the visited states and the events that lead from one state to the next. Events change the system state by altering state variable values. All events and state changes are the result of active elements inside a discrete event system, which we will call **actions** in the following. Actions might correspond to a transition in a Petri net model class or a server of a queuing system. They describe possible **activities** that might become **enabled**, start, take some time to complete, and are finally **executed** resulting in an event with its corresponding state change. The dynamics of an action depend on the current system state, i.e., the state variable values. An action might for instance only be enabled if one state variable is in a certain range. Interaction and causal dependency between actions is thus possible because their execution depends on state variables and changes them as well.

In many systems there are actions that actually model classes of state changes, i.e., which lead to different possible activities for one state during the evolution of the dynamic behavior. An example is transitions in colored Petri nets. Hence it is not sufficient to talk of actions which are enabled or not; we need to compute the set of enabled modes of the action for a state, and to decide which one may be executed. We thus distinguish between **actions** and **action modes**. One of the latter corresponds to a specific enabling instance of an action. A pair of an action together with one of its action modes is called an **action variant**. This distinction is not necessary for model classes without this property, like Automata. In this simple case there is exactly one action mode if the action is enabled. We will therefore use the term **action** for both terms in this case. During the evolution of the dynamics of an SDES it might be necessary for the correct computation to store internal states of an action mode. An evaluation algorithm then works on an extended model state that includes both the values of state variables and internal states of actions. This issue is covered in detail together with the definition of the dynamics later on.

During an actual evaluation of the dynamic behavior of an SDES there are several model classes in which the same action (or modes of it) can be enabled concurrently with itself. This is, e.g., the case for queues with infinite server semantics. To capture this important property, we need to distinguish not only executable action modes, but different concurrent executions of them as well. We denote one individual running action variant with the term **activity** in the following. Such an activity models an enabled action variant, and contains the remaining delay until execution or the planned execution time as well. If it becomes disabled due to a state change, the activity ceases to exist.

The time between two subsequent **action executions** (or more exact **activity** or **action variant executions**) depends on the times that the ongoing activities have been enabled. Whenever an activity becomes newly enabled, a **delay** is sampled from the delay distribution of the action. When the time is used up, the activity is executed, provided there is no other activity scheduled for the same point in time, which disables the other one by its prior execution. When we take a snapshot of the system dynamics, the **remaining activity delay** stores the time that still has to elapse for an activity before it may be executed. It is obvious that every activity is an action variant as well; in the formal definition, however, activities have an associated remaining activity delay which action variants naturally do not have. The time that is spent in one individual state is called the **sojourn time** and might be zero.

2.2 Static Model Definition

A SDES is a tuple

$$\text{SDES} = (SV^\star, A^\star, S^\star, RV^\star)$$

describing the finite sets of **state variables** $SV^\star$ and **actions** $A^\star$ together with the **sort function** $S^\star$. The **reward variables** $RV^\star$ correspond to the quantitative evaluation of the model and are covered in Sect. 2.4. The $\star$-sign will be used consistently to distinguish identifiers of the general SDES definition from the later individual model class definitions. The elements of the tuple are explained in the following.

$S^\star$ is a function that associates an individual **sort** to each of the state variables $SV^\star$ and action variables $Vars^\star$ in a model (see later). The sort of a variable specifies the values that might be assigned to it. In the following we assume that a type system is implicitly used that allows standard operations and functions on sorts. We do not elaborate on a more formal foundation of types here. Because of the nature of the SDES model classes with well-known basic types it is usually obvious how constants, variables, terms, and formulas are constructed and evaluated. We denote by $\mathcal{S}^\star$ the set of all possible sorts.

$$S^\star : \left(SV^\star \cup Vars^\star\right) \to \mathcal{S}^\star$$

$SV^\star$ is the finite set of n state variables, $SV^\star = sv_1, \ldots, sv_n$, which is used to capture states of the SDES. A state variable sv_i usually corresponds to a passive element of the SDES, like a place of a Petri net or a queue of a queuing model.

For the later definition of state-dependent properties we denote with Σ the set of all theoretically possible states of a certain SDES, which contains all associations of values to each state variable allowed by the sort. However, not all of these states need to be actually reachable.[1]

$$\Sigma = \prod_{sv \in SV^\star} S^\star(sv)$$

State variables $sv \in SV^\star$ have the following attributes.

$$sv = \left(Cond^\star, Val_0{}^\star\right)$$

There are cases in which not all values that belong to a sort of a state variable are actually allowed. Think, e.g., of a buffer with a limited capacity. It would be possible (but not straightforward) to specify this as a condition of an action. The **state condition** $Cond^\star$ is a boolean function that returns for a state variable in a specific model state whether it is allowed or not.

$$Cond^\star : SV^\star \times \Sigma \to \mathbb{B}$$

$Val_0{}^\star$ is a function that specifies the **initial value** of each state variable, which is necessary as a starting point for an evaluation of the model behavior. The associated value obviously needs to belong to the sort of the corresponding state variable.

$$\forall sv \in SV^\star : Val_0{}^\star(sv) \in S^\star(sv)$$

and it is required to fulfill the state condition

$$\forall sv \in SV^\star : Cond^\star\left(sv, Val_0{}^\star(sv)\right) = \text{True}$$

$A^\star$ denotes the set of **actions** of an SDES model. They describe possible state changes of the modeled system. An action $a \in A^\star$ of an SDES is composed of the following attribute functions.

---

[1] The product symbol in the equation denotes the cross-product over all sets.

$$a = (Pri^\star, Deg^\star, Vars^\star, Ena^\star, Delay^\star, Weight^\star, Exec^\star)$$

$Pri^\star$ associates to every action a global **priority**. The priority is used to decide which action is executed first if there are several activities that are scheduled to finish at the same point in time. Actions with numerically higher priorities complete first.

$$Pri^\star : A^\star \to \mathbb{N}$$

The **enabling degree** $Deg^\star$ of an action specifies the number of activities of it that are permitted to run concurrently in any state. This is for instance used to capture the difference between **infinite server** and **single server** semantics, for instance, of Petri net transitions or in queuing systems. Positive natural numbers including infinity are allowed as values.

$$Deg^\star : A^\star \to \{\mathbb{N}^+ \cup \infty\}$$

Actions $a \in A^\star$ may be composed of several internal actions with different attributes in some SDES model classes. In other examples, actions may contain individual variants or modes. To capture this, the **action variables** $Vars^\star$ define a model-dependent set of variables $Vars^\star(a)$ of an action a with individual sorts. One setting of values for these variables corresponds to an action mode *mode*. This is, e.g., equivalent to a **binding** in a colored Petri net.

For model classes with exactly one action mode per action, the set $Vars^\star$ is empty. The actual sorts need to be defined in the model class and are specified formally by $S^\star : Vars^\star(a) \to \mathcal{S}^\star$.

Any one of the possible associations *mode* of values to the action variables $Vars^\star(a)$ of an action a is called an **action variant** and formally defined as a mapping

$$\forall var^\star \in Vars^\star(a), mode(a) : var^\star \to S^\star(var^\star)$$

The set of all action modes of an action a is denoted by $Modes^\star(a)$ and defined as

$$\forall a \in A^\star : Modes^\star(a) : \{mode(a)\}$$

Many attributes depend on an action a together with one of its corresponding modes $mode \in Modes^\star(a)$. To simplify notation, any possible pair of action and action mode is called an **action variant** and written as v. The set of all possible action variants AV is defined as

$$AV = \{(a, mode) \mid a \in A^\star, mode \in Modes^\star(a)\}$$

The following attributes of actions are defined on individual action variants rather than the action itself. In many model classes, there are no action variables and actions, thus contain only one action mode and variant. In those cases it is not necessary to distinguish between an action and its modes. We will then just write Attribute(a) instead of the complete Attribute $(a, mode(a))$.

Action variants may only start and proceed over their delay under certain conditions until execution. If these conditions hold in a state, we say the action variant is **enabled** in it. The value of the boolean **enabling function** $Ena^\star$ of an action variant returns for a model state if it is enabled or not.

$$Ena^\star : AV \times \Sigma \to \mathbb{B}$$

An action is informally called enabled in a state if at least one of its variants is enabled in it. It should be noted that the enabling degree of an action is allowed to be positive in a state even if it is not enabled in it. On the other hand, an action variant v that is formally enabled ($Ena^\star(v, \cdot) = \text{True}$) may not be effectively enabled because the enabling degree of the action in the state is zero. Definitions and algorithms for the dynamic behavior observe these special cases.

The **delay** $Delay^\star$ describes the time that must elapse while an action variant is enabled in an activity until it finishes. This time is in most cases not a fixed number, but a random variable with positive real values. $Delay^\star$ thus defines the probability distribution function for this random time.

$$Delay^\star : AV \to \mathcal{F}^+$$

Some background on distribution functions and the set $\mathcal{F}^+$ is given in Sect. 1.4.

The **weight** $Weight^\star$ of an action variant is a real number that defines the probability to select it for execution in relation to other weights. This applies only to cases in which activities with equal priorities are scheduled for execution at the same instant of time. An example are firing weights of immediate transitions in Petri nets. The calculation of the individual execution probabilities is explained in more detail in the subsequent section, together with the behavior of an SDES.

$$Weight^\star : AV \to \mathbb{R}^+$$

$Exec^\star$ defines the state change that happens as a result of an action variant execution (i.e., the finishing and execution of the activity) and is called **execution function**. As actions change the state, $Exec^\star$ is a function that associates a destination state to a source state for each action variant. This function does not need to be defined or have a useful value for pairs containing a variant that is not enabled in the respective state.

$$Exec^\star : AV \times \Sigma \to \Sigma$$

The given definition of an SDES will be used to map other well-known modeling formalisms into one common framework in the subsequent chapters, based on which, e.g., a set of analysis algorithms can be used transparently. In the individual definitions of how the model classes can be mapped into (or described as) an SDES, elements and attributes of the model class are used to specify the corresponding SDES elements. One example is the mapping of Petri net places to SDES state variables.

2.3 Dynamic Behavior of Stochastic Discrete Event Systems

This section formally defines the dynamic behavior of an SDES, the **stochastic process** that an SDES model describes. Some definitions for the behavioral specification are introduced first.

A **state** σ of an SDES captures a snapshot of all local states; it thus is a mapping of values (of correct sorts) to all state variables. It is obviously contained in the set of all possible states Σ.

$$\sigma \in \Sigma$$

The ith element of σ contains the value of the state variable sv_i in a state and is denoted by $\sigma(sv_i)$ with $\sigma(sv_i) \in S^*(sv_i)$.

For a description or an analysis of the dynamic behavior, things are greatly simplified if the complete information that is necessary for the further behavior is captured in only one current state. In addition to the state variable values stored in σ we need to keep track of internal states of actions. A **complete state**[2] $cs \in CS$ thus contains a (conventional) state of the state variables and the internal states of all actions:

$$CS = \{(\sigma, as) \mid \sigma \in \Sigma, as \in AS\}$$

and AS as defined below. The set of all possible complete states is denoted by CS.

The timing semantics of an SDES can be defined based on the assumption of a **remaining activity delay** (RAD for short), which measures the time that has still to elapse before an activity is executed. This term is similar to the **remaining firing time**, e.g., used in the area of Petri nets. The **action state** describes a set of enabled actions and action modes together with their RAD. Following these considerations, an element of an action state is a 3-tuple containing action a, action mode $mode$ and the remaining activity delay RAD. Such a tuple is the formal representation of an activity, which can obviously be considered as an action variant together with a remaining delay as well.

$$as \subseteq \{(a, mode, RAD) \mid (a, mode) \in AV, RAD \in \mathbb{R}^{0+}\}$$

An element of an action state is called an **activity**. Activities are the entities that are scheduled to happen at some instant of time due to the enabling of action variants. Variants that are not scheduled for execution or are simply not enabled cannot be contained in the corresponding action state. There might be action variants that are enabled in a state of the state variables, but are not contained in a corresponding action state because of concurrency restrictions of an action (as defined by the enabling degree).

---

[2] Also termed **augmented state** in the literature.

The notion of action states will be used for the formal definition of the dynamics of the stochastic process underlying an SDES later. The set of all action states is denoted by $AS = \{as\}$. Please note that the same action variant can be enabled multiply in a state, and thus might be contained several times with the corresponding RADs in the set as. The associated remaining activity delays RAD might be equal as well as a special case, thus requiring as to be formally interpreted as a multiset (bag).

It should be noted that all sets as are finite, because the actual enabling degree in any state of the model is bounded. The set of pairs is created and updated such that for enabled action modes the remaining delays are stored, which are finite nonnegative real values. Using ∞ formally as a delay, as it is sometimes done in the literature, is not necessary because disabled action modes have no associated element in the set of pairs.

2.3.1 Rules for a Behavioral Definition

After the introduction of states of an SDES model, the dynamic behavior can be specified. This is informally done by the following rules; a more thorough definition is given in the subsequent section.

The future behavior of an SDES model at a certain state and time depends in general on the complete history. Such a way of describing (and analyzing the model by any real implementation) would, however, not be reasonable to do. It is more convenient both for the definition as well as an algorithm if we are able to describe the future behavior by using only the current state. This requires to keep information in such a necessary complete state that contains information about ongoing activities and their attributes, like the execution mode and delay. By doing so, it is possible to describe the behavior with only a few rules below. The same principle is used to formally define the behavior. It influences the resulting kind of stochastic process (cf. Sect. 2.3.2).

Enabling Rule
: An action variant v is enabled in a state if its enabling function $Ena^\star$ evaluates to True in the state, and if for every state variable of the SDES the prospective future value after an execution of v fulfills the state condition $Cond^\star$. We call an action enabled in a state if any one of its variants is enabled in it.

Initial State Rule
: The initial complete state of the SDES model is given by the initial state variable values $Val_0{}^\star$ which are usually given as part of the model. In addition to that, the action states are initialized such that for every action with enabled modes, as many of them are stored as specified by the enabling degree. The selection of a certain mode

out of all possible ones is done with equal probabilities. The initial remaining activity delays are sampled from the corresponding delay.

Sojourn Time Rule   When the SDES is in a complete state, i.e., all ongoing activities have been specified, the model time passes. During such a period all remaining activity delays (RAD) decrease with equal speed, until (at least) one of them reaches zero. Please note that the case in which there is already an activity with $RAD = 0$ is a special case that does not need to be treated separately. The first activity for which the remaining delay reaches zero is executed; this behavior is often called a *race policy*. The scheduling rule defines in more detail how the activity that is executed first is selected in the general case.

Scheduling Rule   If there is more than one activity that is scheduled to finish at the same time, because their remaining activity delays reach zero together, a decision needs to be made which activity is executed first (although at the same model time, if they do not disable each other).

First of all, the activity with the highest priority is selected. If this still does not solve the conflict, meaning that there are activities with equal priorities reaching a zero RAD, a probabilistic choice is applied. The relative probability of an activity to be selected is given by its weight *Weight**.

Execution Rule   Execution of an activity (or an action variant) marks an event and changes the values of the SDES state variables according to the execution function *Exec**. The executed activity itself and all activities that are not enabled any more due to the state change are removed from the set of activities of the destination state. The action delays are updated according to the elapsed sojourn time for the remaining ones.

In a second step, the enabling degrees of all enabled actions are considered to update the set of activities of the destination state. If the number of activities of an action already contained in the set is bigger than the allowed enabling degree, the exceeding ones are selected with equal probability and removed. In the case that the enabling degree is bigger than the number of existing activities, action variants are selected with equal probability to form new activities. The remaining delay of every new activity is sampled from the delay distribution of the action.

2.3.2 The Stochastic Process Defined
by a Stochastic Discrete Event System

Section 2.3.1 gave informal rules about how an SDES evolves from state to state by the execution of subsequent actions. For the evaluation of an SDES we are mainly interested in its stochastic behavior over time, which is commonly abstracted by a **stochastic process**. Such a process can be viewed and defined as a family of random variables $\{X(t), t \in T\}$ as recalled in Sect. 1.4. Throughout this section the stochastic process of an SDES is formally defined and its evolution over time is specified in detail.

Different views on the behavior of an SDES are meaningful for the various usages of the model: for the later evaluation of performance measures we are mainly interested in the evolution of state variables over time and actions that lead to state changes. For a complete definition of the dynamics of an SDES, however, the stochastic process also needs to keep track of action variables, concurrently enabled action modes, and their remaining activity delays. Hence, we define different stochastic processes in the following, starting with the most detailed process that used to define the behavior of an SDES formally. A simplified view on that process is defined based on it later on, which can be used for the evaluation of quantitative measures of a model.

For the definition of the stochastic process given by (or underlying) an SDES, the natural choice would be a continuous-parameter process, where the parameter t is interpreted as the time. However, as we allow action delays to be zero, there might be several action executions at the same point in time. This may lead to ambiguities because the causal ordering among actions is often important despite their execution at the same time.

We therefore define the stochastic **complete process** that underlies an SDES as a discrete-parameter process with three parameters

$$CProc = \left\{ \big(cs(n), \theta(n), ce(n)\big), n \in \mathbb{N} \right\}$$

where n is an index variable that numbers consecutive elements of the process. The nth complete state of the model is denoted by $cs(n)$. $\theta(n)$ specifies the **state sojourn time**, i.e., the time that the model spends in state $cs(n)$ before **event** $ce(n)$ happens, changing the model state to the subsequent state $cs(n+1)$. The model time at which the nth event $ce(n)$ happens is given by the sum of all state sojourn times passed before, $\sum_{i=0}^{n} \theta(i)$. Recall that for a complete definition of the stochastic process $CProc$ the remaining activity delays of all enabled action variants need to be tracked. They are captured in the second part of the complete state $cs(n)$ for state n. To be specific, the remaining times are stored that correspond to the point in time just after event $ce(n-1)$ has happened. The initial values of the parameters are set according to the initial state of the SDES model, which is described in detail below.

$CProc$ is called complete process of an SDES because all events (activity executions) of the SDES are considered in their execution sequence, and the

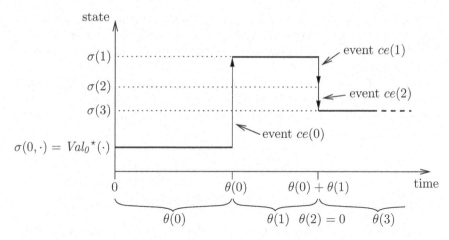

Fig. 2.1. State, event, and sojourn time examples of the complete process

full information about complete states and remaining activity delays is captured. Figure 2.1 visualizes example relations between the elements. Events $ce(1)$ and $ce(2)$ happen at the same time in the example; the sojourn time in state $cs(2)$, $\theta(2)$, is thus zero.

The elements of the complete process *CProc* are defined as follows for every index $n \in \mathbb{N}$. A state cs is a **complete state** of the SDES as defined in Sect. 2.3, $cs(n) = \big(as(n), \sigma(n)\big) \in CS$. It contains a value for each state variable in $\sigma(n)$, and a possibly empty set $as(n)$ of activities (3-tuples containing action, action mode, and remaining delay). This is necessary to store which and how many action variants are concurrently enabled in a complete state.

The **state sojourn time** θ is defined as $\theta(n) \in \mathbb{R}$, and every event $ce(n)$ equals the executed action variant.

$$ce : n \to AV$$

It should be noted that the execution of an action variant maps from a (simple) source state to a (simple) destination state, while the states of the complete process are complete states of the SDES.

It now remains to define in detail how the values of the elements of the complete process change over time. This follows directly by applying the rules that govern the behavior of an SDES given textually in Sect. 2.3.1 to the formal definition of the complete process.

The evolution of the stochastic process is defined in an iterative manner: we first specify how the initial complete state of the process $cs(0)$ is set up. A number of subsequent equations are given afterwards, which define $\theta(n)$, $ce(n)$, and finally $cs(n + 1)$ based on the knowledge of $cs(n)$. The stochastic process for an SDES is thus completely defined, provided that it does not exhibit improper behavior. This is discussed at the end of this section.

The reader should keep in mind that the following is intended as a formal definition of the process dynamics and not as an actual algorithm. Although it could be used as one, an efficient software implementation should exploit model properties.

The set of enabled action modes $Enabled(a, \sigma)$ of an action a in a state σ is defined as a prerequisite. It requires a variant of a to be enabled in state σ, and the condition function to hold in the state that would be reached by executing it.

$$\forall a \in A^\star, \sigma \in \Sigma : Enabled(a, \sigma)$$
$$= \{v = (a, mode) \in AV \mid Ena^\star(v, \sigma) = \text{True} \qquad (2.1)$$
$$\wedge \; \forall sv \in SV^\star : Cond^\star(sv, Exec^\star(v, \sigma)) = \text{True}\}$$

The enabling degree $Deg^\star(a)$ of an action specifies a maximal number of concurrent activities that may be running in parallel. For actions with different modes, we need to differentiate between the enabling degrees of the individual modes as well. Remember that one mode (and thus one action variant) corresponds to a setting of values to the action variables. There might be several of them enabled in a state, and it is moreover possible that one enabled variant can be activated in parallel to itself.

Think of a colored Petri net example as shown in Fig. 2.2 with numbers as tokens in places P1 and P2. Transition T1 is enabled under the bindings (i.e., has action variants for) $x = 1, y = 3$ and $x = 2, y = 3$. The enabling degrees of the two action variants are 2 and 1, respectively, because there is only one token 2 and two 1 in P1. In the general framework of SDES, the **enabling degree of an action variant** $VDeg^\star$ equals the number of times the variant could be executed subsequently in a state, i.e., the number of times it is concurrently enabled with itself. This value must be finite in any state, even for actions with an unbounded number of servers. An infinite number of concurrent activities would not make sense in a finite model.

The enabling degree of an action variant v returns a natural number for a state.

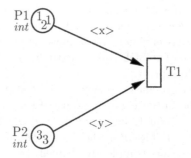

Fig. 2.2. Colored Petri net example for enabling degrees

$$VDeg^\star : AV \times \Sigma \to \mathbb{N}$$

It is not considered as an integral part of the SDES specification and thus not specified in Sect. 2.2, because the actual value in a state could be derived by checking how many times the variant can be executed until it is not enabled any more.

The allowed degree of concurrency of any action variant is of course always bounded by the overall enabling degree $Deg^\star(a)$ of the action a. The minimum of action degree and enabling degree of action variant must therefore be used in all formulas. To avoid writing $\min\big(Deg^\star(a), VDeg^\star(v,\sigma)\big)$ in every place where the degree of concurrency is used, we define the enabling degree of an action variant accordingly, taking the action degree already into account.

$$\forall v = (a, mode) \in AV, \sigma_0 \in \Sigma :$$
$$VDeg^\star(v, \sigma_0) \;=\; \min\big(n, Deg^\star(a)\big) \quad \text{if } \exists \{\sigma_0, \sigma_1, \ldots, \sigma_n\} \subseteq \Sigma$$
$$\text{with } \forall i \in \{0 \ldots n-1\} : mode \in Enabled(a, \sigma_i)$$
$$\wedge \quad \forall i \in \{0 \ldots n\} : \sigma_{i+1} = Exec^\star(v, \sigma_i)$$
$$\wedge \quad mode \notin Enabled(a, \sigma_n)$$

This is, however, a very inefficient approach for most model classes. Hence, an individual $VDeg^\star$ definition is given in each of the model class related chapters.

We require in the following that the enabling degree of an action must either be one or infinity if there is a state in which there is more than one of its action modes enabled. The possible combinations of enabling degrees of actions and action variants would otherwise become quite complex and hard to specify and analyze. There has been no restriction whatsoever for the model classes considered so far.

$$\forall a \in A^\star : \big(\exists \sigma \in \Sigma \mid |Enabled(a, \sigma)| > 1\big) \to \big(Deg^\star(a) \in \{1, \infty\}\big)$$

The Initial State of the Complete Process

The **initial complete state** of the complete process $cs(0) = \big(\sigma(0), as(0)\big)$ comprises the initial state of the state variables as well as the initial action state. The initial state of the state variables $\sigma(0)$ is directly given by the SDES definition, as specified by the modeler.

$$\forall sv_i \in SV^\star : \sigma(0)(sv_i) = Val_0{}^\star(sv_i)$$

The initial action state $as(0)$ is a multiset containing tuples such that for every action a with enabled action variants there is a maximum of them given by the degree of concurrency for the variant. The overall degree of concurrency of the action $Deg^\star(a)$ must not be exceeded as well. The initial remaining activity delays are set according to the corresponding delay distribution. The following equation defines $as'''_{a,mode}$, the set of activities for each action variant,[3] of which the union is taken resulting in the initial action state below.

$$\forall v = (a, mode) \in AV :$$
$$as'''_v(0) = \{(a, mode, RAD) \mid mode \in Enabled(a, \sigma(0)), RAD \sim Delay^\star(v)\}$$
$$\text{such that } |as'''_v(0)| = VDeg^\star(a, mode, \sigma(0))$$

Selection of the action modes is done with equal probabilities[4] among the enabled ones if necessary. This is done during the construction of the action state $as'''_a(0)$ of every action a, over which the union is finally taken resulting in $as(0)$.

$$\forall a \in A^\star :$$

$$as'''_a(0) = \begin{cases} \emptyset & \text{if } |Enabled(a, \sigma(0))| = 0 \\ \bigcup_{v=(a,\cdot)\in AV} as'''_v(0) & \text{if } Deg^\star(a) > 1 \\ as'''_v(0) & \text{otherwise; select } v = (a, \cdot) \text{ randomly:} \\ & P\{v\} = \frac{1}{|Enabled(a,\sigma(0))|} \end{cases}$$

$$as(0) = \bigcup_{a \in A^\star} as'''_a(0)$$

As this definition is very similar to the one below for the general definition of $as(n)$, any actual algorithmic implementation should instead use the corresponding steps after proper initialization of the temporary items defined later.

Iterative Derivation of Complete States

Having defined $cs(0)$ based on the SDES definition, we assume in the following that $cs(n) = (\sigma(n), as(n))$ is known, and define how $\theta(n)$, $ce(n)$, and $cs(n+1)$ are defined by construction.

With the knowledge of all remaining activity delays RAD of the ongoing activities from $as(n)$, the state sojourn time $\theta(n)$ of state n is simply the minimum of all RAD values.[5]

$$\theta(n) = \begin{cases} \min_{(\cdot,\cdot,RAD)\in as(n)}(RAD) & \text{if } as(n) \neq \emptyset \\ \infty & \text{otherwise} \end{cases}$$

---

[3] The prime notation has been chosen in correspondence to similar temporary items for the general case n below.

[4] A different approach would select action variants v that are multiply enabled (i.e., with $VDeg^\star(v, \cdot) > 1$) with a higher priority. This would, however, make the selection even more complex and is thus avoided here.

[5] In the case of $as(n) = \emptyset$, the state $cs(n)$ is a *dead state* and the process comes to a stop. The end of the section discusses this case. We assume the sojourn time to be infinity then.

The value of $\theta(n)$ specifies the time that is spent in state $cs(n)$. Based on it we can now define which activity will cause the next event $ce(n)$ by looking for the activity with $RAD = \theta(n)$, the shortest time. Things become a bit more complex if there are several activities for which this equation holds. Priorities and weights are used to resolve the conflict, as it has been described in Sect. 2.3.1.

We first define the set $as^0(n) \subseteq as(n)$ such that it contains all activities with remaining activity times equal to the state sojourn time $\theta(n)$.

$$as^0(n) = \{(\cdot, \cdot, RAD) \in as(n) \mid RAD = \theta(n)\}$$

If $as^0(n) = \emptyset$, there are no enabled activities in the state. The process is then completely described and we do not need to follow the iterative definition further. It depends on the type of analysis and applied algorithm if this can be accepted or hints to a modeling error. We assume in the following that there are enabled activities. To cover the case $|as^0(n)| > 1$ when more than one activity is scheduled to complete at $\theta(n)$, we further restrict the set $as^0(n)$ to activities with the highest priority $\pi(n)$.

$$\pi(n) = \max_{(a, \cdot, \cdot) \in as^0(n)} \left(Pri^\star(a)\right)$$
$$as^\pi(n) = \{(a, \cdot, \cdot) \in as^0(n) \mid Pri^\star(a) = \pi(n)\}$$

If the number of activities contained in $as^\pi(n)$ is still bigger than one, weights are used to perform a probabilistic choice between the remaining activities. Any one of the activities $(a, mode, \theta(n)) \in as^\pi(n)$ can be selected with the associated weight as the relative probability.

$$ce(n) = (a, mode) \text{ with } (a, mode, \cdot) \in as^\pi(n),$$
$$P\{ce(n) = (a, mode)\} = \frac{Weight^\star(a, mode)}{\sum_{(a_i, mode_i, \cdot) \in as^\pi(n)} Weight^\star(a_i, mode_i)}$$

An algorithm does of course not need to execute all of the steps described above to select the next event, if one of the sets $as(n)$ or $as^0(n)$ already has only one element.

After the selection of the activity $ce(n)$ to be executed, we are ready to define how the subsequent state $cs(n{+}1)$ is constructed. There are two steps to this task, which correspond to defining the two elements $\sigma(n{+}1)$ and $as(n{+}1)$. The state $\sigma(n+1)$ of the SDES state variables can easily be constructed from the current state by applying the execution function of the activity $ce(n)$, because $\sigma(n), \sigma(n+1) \in \Sigma$, and $Exec^\star(v) : \Sigma \to \Sigma$.

$$\sigma(n+1) = Exec^\star\big(ce(n), \sigma(n)\big)$$

The derivation of the subsequent action state $as(n{+}1)$ is more complex and carried out in the following steps. The set of enabled action modes as defined in the beginning is used. When applied to the state of interest $\sigma(n+1)$, we get the set of enabled action modes for every action a as $Enabled\big(a, \sigma(n+1)\big)$.

As a next step we define how the remaining activity delays are updated for ongoing activities. We consider activities that have been active in state $\sigma(n)$ and are still enabled in $\sigma(n+1)$ first. The RAD is decreased by the state sojourn time $\theta(n)$ for them, while the disabled activities in $\sigma(n+1)$ as well as the executed event $ce(n)$ do not have to be considered any more. The elements are kept in different multisets $as'_{a,mode}$ for every action variant $(a, mode) = v$.

$$\forall v = (a, mode) \in AV \ :$$

$$as'_v(n+1) = \big\{\big(a, mode, RAD - \theta(n)\big)$$

$$\mid (a, mode, RAD) \in as(n)$$

$$\wedge \ mode \in Enabled\big(a, \sigma(n+1)\big)\big\} \setminus \{\,(ce(n), 0)\,\}$$

Now after having specified the continuing activities, it remains to update the action state by considering the enabling degrees of all actions and action variants. Recall that the enabling degrees specify the maximum number of concurrent activities in a state σ. In the case that the updated action state $as'_a(n+1)$ contains more activities, a probabilistic equal choice is necessary to reduce the number of activities accordingly.

$$\forall v = (a, mode) \in AV \ :$$

$$as''_v(n+1) \subseteq as'_v(n+1) \text{ with}$$

$$|as''_v(n+1)| = \min\big(|as'_v(n+1)|\,, VDeg^\star(v, \sigma(n+1))\big)$$

$$\mathrm{P}\{(a, mode, \cdot) \in as''_v(n+1)\} = \min\left(\frac{VDeg^\star(v, \sigma(n+1))}{|as'_v(n+1)|}, 1\right)$$

The opposite case is of course also possible: There are enabled action modes of an action a (i.e., $Enabled(a, \cdot) \neq \emptyset$), but the number of activities of that action that are already contained in the action state is smaller than the allowed enabling degrees of action and variants. The action state then needs to be extended by adding more activities to fully exploit the possible degree of concurrency. The following equation defines as'''_v, the set of activities that are added to as''_v in order to form the updated action state. The selection of one of the enabled action modes to be an element of as'''_v is again done by a probabilistic equal choice among the enabled modes. The initial remaining activity delay for every new activity is sampled from the corresponding delay distribution. In the special case that there is no enabled action mode of the action, the set is obviously empty.

$\forall v = (a, mode) \in AV :$

$\quad as_v'''(n+1) = \{(a, mode, RAD) \mid mode \in Modes^\star(a), RAD \sim Delay^\star(v)\}$

with

$$|as_v'''(n+1)| = \begin{cases} 0 & \text{if } |Enabled(a, \sigma(n+1))| = 0 \\ VDeg^\star(v, \sigma(n+1)) - |as_v''(n+1)| & \text{otherwise} \end{cases}$$

$$\text{and} \quad P\{(a, mode, \cdot) \in as_v'''(n+1)\} = \frac{1}{|Enabled(a, \sigma(n+1))|}$$

The combination of the different action states for the variants realizes a probabilistic choice for the case $Deg^\star(a) = 1$. It should be noted that only one of the sets $as_v'''(n+1)$ in the second line of the case differs from the empty set if $Deg^\star(a) \neq \infty$ because of the restriction introduced on p. 29. This restriction is necessary for the equations above to work properly.

$$as_a'''(n+1) = \begin{cases} \emptyset & \text{if } |Enabled(a, \sigma(n+1))| = 0 \\ \displaystyle\bigcup_{v=(a,\cdot)\in AV} as_v'''(n+1) & \text{if } Deg^\star(a) > 1 \\ as_v'''(n+1) & \begin{array}{l} \text{otherwise; select } v = (a, \cdot) \text{ randomly:} \\ P\{v\} = \frac{1}{|Enabled(a,\sigma(n+1))|} \end{array} \end{cases}$$

All necessary building blocks are now available through the definitions above. Together they make up the next complete state $cs(n+1)$, which is defined accordingly.

$$cs(n+1) = \left(\sigma(n+1), as(n+1) = \bigcup_{a \in A^\star} as_a'''(n+1) \cup \bigcup_{v \in AV} as_v''(n+1)\right)$$

The section has shown how the step from a state n to the subsequent one $n+1$ is defined. The iterative application of the definitions thus completely defines the stochastic process underlying an SDES.

Process Types and the Simplified Process

The complete process contains in each of its complete states all the information that is needed to decide about the future of the process. The state sojourn times are, however, not memoryless in general, leading to a semi-Markov process. Part of the information that is kept in the complete state describes the remaining activity delays. As they specify continuous time values, the complete process can be characterized as a *continuous-state semi-Markov process* or *generalized semi-Markov process* (GSMP). The variable over which the process evolves counts discrete event executions, which makes the process a *time-homogeneous discrete-time GSMP* [143].

If we are interested in a view onto the basic system state as it changes over time, we can define a corresponding **simplified process** *SProc* as a continuous-time stochastic process

$$SProc = \left\{ \left(\sigma(t), SE(t) \right), t \in \mathbb{R}^{0+} \right\} \text{ with } \sigma(\cdot) \in \Sigma$$

where the state of the state variables at time t is given by $\sigma(t)$, and the (possibly empty) set of events that are executed at time t by $SE(t)$. This process is completely determined by *CProc* following

$$\sigma(t) = \sigma \quad \text{iff} \quad cs \left(\min_{x \in \mathbb{N}} t \leq \sum_{i=0}^{x} \theta(i) \right) = (\sigma, \cdot)$$

and

$$SE(t) = \left\{ ce(n) \,\middle|\, t = \sum_{i=0}^{n} \theta(i) \right\}$$

The simplified process does not visit states of the complete process with zero sojourn time, it is right-continuous in the states where some time is spent. As there might be several executions of activities at one point in time, the events of an instant t are stored as a multiset. It is not possible in general to derive the correct ordering of events from this description. The simplified process does, however, still contain all information necessary to derive quantitative measures from it, as explained in detail in Sect. 2.4. As the sets of ongoing activities and their remaining delays are neglected in this definition, it is not possible to foresee the next state and event. The future of the state process could therefore be defined only based on the knowledge of its complete history. The process is thus not semi-Markov, but can be characterized as a *continuous-time stochastic process* due to its definition over a continuous time variable t.

There are special cases of models in which the underlying stochastic process belongs to a more restricted class. If in a model all delays are either exponentially distributed or zero, the sojourn times in the process states become memoryless, resulting in a *Markov process*. It is then not necessary to store the remaining activity delays because of the memoryless property of the exponential distribution, which means that the underlying process can be described as a *Markov chain*.

If we allow one action variant with a nonexponentially distributed (or immediate) delay executable in every reachable state, the process is not memoryless any more. It is, however, sufficient for the description of the process to capture the remaining activity delay of the nonexponential action variant in a so-called *supplementary variable* [130]. Such a continuous variable is added to the otherwise discrete state space. The whole information necessary for the further evolution of the stochastic process is then captured in the augmented state, making it a Markov process. For the derivation of the associated state equations it is not important whether the supplementary variable stores the remaining delay or the elapsed enabling time.

An alternative view is to observe the stochastic process only at instants of time when it is in fact memoryless. This will always happen eventually in a model with the restriction described earlier, because an enabled nonexponential action variant is executed at some time. In states without such an enabled action variant the process is memoryless anyway. Such a process is referred to as a *Markov regenerative process* or *semi regenerative process* [71]. It is of special interest due to the existence of numerical analysis techniques, see Sect. 7.3.3.

Restrictions and Special Cases

After the general definition of the stochastic process underlying an SDES, some notes on useful restrictions and possible problems follow. They stem from the consideration that most evaluation algorithms, both numerical/analytical or by simulation, follow in principle the iterative construction of the stochastic process as defined earlier. The absence of improper dynamic behavior is also a prerequisite for the applicability of limit theorems for the calculation of quantitative measures [150].

To avoid algorithms running forever without significant progress in model time, the stochastic process underlying an SDES is not allowed to exhibit an infinite number of events inside a finite interval of time. This would also lead to a finite lifetime of the stochastic process, which is often not desired, especially for the definition of the continuous-time processes above. There are several ways in which such a problem may occur. One obvious case is called *absorption into the set of vanishing states*, and happens when there is a set of vanishing states which is entered once and can not be left any more. A less problematic case occurs when there is a loop of vanishing states in the reachability graph, which eventually is left to a tangible state. Algorithms that use an on-the-fly elimination of vanishing states need to treat this case separately.

Another case with an infinite number of events in a finite interval of time is not so obvious. Even if the model time increases during the evolution of the model, the sojourn times in subsequent states might become smaller and smaller such that the model time has an accumulation point [150]. This case is called *explosion* in the literature. Absence of an explosion is also called *nonZeno*, e.g., in the field of timed automata after the philosopher's imaginary race between Achill and a turtle.

Finiteness of the lifetime might not be a technical problem in the case of a transient analysis, as long as the considered transient interval is contained in the lifetime. However, in the definitions above, we implicitly assumed an infinite lifetime to simplify definitions. It should be noted that a much bigger problem than explosion or absorption into the set of vanishing states is in practice in the existence of *dead states*, i.e., in which no action is enabled and the complete process would come to a stop and be undefined from that point on.[6] We assume in the following that models do not exhibit the kinds of problems mentioned earlier. Algorithms that depend on these restrictions are often able to test for them and exit with appropriate warnings whenever necessary.

2.4 Measuring the Performance of SDES

Although SDES models can be useful on their own (like for documentation purposes), the main application of stochastic models aimed at in this text is quantitative evaluation. The SDES model itself as defined in the earlier sections describes through its semantics how a system under investigation evolves over time. To define what we want to know about the model and thus the system itself, some kind of measure needs to be specified as well. Examples could be the number of customers in a waiting room, the throughput of a communication system, or other issues of performance and dependability. The measures depend on the dynamic evolution of the SDES and are thus formally defined based on the stochastic process.

The formal notion of such a measure is called **reward variable** in the following in accordance with the relevant literature. It is named *reward* because there might be any kind of positive bonus or negative penalty associated with elements of the stochastic process. Such a reward variable is merely a function of a stochastic process that returns a real value. This is why the general type of evaluation we are interested in here is coined *quantitative* evaluation. It should be noted that the reward values are unitless just like the numerical model attributes. Their interpretation is completely in the hands of the modeler and needs to be consistent throughout model and rewards to avoid misinterpretations.

Two types of elements of such a reward variable have been identified in the literature. This was based on the basic observation that the stochastic process of a discrete event system remains in a state for some time interval and then changes to another state due to an activity execution, which takes place instantaneously. For the efficient computation and user-friendly specification of reward variables, their building blocks should be associated to process elements like states and state transitions. The natural way of defining a reward variable thus includes **rate rewards**, which are accumulated over time in a state, and **impulse rewards**, which are gained instantaneously at the moment of an event, i.e., an activity execution.

The definition of a reward variable with rate and impulse rewards is, however, not sufficient to completely describe what the modeler is interested in. If the number of vehicles driving over a crossing should be evaluated in an

---

[6] The formal definition of the complete process and the simplified process are valid even in the presence of dead states; the final state then lasts forever.

SDES model of a traffic system, the observation time must obviously be specified as well. An interval with fixed length or the average behavior over an infinitely long period are examples. They are closely related to **transient** and **steady-state** evaluation. In addition to that, one might be interested in a mean value over an interval, leading to a division by the interval length. There are a number of possible combinations, which are covered extensively in Sanders and Meyer [284]. Not all of them are meaningful for a quantitative evaluation; the most important ones are defined for the SDES setting in detail later.

We have seen so far that a reward variable is a function of a stochastic process and is computed for some specific time setting. Definitions of reward variables are usually based on one sample path of the system, which would only reflect a randomly selected one. However, as the process is stochastic, there are many different possible outcomes of the dynamic model evolution. The full information about a reward variable would thus require to analyze every possible process instance, and to compute a probability distribution function of the reward variable value from them. Many analysis algorithms are not capable of delivering this information, which is fortunately not necessary in most cases. The term **reward variable measure** is thus used as a derived value like the expectation or a quantile of an actual reward variable.

The detailed information about quantitative measures of SDES given below is structured as follows. The next section describes how reward variables are formally specified in the SDES framework. The semantics of the reward variables in terms of the stochastic process is shown in the subsequent section.

However, this is only an abstract description just as the unified SDES definition itself. For every actual model class like Automata or Petri nets, a definition of specific reward variables is given in the respective sections. Their formal relationship to SDES reward variables is defined there as well, which allows to interpret the measures on the SDES level just as the model parts itself. Typical model-class specific reward variables are listed, and application examples can be found in Part III.

2.4.1 Reward Variable Specification

The finite set of reward variables $RV^\star$ of an SDES is a part of its definition (compare p. 19). This underlines the view that the definition of performance measures is an integral part of a model description. Every element $rvar^\star \in RV^\star$ specifies one reward variable and maps the stochastic process to a real value. How this is done, i.e., the actual semantics of the reward variable definition, is described in the subsequent Sect. 2.4.2.

$$\forall\, rvar^\star \in RV^\star : \quad rvar^\star : CProc \to \mathbb{R}$$

More complex quantitative measures can be constructed on a higher level in a model using the reward variable results and combining them in arithmetic expressions. This is for instance necessary if a measure of interest is calculated as a nonlinear function of rate and impulse rewards. This is not specified inside the reward variable parameters. We do not go into the details how this can be done, because it appears as a technical question.

Each variable $rvar^\star$ is further defined by a tuple that contains specific information about how the value of the variable is derived from the stochastic process. The elements are defined and explained below.

$$\forall rvar^\star \in RV^\star : \quad rvar^\star = (rrate^\star, rimp^\star, rint^\star, ravg^\star)$$

The reward that is gained in a state of the model over time is defined by the **rate reward** $rrate^\star$. It returns a real number for every SDES state of the model, which is the rate of reward collected per model time unit.

$$rrate^\star : \Sigma \to \mathbb{R}$$

Activity executions may lead to gaining a reward instantaneously. This is specified by the **impulse rewards**, which are denoted by $rimp^\star$. Each one returns a real number for a state change, depending on the action variant that completes. Only action variant information is used for the activity, because the remaining firing time of an executed activity is always zero.

$$rimp^\star : AV \to \mathbb{R}$$

Similar definitions in the literature also allow impulse rewards to depend on the state before the completion of action. This was intentionally left here to simplify definitions and because it is not necessary for the application examples considered. It would, however, be simple to add this possibility to the SDES framework. Rate and impulse rewards are together often called a **reward structure** in the literature.

The model time interval during which we want to observe the stochastic process is given by the **observation interval**, $rint^\star$. It is a closed interval of positive real values, including infinity. The latter is used to denote cases in which the limiting behavior of the stochastic process should be analyzed (explanations are given below).

$$rint^\star : [lo, hi] \quad \text{with} \quad lo, hi \in \left(\mathbb{R}^{0+} \cup \{\infty\}\right) \wedge lo \le hi$$

The interval definition is already sufficient to differ between **instant-of-time** and **interval-of-time** measures. If $lo = hi$, the instant of time is given by both values, while an interval is obviously specified if $lo < hi$.

The last element $ravg^\star$ of a reward variable definition stores the information whether the resulting measure should be computed as an **average over time** or **accumulated**. If the boolean value is True, the variable is intended as an average measure.

$$ravg^\star \in \mathbb{B}$$

The reward variable *measure* itself is not kept as a part of the model definition. It is in general very complex (and often not needed, as for the example applications given in Part I) to solve for the reward variables in distribution. Hence only **expected values** of reward variables (expected accumulated reward, expected average reward, expected instantaneous reward) are considered in this text.

Only a few of the many possible types of reward variables are commonly used in the literature and in this work. A selection of typical variables can be expressed in the SDES framework as follows.

Instant-of-time measures analyze the value of a reward variable at a certain point in time t. This is often referred to as *transient analysis*. The instantaneous reward at time t is computed if we set the reward variable parameters as follows:

$$rint^\star = [t,t]\,; \quad ravg^\star = \text{False}$$

Often the algorithmic implementation requires to compute the values of the reward variable for all arguments inside the interval $[0, t)$ as a prerequisite to compute it for t. The evolution of the reward variable value over this interval can be a valuable information in addition to the numerical value that needs to be computed. Software tools can graphically show the value of the reward variable over time. The underlying analysis is, however, of the instantaneous type, despite the information about an interval.

Interval-of-time analysis of a reward variable asks for the accumulated reward over a certain time period. The left boundary of the interval is usually zero, because one could otherwise set the initial state of the model to an appropriate value for the different starting time in many cases. The accumulated reward is normally different from zero and does not sum up to a finite value over an infinite time period. Therefore, the normal case for an interval-of-time analysis has a finite right interval boundary. This case (accumulated reward until t) is captured by a parameter setting

$$rint^\star = [0, t]\,; \quad ravg^\star = \text{False}$$

Many software tools allow only a starting time of zero for an interval-of-time analysis. If a different time is needed, one can in this case obviously execute two analyses and subtract the results as

$$rvar^\star_{[lo,hi]} = rvar^\star_{[0,hi]} - rvar^\star_{[0,lo]}$$

A slightly different type of analysis can be done if the average reward over the transient time interval needs to be computed.

This case is called **time-averaged interval-of-time** or just **average reward until** t, and is simply specified as

$$rint^\star = [0, t]; \quad ravg^\star = \text{True}$$

Steady-state analysis derives the mean value of the reward variable after all initial transient behavior is left behind. The usual way to specify (and to compute) this value is to set the following reward variable parameters

$$rint^\star = [0, \infty]; \quad ravg^\star = \text{True}$$

That is, the reward is accumulated from time zero to infinity and is averaged over (divided by) the interval length. It should be noted that the expected time-averaged limiting value is the same as the expected limiting value for the instantaneous reward, if both limits exist. The latter would be described by

$$rint^\star = [\infty, \infty]; \quad ravg^\star = \text{False}$$

2.4.2 Derivation of Reward Variables

After the specification of reward variables as described in the previous section, we now formally describe how the reward variable values are defined. This is done based on the simplified process *SProc*. The definition of *SProc* was uniquely defined by the complete process *CProc*, and we may thus use it instead of the latter. Recall that *SProc* is characterized at time t by the state $\sigma(t)$, and the set of activities $SE(t)$ that are executed.

For the definition we first introduce an intermediate function $R_{inst}^\star(t)$. This value can be interpreted as the instantaneous reward gained at a point in time t. It is a generalized function in that it contains a Dirac impulse Δ if there is at least one impulse reward collected in t.

$$R_{inst}^\star(t) = \underbrace{rrate^\star(\sigma(t))}_{\text{rate rewards}} + \underbrace{\Delta \sum_{se \in SE(t)} rimp^\star(se)}_{\text{impulse rewards}}$$

Depending on the type of reward variables $rvar^\star \in RV^\star$ of an **SDES**, their values can now be derived from the simplified stochastic process by using the individual functions $R_{inst}^\star$. Assume for notational convenience

$$lo = \min(rint^\star), \quad hi = \max(rint^\star),$$

we define the reward variable value

$$
rvar^\star(CProc) = \begin{cases} \displaystyle\lim_{x\to lo^-} R_{inst}^{\;\star}(x) & \text{if } lo = hi < \infty \wedge \neg ravg^\star \\[3mm] \displaystyle\lim_{\substack{x\to lo^-\\y\to hi^+}} \int_x^y R_{inst}^{\;\star}(t)\,dt & \text{if } lo < hi < \infty \wedge \neg ravg^\star \\[3mm] \displaystyle\lim_{\substack{x\to lo^-\\y\to hi^+}} \frac{1}{y-x}\int_x^y R_{inst}^{\;\star}(t)\,dt & \text{if } lo < hi < \infty \wedge ravg^\star \\[3mm] \displaystyle\lim_{\substack{x\to lo^-\\y\to\infty}} \frac{1}{y-x}\int_x^y R_{inst}^{\;\star}(t)\,dt & \text{if } lo < hi = \infty \wedge ravg^\star \end{cases}
$$

$$(2.2)$$

The first case covers *instant-of-time* variables and the second *accumulated interval-of-time* variables. The third and the last one capture variables of the types *averaged interval-of-time* as well as *steady-state*. Averaged interval-of-time measures are obviously undefined for zero length intervals. Please note that the limits are taken from above ($x \to lo^+$) or from below ($x \to lo^-$) accordingly.

An example for the use of infinite values is the steady-state probability of being in a state σ_i, which can be computed as the averaged interval-of-time measure ($ravg^\star = \text{True}$) over time $rint^\star = [0, \infty]$. We use zero impulse rewards and a rate reward that equals one in state σ_i only:

$$
rimp^\star(\cdot) = 0; \quad rrate^\star(\sigma) = \begin{cases} 1 & \text{if } \sigma = \sigma_i \\ 0 & \text{otherwise} \end{cases}
$$

Following the definitions above and after simplifying we reach

$$
P\{SProc \text{ in state } \sigma_i\} = \lim_{hi\to\infty} \frac{1}{hi} \int_0^{hi} rrate^\star\big(\sigma(t)\big)\,dt
$$

$$
= \lim_{n\to\infty} \frac{\sum_{i=0;\ cs(i)=(\sigma_i,\cdot)}^{n} \theta(n)}{\sum_{i=0}^{n} \theta(n)}
$$

where the final equation is based on the complete process *CProc*. Besides that, it shows the basis of how simulation algorithms work for that kind of performance measure. An estimator for the probability to be calculated is the sum of simulation times in which the condition holds divided by the overall simulation time.

This leads to the general question if the limits that have been defined above for infinity exist and which values they take. In fact, if both the limit for instant-of-time and time-averaged interval-of-time exist, they are the same. They differ for instance in a purely deterministic, periodic system, for which an example is given in German [127]). A rigorous treatment of limit theorems for stochastic discrete event systems can be found in Hass [150].

Notes

An integration of different discrete event model classes in one framework is advantageous because it allows the modular extension by new model classes and analysis algorithms. Several software tools and modeling frameworks have been proposed in this context. Following Deavours [82], the different approaches are characterized as follows. *Single formalism, multiple solutions* methods develop different evaluation methods for a single class of models. An appropriate analysis algorithm, e.g., simulation or numerical analysis, can be selected depending on the actual model properties. Integration of several model classes is tackled in *multiple formalism, multiple solutions* approaches. Tools that offer a common user interface to a variety of models and evaluation techniques belong to this class. Please refer to Chap. 12 for a list of related software tools.

A system-theoretical approach to the abstract notation of discrete event systems is followed in the *Discrete Event System Specification* (DEVS [332]). Internal states and the interface to the model environment are thus focused in the description. Coupling and hierarchical composition of models is possible in this framework, which can be used for simulation.

Background on stochastic processes can, e.g., be found in Refs. [71, 143, 170, 305], and a brief overview was given in Sect. 1.4.

Performance evaluation of SDES models is done with the specification and computation of *rewards*. An early reference is [170], in which a reward structure is defined on semi-Markov processes with yield rates and bonuses, which correspond to rate and impulse rewards. Yield rates and bonuses are, however, more complex in that they may depend on the previous state as well as the sojourn time. Possible types of reward variables are characterized in Sanders and Meyer [284]. Reward structures with rate and impulse rewards are defined in a unified way for stochastic activity networks (SANs).

The definition of rewards presented for SDES models is close to the one given in German [130]. Rate and impulse rewards as well as the relations between instantaneous and steady-state measures are considered. Information about the existence of the limits for the result derivation can be found in [130, 150]. Summation of indicator variables over all states is followed for impulse and rate rewards in German [130], which restricts the definition to finite state sets. This is not necessary because the stochastic process is in exactly one state at any time, which is exploited in the definition given here. Another difference is that impulse rewards are not included in instant-of-time measures in this text. The result might otherwise contain a real value and a Dirac impulse.

In an earlier work about stochastic Petri net performance evaluation in the setting of stochastic reward nets [59], rate and impulse rewards are defined on the stochastic process in a mixed way: rate rewards are integrated over time, while impulse rewards are derived by summation. This avoids the use of the Dirac impulse in the definition, but can therefore not be defined straightforward on the stochastic process like it is done here.

Summation for both impulse and rate rewards over indication variables is followed for the definition of reward measures in German [284]. This approach is extended to abstract discrete event system models in Deavours [82] using a reward automaton defined on top of the stochastic process of a model.

3

Stochastic Timed Automata

One of the simplest model classes of stochastic discrete event systems are **stochastic timed automata**, which are explained in this chapter. The **states** and **state transitions** of a discrete event system are explicitly modeled in an automaton. Their level of abstraction is thus identical to the actual set of reachable states and state transitions, which makes them easy to understand and use for simple systems. More complex behavior is however better expressed with one of the model classes that are covered later in this text. The only abstraction of an automaton is in the notion of **events**, which correspond to actions that may happen in several states. One event can thus lead to many state transitions in the full model.

An automaton is said to be **deterministic** or **nondeterministic** depending on whether there are events for which more than one associated state transition starts at one state. Standard automata describe states and state transitions without a notion of time. Delays of actions (interevent times) are associated to events and their associated state transitions in **timed automata**, for which the performance can then be evaluated. In a **stochastic timed automaton**, the interevent times are allowed to be random, and given by a probability distribution function.

Automata are a way of representing *languages*, if we think of the set of events as an alphabet and sequences of events as strings. A string that can be produced by an automaton corresponds to a *trace* of events. However, the relationship between languages and automata is not deepened here, because we are mainly interested in stochastic automata as a way of describing the behavior of a discrete event system by looking at states and events. The interested reader is deferred to the rich body of literature on this topic (e.g. [168]).

Stochastic timed automata as used in this text are introduced informally in the subsequent section. The model class is formally defined and the translation into an **SDES** model is covered in Sects. 3.2 and 3.3. Section 3.4 explains the basic elements of Statecharts in the form that is present in the Unified Modeling Language (UML). They can be seen as an extension of automata model classes and are included here due to their industrial acceptance. The transformation of Statecharts into stochastic Petri nets is the topic of Sect. 3.5. Details are shown for Statechart *states* and *transitions* as well as the interpretation of annotations. An algorithm for the transformation into stochastic Petri nets is explained in Sect. 3.5.3. Finally, some notes are given.

Evaluation methods for SDES models, into which automata are translated, are given in Part II. A small application example of UML Statecharts as an automata variant is presented in Sect. 14.3.2, which is subsequently translated into a stochastic Petri nets with the method explained in Sect. 3.5.

3.1 Informal Introduction

An automaton is an explicit description of system states and events, together with state transitions due to the events. It is a simple model (compared to the other modeling methods considered throughout this work) and very close to the general understanding of discrete event systems on a state/event level. They can be formally defined using sets and relations as it is done in the subsequent Sect. 3.2. A notation by a labeled directed graph, called the **state transition diagram**, is more useful to understand the concept.

Consider for example the automaton shown in Fig. 3.1. The **states** of the system that is modeled with the automaton are the nodes of the graph. Arcs correspond to **state transitions**, indexState transition that are labeled by names of **events** . The behavior that is described by the automaton is the following. The automaton is always in one of the states shown in the graph. The **initial state** of the automaton (x in the example) is denoted by an arc without source state. The events that are inscribed at any one of the outgoing arcs of that current state are called **active events**.

It should be noted that because not all events are active in every state, there are events that are not allowed to occur in some states, and there is

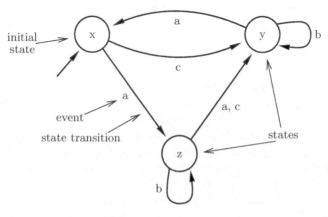

Fig. 3.1. Example of an automaton

no corresponding state transition for them (see e.g. event c in state y). This means that the state transition function, which maps states and events to destination states, does not need to be completely defined for its domain. In some cases an event does not change the state, which is denoted by a **self-loop** (event b in the example). In other cases, two or more events may lead from one source state to the same destination state (like events a and c in state z in the example).

If there are several events active in one state, the model needs to specify which one occurs. For any number of outgoing events, the holding time for each state needs to be defined as well. This is done as follows. Whenever an event becomes active in a state and was not active in the previously visited one, a **clock value** is set for it. This value is randomly selected from the **interevent time** (a probability distribution function) that is defined for each event. The clock values of all active events are decreased with model time until one of them reaches zero. The corresponding event, the fastest one, is the one that occurs and leads to a state transition. If there is more than one event for which the clock value reaches zero, a probabilistic choice with equal probabilities is made to select the event that occurs first. If an event was active in a state and is still so in a consecutive one, its clock value is not changed by the state change of the automaton.

It should be noted that our understanding of Automata is in the following based on stochastic timed extension of *deterministic* automata, i.e., with only one possible state change for an event that happens in a state. This choice is merely done for simplicity of explanation. A nondeterministic automaton for example can have different destination states for one enabled event in a state, which requires a probability distribution to be defined over them. It is possible to capture stochastic timed nondeterministic automata as an SDES by adding intermediate states and subsequent state transitions for the nondeterministic choices that can take place after the execution of an event.

If the set of states of an automaton is finite, it is called a **finite-state automaton**. If the automaton is taken as a model of a discrete event system, and to be stored and analyzed by a computer program as the intention is throughout this work, we require the sets of states to be finite. The basis for our definitions are thus **stochastic timed finite-state deterministic automata**.

The modeler finally needs to express quantitative measures that he is interested in. Typical examples include the probability of being in one of the automaton states or the number of times that one event occurs. All these measures may then be evaluated for a specific point in time after the initial state, in steady state or in other settings as it has been explained in Sect. 2.4.

We adopt a syntax for the specification of performance variables here that is similar to the ones used for Petri nets in later chapters. The modeler may

create a performance measure as a usual arithmetic expression containing the
following model-relevant elementary types:

1. P{<state>, <state>, ... } corresponds to the probability of being in
   one out of a list of specific states of the automaton, which are identified
   by their names. Example for Fig. 3.1: P{z, y}.
2. T{<event>, <event>, ... } computes the number of occurrences for
   all listed events.[1] Example: T{c}

Usage of the simple performance measure types should be quite clear. State-
related issues like the probability of a resource being in a certain state is
simply expressed as a P-type measure with all corresponding automata states
in the list. Simple measures that are related to events like throughput can be
solved in a similar way.

3.2 Model Class Definition

A **Stochastic Automaton** is defined as a tuple

$$A = (\mathcal{X}, \mathcal{E}, f, \Gamma, x_0, G, RV)$$

with the following elements.

The **state space** is a set denoted by $\mathcal{X}$, containing the individual states
$x \in \mathcal{X}$ of the automaton. The actual elements of the state space set are not
important, we just interpret every one of them as one state of the specified
automaton, like an identifier.

$\mathcal{E}$, the **event set**, is the finite set of **events** $e \in \mathcal{E}$ of the automaton. We
require it to be finite in order to apply algorithms that iterate over the set of
events later.

The events that are active (feasible, enabled) in a state are given by the
active event function, which is denoted by Γ. It returns the set of active
events for a state.

$$\Gamma : \mathcal{X} \rightarrow 2^{\mathcal{E}}$$

We say that an event e is active in a state x iff $e \in \Gamma(x)$.

Active events may lead to a change of state. This is described by the **state
transition function** f, which returns the destination state for a source state
and an event. The function does not need to be defined or the result is ignored
for state/event pairs where the event is not active in the state.

$$f : \mathcal{X} \times \mathcal{E} \rightarrow \mathcal{X}$$

x_0 specifies the **initial state** of the automaton.

$$x_0 \in \mathcal{X}$$

---

[1] The notation is chosen for a compact description of performance measures; obviously, P{z, y} = P{z} + P{y} and T{b, c} = T{b} + T{c}.

For every event e that becomes active when a new state is entered, the corresponding clock value needs to be randomly initialized. This is done according to a distribution function, which is specified by the **interevent times**[2] G.

$$G : \mathcal{E} \rightarrow \mathcal{F}^+$$

The set of distribution functions $\{G(e) \mid e \in \mathcal{E}\}$ is sometimes referred to as a **stochastic clock structure** or **timing structure**.

RV specifies the set of **reward variables** of the stochastic automaton. Two elementary types have been proposed in the informal introduction above, namely the probability of being in any one of the states $\{x_i, ..., x_j\} \subseteq \mathcal{X}$ (syntax: $\mathrm{P}\{x_i, ..., x_j\}$) and the number of state transitions due to a set of individual events $\{e_i, ..., e_j\} \subseteq \mathcal{E}$ (syntax: $\mathrm{T}\{e_i, ..., e_j\}$).

A flag *rtype* denotes the corresponding type of each reward variable. Obviously the first type corresponds to a rate reward, while the latter leads to an impulse reward in the SDES framework. The parameter *rexpr* either specifies the state set for the first case or the event set for which the number of occurrences should be computed in the second case:

$$\forall rvar \in RV : rvar = (rtype, rexpr)$$
$$\text{with} \quad rtype \in \mathbb{B}$$
$$\text{and} \quad \begin{cases} rexpr \subseteq \mathcal{X} & \text{if } rtype = \text{True (P-case)} \\ rexpr \subseteq \mathcal{E} & \text{if } rtype = \text{False (T-case)} \end{cases}$$

3.3 Automata as SDES

The SDES definition comprises state variables, actions, sorts, and reward variables

$$SDES = (SV^\star, A^\star, S^\star, RV^\star)$$

which are set as follows to capture a stochastic automaton as defined in the previous section.

There is exactly one state variable sv, whose value is the state of the automaton. The set of state variables $SV^\star$ has therefore only one element.

$$SV^\star = \{sv\}$$

The set of SDES actions $A^\star$ is given by the events of the automaton.

$$A^\star = \mathcal{E}$$

The sort function of the SDES maps the state variable to the allowed values, i.e., the state space of the automaton. As there are no SDES action variables necessary, no sort is defined for them.

$$S^\star(sv) = \mathcal{X}$$

---

[2] See Sect. 1.4 for the definition of $\mathcal{F}^+$.

The set of all possible states Σ is obviously

$$\Sigma = \mathcal{X}$$

The condition function is always true, because all states in $\mathcal{X}$ are allowed.

$$Cond^{\star}(\cdot, \cdot) = \text{True}$$

The initial value of the state variable is directly given by the initial state of the automaton.

$$Val_0{}^{\star}(sv) = x_0$$

Actions of the SDES correspond to events of the stochastic automaton. There is no explicit priority given for them, we thus select w.l.o.g. the value one.

$$\forall a \in A^{\star} : Pri^{\star}(a) = 1$$

Events of an automaton are either active or not, there is no enabling degree other than the standard value of one.

$$\forall a \in A^{\star} : Deg^{\star}(a) = 1$$

Thus the individual enabling degree of any action variant can be set to one in every state as well.

$$\forall \sigma \in \Sigma : VDeg^{\star}(\cdot, \sigma) = 1$$

An action of an SDES for a stochastic automaton is completely described by the attributes of a single event. Action modes or variables are not necessary.

$$\forall a \in A^{\star} : Vars^{\star}(a) = \emptyset$$

Thus there is exactly one action mode per event, $|Modes^{\star}(a)| = 1$, and we can omit the action mode and variant in the following.

An event of an automaton is enabled in a state if it is contained in the active event set for that state.

$$\forall a \in A^{\star}, \forall \sigma \in \Sigma : Ena^{\star}(a, \sigma) = \begin{cases} \text{True} & \text{if } a \in \Gamma(\sigma) \\ \text{False} & \text{otherwise} \end{cases}$$

The delay of an SDES action is distributed as specified by the interevent time for the corresponding event.

$$\forall a \in A^{\star} : Delay^{\star}(a) = G(a)$$

No weights of events influencing their execution probabilities are defined for a stochastic automaton. We thus assume an equal weight of 1 to resolve cases in which two events are scheduled for execution at the same time.

$$\forall a \in A^{\star} : Weight^{\star}(a) = 1$$

If an enabled action is executed, i.e., an active event happens, the corresponding state change is directly given by the state transition function of the automaton.

$$\forall a \in A^\star, \forall \sigma \in \Sigma : Exec^\star(a,\sigma) = f(\sigma,a)$$

The set of automata reward variables RV can be converted into the set of SDES reward variables with the following simple rules. The two elements $rint^\star$ and $ravg^\star$ of an SDES reward variable specify the interval of interest and whether the result should be averaged. They correspond to the type of results the modeler is interested in and are not directly related to the model. Hence they are set according to the type of analysis (algorithm) and not considered here.

One SDES reward variable is constructed for each automaton reward variable such that

$$RV^\star = RV$$

and

$$\forall rvar \in RV^\star : \begin{cases} \begin{rcases} rrate^\star(x_i) = \begin{cases} 1 & \text{if } x_i \in \mathcal{X}' \\ 0 & \text{otherwise} \end{cases} \\ rimp^\star = 0 \end{rcases} & \text{if } rvar = (\text{True}, \mathcal{X}') \\[4ex] \begin{rcases} rrate^\star = 0, \\ rimp^\star(e_i) = \begin{cases} 1 & \text{if } e_i \in \mathcal{E}' \\ 0 & \text{otherwise} \end{cases} \end{rcases} & \text{if } rvar = (\text{False}, \mathcal{E}') \end{cases}$$

Thus a rate or impulse reward of one is earned either per time unit as long as the model spends time in a state $x \in \mathcal{X}'$ in the first case, or whenever one of the specified events $e_i \in \mathcal{E}'$ occurs for the latter.

3.4 UML Statecharts

The UML is a collection of semiformal modeling languages for specifying, visualizing, constructing, and documenting models of discrete event systems and of software systems. It provides various diagram types allowing the description of different system viewpoints. Static and behavioral aspects, interactions among system components, and implementation details are captured. UML is very flexible and customizable because of its extension mechanism with so-called *profiles*. A profile for a special application domain maps aspects from the domain to elements of the UML meta model. The *UML Profile for Schedulability, Performance, and Time* [255] is an example. The term **Real-Time UML** (RT UML) is used in the following to denote the UML 2.0 [256] in combination with the mentioned profile.

UML defines 13 types of diagrams, which can be divided into structural and behavioral ones. Structural diagrams (like class diagrams) are used to model the logical and architectural structure of the system. Behavioral diagrams (like sequence charts) describe system dynamics and thus include or may be enhanced by timing information. The latter type is thus important when dealing with quantitative modeling and analysis of systems.

Among the behavioral diagrams we consider UML Statecharts (UML-SC) as an appropriate basis for modeling stochastic discrete event systems and their behavior. Other behavioral diagram types such as sequence charts are not as easily usable: they originally describe only one trace (sequence) of the behavior, and are thus better suited for the specification of usage and test cases. Collaboration diagrams as another example focus on the interactions and not the states.

UML-SC follow the idea of states and state transitions like an automata model. They can thus be used for specifying possible sequences of states that an individual entity may proceed through its lifetime. This type is called *Behavioral Statechart* [256], which is a variant of statecharts as defined by Harel [153, 154]. Statecharts in their various forms are widely accepted in the industry. The software tool StateMate [153, 155] has set a de-facto standard and is used, e.g., for the model-based verification of signaling systems [77].

Model elements include *states*, different *pseudostates*, and *transitions*. A simple UML-SC example with basic elements is sketched in Fig. 3.2. A **state** (A and B in the figure) models a situation during which some condition holds. When a state is entered as a result of a transition, it becomes *active*. It becomes *inactive* if it is exited as a result of a transition. Every state may optionally have one of each so-called *entry*, *exit*, and *do* actions or activities, like for example state A in Fig. 3.2. Whenever a state is entered, it executes its *entry* action before any other action is executed. A *do* activity occurs while the UML-SC is in the corresponding state, and might be interrupted if the state is left. The *exit* action is executed before the state is left due to an outgoing transition.

A **transition** causes the exit of a source state and leads to a target state. In Fig. 3.2 a transition leads from state A to state B. Transitions may be inscribed with optional **event**, **guard expression**, and an **action list**. An event

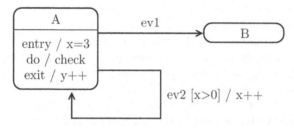

Fig. 3.2. Basic UML Statechart elements

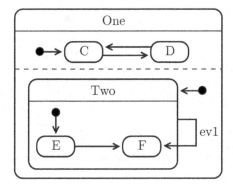

Fig. 3.3. Composite state example

triggers the transition, but it may only be taken if the guard evaluates to true. The transition from A to B in the figure is triggered by event ev1. **Events** may be of the types *Signal* (asynchronous communication), *Call* (synchronous communication), relative or absolute *Time* (for deadline and others), or simply a *Change* of a variable value. A guard is a boolean expression that may depend on variables. The *self-transition* from state A to itself in Fig. 3.2 is taken when event ev2 occurs, provided that the guard x>0 is true. If there is an action list attached to a transition like x++ in the example, it is executed while the transition takes place.

A state of a UML-SC may either be a simple automaton state (i.e., a *simple state* like A in Fig. 3.2) or contain internal states and transitions. In that case it is called **composite state**; Fig. 3.3 gives an example.[3] The contents of the states thus allow hierarchical modeling, hiding the inner behavior. A state of a composite state is called its **substate**, E is for instance a substate of Two.

Only one state is active at any time in a classic automaton, which makes the description of concurrent activities cumbersome. State Two is either in substate E or F if it is active itself. UML-SC have the ability to model different activities within one object with *orthogonal states*.[4] The concurrent state machines inside such a state are called **regions** and divided by a dashed line. State One in Fig. 3.2 contains two regions, the lower one is again hierarchically refined. The complete state of such an orthogonal state is a subset of the cross product of the states contained in the regions. A state may also contain only one region with its states and transitions,[5] state One in Fig. 3.2 is an example.

In a hierarchical UML-SC model, all composite states that directly or indirectly contain an active simple state are active as well. The complete state of such a model is thus represented by a tree of states, and the tree structure is given by the substate and composition relations. Complex nested UML-SC

---

[3] The black dots depict initial states and are explained later.
[4] Called AND-states in [93].
[5] It is then called nonorthogonal [256] or OR-state [93].

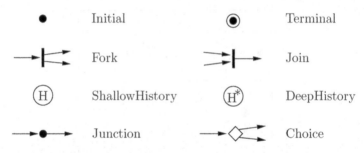

Fig. 3.4. Graphical notation of pseudostates

models may be visually simplified by hiding detailed substates in *submachines*. This is however only a drawing alternative, and has no actual semantics. Transitions that cross the border of such a state could not be differentiated from the ones that start from the enclosing state. Entry and exit points (see below) are used to visualize this difference.

A **pseudostate** is a node in the state machine graph, which is transient because it is left immediately. They can be used to express decisions or paths that accumulate state transitions. The following pseudostates are defined in a UML-SC [93, 256] and depicted in Fig. 3.4:

Initial	This pseudostate belongs to a composite state, and has one state transition (the *default transition*) that leads to the default state of the composite. It is depicted by a small filled circle.
Terminal state	A transition to it represents the end of activity in the region. A composite state is however only completed when all its regions have reached their final states.
Fork	These pseudostates are used to split transition paths into several transitions that lead to individual regions. In other words, all outgoing transitions are taken concurrently, and lead to individual initial states in the concurrent regions.
Join	Have the opposite function, and merge transitions coming from different regions.
ShallowHistory	This one is shown as a letter H in a circle, and a transition to it from the surrounding state means that the initial state equals the most recent local substate that has been active before the composite state was exited. It thus stores the last state over periods in which the state itself is not active.
DeepHistory	Similar to ShallowHistory, but restores the whole composite state with all of its contained substates.
Junction	Only serve to connect subsequent transitions inside one region to a path. They can however be used to merge and split transition paths as well, with individual guards on outgo-

ing transitions in the latter case. Their connected transitions belong to the same region in difference to a Join or Fork.

Choice   This type of vertex is used to implement dynamic conditional branches. The guards of their outgoing transitions may depend on results that have been completed during the same path execution. Junction states differ from this behavior, because their path transitions are all considered in one execution step.

EntryPoint   Entry points of a state machine, which can be accessed from outside.

ExitPoint   Leaves the surrounding composite state whenever it is reached in any region. Entry and exit points symbolize state-border crossing transitions of a submachine, and therefore do not add to the behavioral semantics and are omitted in Fig. 3.4. Both serve as interfaces of a submachine, and are depicted as circles on the border of the submachine; ExitPoints with a cross.

Transition execution happens after the exit action of the previous state has been finished, and before the entry action of the destination state starts. In the case of nested states, exit actions are executed bottom-up (starting from inside), and vice versa for entry actions. Complex transitions can be modeled with a sequence of transitions and pseudostates. The simplest case is a sequence with Junctions to connect them. Junctions may have several incoming and outgoing transitions. In the latter case, all outgoing transitions must be guarded such that at most one of them can be taken. The transition path is not executed as long as there is no guard fulfilled. All guards of a transition path are evaluated before any of the corresponding actions is performed. Choice points can be used if this is not the intended behavior: the actions of transitions on the path before the choice point are executed before the guard is evaluated. It is thus required that one of the guards of the outgoing transitions of a choice point must be true, because otherwise a dead end is reached. More details on transition execution semantics can be found in [93, 256].

The UML Profile for Schedulability, Performance, and Time enables advanced annotation of timing and performance information within the behavioral UML diagrams. Profiles are used to specialize UML for a restricted area of application. The SPT profile is intended to standardize the expression of issues related to timing and performance within a UML model. It provides a set of *stereotypes* and *tagged values* specializing UML without violating its existing semantics. In Fig. 3.5, such items are attached to the corresponding model items by dashed lines.[6]

---

[6] The enclosing apostrophes of values are omitted in this figure as well as all following UML-SC material, in order to save space and improve readability. An example would otherwise read RTduration=('exponential','36')

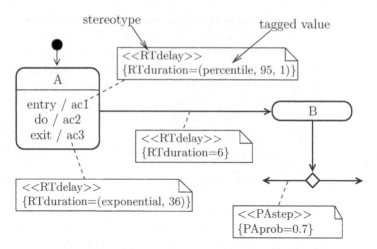

Fig. 3.5. UML Statechart with SPT inscriptions

Table 3.1. Example stereotypes and tagged values

Stereotype	Tagged value	Meaning
RTdelay	RTduration = 6	Fixed duration of 6
RTdelay	RTduration = (exponential, 36)	Exponentially distributed delay with mean 36
RTdelay	RTduration = (percentile, 95,1)	Delay that is distributed such that the random variable value is smaller than 1 with probability 95%
PAstep	PAprob = 0.7	The probability of the corresponding choice is 70%

Stereotypes indicate the type of a model element. Tagged values contain a property name and a value, and thus assign information about the specific instance of a stereotype. <<RTdelay>> is an example for a *stereotype*. It is mapped to an *activity* or a state transition. Its *tagged value* RTduration specifies the duration of the activity or transition. A detailed description of the numerous stereotypes and associated tagged values is outside the scope of this work; the reader is referred to [93, 255].

Table 3.1 lists some stereotypes and tagged values that can be used for the specification of time and probability values. We assume model time to be unit-less here, although one may also use time units in the specification. From the later examples it becomes obvious that the existing stereotypes are not yet perfect for a natural description of performance-related issues. They have been used here because they are available in the current UML definitions;

however, it is expected that improvements of the description will be proposed and adopted in the future.

A SDES interpretation of Statecharts is intentionally left here, although it should be possible in principle based on a clear definition of the Statechart semantics [155]. The basic behavior would be very similar to the one of automata due to their equivalences. Pseudostates would lead to intermediate vanishing states. Orthogonal composite states are no problem because actions in SDES models are concurrent, and the SDES state may be composed of different variable states. However, through the translation of UML-SC models into stochastic Petri nets as shown in Sect. 3.5, a performance evaluation is indirectly possible without an explicit SDES interpretation.

3.5 Transformation of Statecharts into Stochastic Petri Nets

The UML [256] including its profiles has gained increasing acceptance as a specification language for modeling real-time systems [93]. For this application area it is especially important to enable quantitative predictions in an early design stage, because timeliness and dependability must be ensured. However, UML models are mainly intended for a structural and functional description of systems and software. Hence the models are not directly analyzable in the sense of a quantitative performance evaluation. An approach to transform one model class of UML into stochastic Petri nets for their later evaluation is therefore proposed in the following.

We consider the UML in combination with its *Profile for Schedulability, Performance, and Time* (SPT, [255]), which has been introduced as a specification language for the design of real-time systems recently. The derivation of quantitative measures from these models is an open research issue. Two main strategies exist to retrieve performance measures from UML models. The direct way requires the development and application of an analysis method that operates directly on the UML specification. The second, indirect way consists of mapping the UML specification to an established performance model such as a stochastic Petri net or a queuing network model. Quantitative measures can then be obtained by applying existing analysis methods and tools for the chosen performance model. We consider the indirect way in the following, because in this case a reuse of established knowledge for the analysis of the model is possible. Furthermore there are numerous software tools available that support the quantitative analysis of the resulting models (see Chap. 12).

Both strategies have in common that quantitative system aspects such as frequency, delay, or service execution time have to be specified in the UML model, as well as performance measures. The mapping of UML into a performance model requires rules that specify how certain UML fragments have to be interpreted in the performance model context. In the resulting performance model, the semantics of the model have to be preserved. It is required that the timing behavior from the UML is transfered equivalently. Finally, the results of the performance evaluation need to be interpreted in terms of the original UML model.

A requirement for the modeling of real-time systems is to specify deterministic and even more general timing behavior in addition to the analytically simple exponential case. Our goal is to translate UML-SC annotated following the SPT profile into stochastic Petri nets with nonexponentially distributed firing times.

Only a reduced set of UML-SC elements is considered for now [308]. States, transitions, and Choice pseudostates were selected, allowing to model and evaluate a wide range of quantitative issues including the case study presented in Sect. 14.3.2. The translation of other UML-SC elements like composite states and the remaining pseudostates are the subject of current work [307].

3.5.1 States

Figure 3.6 depicts a simple UML Statechart example with states A and B as well as a transition and some annotations.

Time may be consumed within each state during the execution of the optional *entry*, *do*, and *exit* activities (ac1 ... ac3 in the example). Elapsing time is modeled in a Petri net by a token in a place with a subsequent timed transition. A corresponding example of how the simple example is translated is shown in the lower part of Fig. 3.6. The exponential transition Tenter_A represents the *entry* activity ac1 with an exponential distribution with parameter $\lambda = 1/36$ such that the mean delay is 36. This is due to the

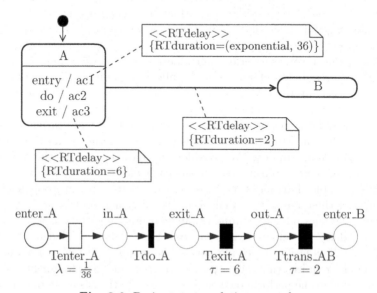

Fig. 3.6. Basic state translation example

RTdelay specification of RTduration = (exponential, 36). The resulting
transition connects place enter_A (entering state A) to place in_A (in state A).

The *do* activity ac2 has no associated time specification and is therefore
assumed to require no time. It is represented by the immediate transition
Tdo_A. The deterministic PN-transition Texit_A represents the *exit* activity
ac3 with the constant delay of 6 time units. The place out_A in the SPN represents the point when the UML-SC state A has been left, after the *exit* activity
is finished. The destination state B has not been entered at this point. This
happens when the deterministic transition Ttrans_AB fires, which represents
the state change between A and B. The timing annotations are used for the
Petri net transition times.

Regardless if there are optional activities specified for a state, we always
follow the logical and temporal order of the optional activities. An immediate
transition is created in the Petri net translation for each activity that does
not exist or for which no time is specified.

Thus the translation of a state X always results in a Petri net fragment
containing places and transitions in the following order: place enter_X →
transition Tenter_X (*entry* activity) → place in$X$ → transition Tdo_X (*do* activity) → place exit_X → transition Texit_X (*exit* activity) → place out_X.
From the exit place there are appropriate transitions for every possible subsequent UML-SC state.

3.5.2 Transitions

The transitions in an UML-SC may consume time and are translated into corresponding Petri net transitions. The naming convention is as follows: For a
UML-SC transition from state A to state B the resulting Petri net transition
is named Ttrans_A_B. It connects the places out_A and ent_B (see Fig. 3.6).
UML-SC transitions without any timing annotation are considered to not consume any time. They are translated into immediate transitions in the Petri
net model. The SPT profile provides the <<RTdelay>> stereotype with its
tag RTduration [255] to express fixed delays. The tag is of type RTtimeValue.
A UML-SC transition with a constant delay such as the one connecting states
B and C in Fig. 3.7 is mapped to a deterministic Petri net transition with the
given delay.

UML-SC transitions may have an exponentially distributed delay like the
one connecting states A and B in Fig. 3.7b. We translate such a UML-SC transition into an exponential transition in the Petri net as shown. The parameter
of the exponential distribution, the firing rate λ, is set such that the mean
delay equals the given time, $\lambda = 1/$specified delay.

Quantiles are another important way to express incomplete knowledge of a
delay distribution. We propose an extension of the RTtimeValue syntax similar to the PAperfValue. A percentile construct is introduced (percentile,
<percentage>, <time value>). This enables for example the specification
of a UML-SC transition with a delay of 5 time units in at most 75% of all

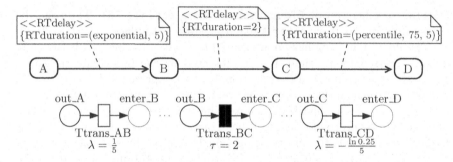

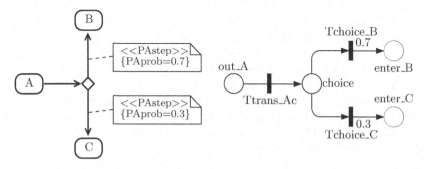

Fig. 3.7. Translation of Statechart transitions

Fig. 3.8. Choice pseudostate and its translation

cases: `RTduration = (percentile, 75, 5)` as in the rightmost example in Fig. 3.7. We assume that in these cases the delay is exponentially distributed. It is then possible to calculate the parameter λ of the resulting exponential transitions in the stochastic Petri net for a percentage p via its distribution function: $F(x) = 1 - e^{-\lambda x} = P\{X \leq p\}$ and thus $\lambda = -\ln(1-p)/x$. For the example in Fig. 3.7, this results in an exponential transition `Ttrans_CD` with rate $\lambda = -\ln 0.25/5 \approx 0.2773$.

Probabilistic Choice

In addition to states, transition and time annotations, another important element of stochastic discrete event systems are nondeterministic choices. The *choice* pseudostate serves this function in UML-SCs. The left side of Fig. 3.8 shows an example: after state A is left, there is a choice between transitions to states B or C. Choices are depicted as diamonds.

For the performance evaluation of a system, probabilities must be attached to the different outcomes of a nondeterministic choice. We follow a proposal to consider a UML-SC transition as a special kind of a `<<PAstep>>` of the SPT [239]. The `PAprob` tag is used to express the relative probability of each path, just like the firing weights of immediate transitions in a Petri net (into which they are in fact translated). It should be noted that in UML-SCs a guard and actions may be attached to the transitions connected to a choice pseudostate. Guards may disable paths that can not be taken, and could thus be translated into Petri net transition guards.

Figure 3.8 shows the proposed translation of the choice pseudostate. The exit action of state A leads to the pseudostate. In the shown case there is no annotation for the exit action, thus an immediate transition Ttrans_Ac is created. The c denotes the choice pseudostate here. A token in place choice marks the pseudostate itself. Such a UML-SC pseudostate naturally maps to a vanishing state in the resulting Petri net. From the choice place there is one immediate transition for every possible choice in the UML-SC model, namely

---

TRANSLATESTATECHART (UML-SC)

Input: Statechart model UML-SC
Output: stochastic Petri net model SPN

(∗ Translate UML-SC states ∗)
for $\forall s \in States$(UML-SC) **do**
      create places enter_s, in_s, exit_s and out_s
      create transition Tenter_s connecting enter_s and in_s
      set $delay($Tenter_s$) =$ RTduration(s_{entry})
      create transition Tdo_s connecting in_s and exit_s
      set $delay($Tdo_s$) =$ RTduration(s_{do})
      create transition Texit_s connecting exit_s and out_s
      set $delay($Texit_s$) =$ RTduration(s_{exit})

(∗ Translate UML-SC choice pseudostates ∗)
for $\forall s \in$ Choices(UML-SC) **do**
      create place choice_s
      for $\forall t \in$ Outgoing transitions from s **do**
            let d denote the destination state of transition t
            create transition choice_d between choice_s and enter_d
            set $weight($choice_d$) =$ PAprob(t)

(∗ Translate UML-SC transitions ∗)
for $\forall t \in$ Transitions(UML-SC) **do**
      let s and d denote the source and destination states of t
      if $d \in States$(UML-SC) **then**
            create transition Ttrans_sd connecting out_s and enter_d
            set $delay($Ttrans_sd $=$ RTduration(t)
      else create transition Ttrans_sc connecting out_s and choice_d
            set $delay($Ttrans_sc $=$ RTduration(t)

return SPN

Algorithm 3.1: Translation of a UML Statechart into a stochastic Petri net

Tchoice_B and Tchoice_C for the example. The weights of the resulting conflicting immediate transitions are simply set according to the probabilities that have been specified with the PAprob tags.

3.5.3 A Transformation Algorithm

An algorithmic formulation of the simple translation rules from a UML-SC UML-SC into a stochastic Petri net model SPN that have been informally explained above is presented.

The algorithm depicted in Fig. 3.1 lists the necessary steps of the translation. All basic UML-SC elements are translated one by one and connected as necessary. The resulting stochastic Petri net contains transitions for all optional actions and state transitions in the UML-SC. Because many of them are empty or not specified in detail using a performance inscription, a big part of the resulting transitions are immediate and do not contribute to the model behavior. These sequences of immediate transitions are not wrong, but unnecessary and clutter the model (see Sect. 14.3.2 for an example).

A trivial simplification based on a check of the model structure can remove these elements: For each pair of places p_i and p_k, which are connected by one immediate transition t such that t is the only transition in the post- or preset ($p_i^\bullet = {}^\bullet p_k = t$), the transition t is deleted and the two places p_i and p_k are merged.

A prototype of the translation algorithm has been implemented in the software tool TimeNET, which is covered in Sect. 12.1. An Open Source modeling environment is used for the creation of UML-SC models. The translation algorithm generates a Petri net model in the TimeNET format.

Notes

Sections 3.4 and 3.5 are based on joint work for which first results have been published in [307–309].

Introductions and overviews of automata are given in [42, 168], while automata extended by time are covered in [8, 9, 42]. Background on stochastic automata and their underlying stochastic process is contained in [42, 143]. Their numerical analysis is, e.g., covered by [78, 298]. Stochastic automata models of parallel systems are analyzed efficiently using a Kronecker algebra approach in [267].

Further literature about Statecharts include [153–155]; the UML [256] contains Statecharts as one of the diagram types. Their use for real-time system modeling is covered in [93].

There are several approaches aiming at a quantitative analysis of annotated UML diagrams, mainly in the area of software performance evaluation. Merseguer et al. present a systematic and compositional approach [26, 239, 240]. This evaluation process includes the translation of extended UML diagrams into labeled generalized stochastic Petri net (GSPN) modules and finally the composition of the modules into a single model representing the whole system behavior [232]. Only exponentially distributed times are taken into account and the resulting Petri net is thus a labeled GSPN. King and Pooley [204,205,268] are also working on the integration of performance evaluation into the software design process based on UML. Again GSPNs are used for the performance evaluation. An intuitive way of mapping UML-SC into GSPNs is introduced. A state in the UML-SC is represented as a place in the GSPN and state transitions in the UML-SC are represented as GSPN transitions. The resulting GSPNs are composed based on UML collaboration diagrams. Hopkins et al. propose the introduction of probabilistic choice and stochastic delay into UML [169] in this context.

Lindemann et al. presented an approach for the direct generation of a generalized semi-Markov process (GSMP) from an annotated UML state diagram or activity diagram in [225]. The diagrams are enhanced by specifying deterministic and stochastic delays. No intermediate model is used. The verification of timed UML Statecharts against time-annotated UML collaboration diagrams is proposed in [208], based on a translation of UML Statecharts into timed automata. UML statecharts are extended by discrete-time distributions to enable probabilistic model checking in [181]. Markov decision processes are used as the basis, but timing information is not explicitly used.

4

Queuing Models

This chapter deals with another classic model type of SDES, namely queuing models. Their underlying idea comes from the everyday experience of requiring some sort of service, which is restricted in a way that not all concurrent requests can be answered at the same time. We are forced to wait for an available teller at a bank, a seat in a cafeteria, or when crossing the street. When it is our turn, we block the service for others that are waiting and release the server for the subsequent customer afterwards. The same sort of behavior is frequently found in technical systems, in which we are mainly interested in this text. Communication packets queue at a switch, data base access requests are performed one at a time, or parts pile up in front of a bottleneck machine.

The main reason for not having enough servers for every possible request is that installation and maintenance of servers cost money, and it is therefore not economically reasonable to have too many of them. On the other hand, customers often request service in an unknown and stochastic pattern, which means that an infinite number of servers would be required in theory to fulfill all requests in parallel.

Common to all of the mentioned examples are the basic elements of a queuing system: customers arrive, wait in a queue, are being served, and finally leave the system. The queue length is restricted in practice, and there might be several servers providing service concurrently. A **queuing system** (QS) is a model in which only one combination of a queue and corresponding server(s) are of interest, together with their customer arrivals and departures. A generalization of this simple model may contain several individual QS, and is called a **queuing network** (QN). In the latter case obviously some additional issues are of importance, for instance the queue that a customer enters after leaving another one or whether he leaves the model altogether.

Queuing theory covers models of this type and has been very successful in creating formulas and algorithms for the computation of performance measures from them. Typical questions are the number of waiting customers, their waiting and service time, throughput and utilization of servers, or the probability for a customer to be blocked or lost because of a full queue.

In comparison to the level of abstraction of other discrete event models covered in this text, queuing models can be seen between automata and Petri nets. The description is obviously on a much higher level w.r.t. automata because the reachability graph of a small QS (and thus a behaviorally equivalent automaton) might already be quite complex. Petri nets are a more powerful description technique because QN models can in principle be transformed into one of them,[1] but not vice versa. The biggest difference is that synchronization between customers can be described by Petri nets, but not with a classic queuing model.

The chapter is organized as follows: simple queuing systems as well as networks of queues are informally introduced in the section below. A more formal definition of standard QN is given in Sect. 4.2, and their mapping to an equivalent SDES is shown in Sect. 4.3. Final notes discuss more sophisticated QN and point to literature relevant to the topic.

4.1 Informal Introduction

Figure 4.1 shows an example of a simple QS. The way of drawing a queuing system is not formally defined, but the one used in the figure is a very popular one. Each customer arrives from the left in the figure and is stored in the queue, if there is an available slot. Customers may of course model any entity of interest in a queuing model, such as a vehicle at a gas station, a passenger waiting at a ticket counter, a document in an office workflow, or a communication packet of a FTP data transfer.

After some waiting time the customer receives service from a server. This requires some time as well. The customer leaves the model after the service is finished. Depending on the type of queue, its waiting slots, and the number of servers in the model, the queue may contain empty slots, waiting customers, and the ones currently receiving service.

The following details need to be specified for a complete description of a queuing system QS:

- The **arrival pattern** of customers describes the times at which new customers enter the model. The individual arrival times are usually unknown, which leads to a stochastic model for this item. The interarrival time between two successive customer arrivals is thus given by a probability distribution function. Many simple models assume that the stochastic arrival behavior does not change over time, i.e., it is a time-independent stationary process.
- A restriction on the number of waiting slots is given by the **system capacity** or storage capacity. It should be noted that customers being served

---

[1] There are, however, extensions of queuing networks with quite complex behavior, for which this might only be possible with a colored Petri net. Some of the extensions are mentioned in the final notes of the chapter.

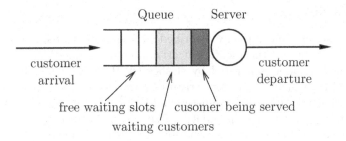

Fig. 4.1. A queuing system

also occupy one waiting slot, although they are not actually waiting. The amount of waiting room may also be unlimited.

– The selection of the next customer to be served if there is an idle server is given by the **queue discipline**. Fair queuing for humans usually means that the one who came first will be selected (first come, first serve; FCFS or FIFO). Other patterns are obviously possible, and make sense in many technical systems. Random order or last come, first served (LCFS or LIFO) are two of them. In addition to that, certain customers may have a higher priority and are thus served before the other ones.

– There is only one server working inside one queuing system as the simplest case; any natural **number of servers** is however possible. There are even cases in which an infinite number of servers is assumed to be available; in practice this means that every single customer receives service in parallel, and that there is no waiting at all.

– The service itself is a process that is most importantly described by the time that it takes for one customer. The **service pattern** is similar to the arrival pattern in that it is not necessarily known a priori, and therefore described by a probability distribution function.

– An additional issue about the arrival of new customers is the question whether customers leave the model completely and arrive at the queue from an unrestricted population. This case is called an **open queuing system** as opposed to a **closed** one, where customers come from a pool of fixed size. In that case customers may obviously only arrive at the queue if they are not all waiting or receiving service, and the **population size** needs to be known.

Following Cassandras and Lafortune [42], arrival and service processes are *stochastic model* parts of the QS, number of servers, and waiting slots *structural parameters*, while the details of customer acceptance, selection, and service differentiation are summarized as *operating policies*.

In a Markovian queuing model, stochastic patterns are time-independent and described by exponential distributions. The interarrival rate is denoted by λ, and the service rate (or departure rate) by μ.

Kendall's Notation

Queuing systems are commonly described with an abbreviated notation. The individual settings of the mentioned attributes for a queue are given in a standardized manner, separated by slashes (A/B/m/K/N/Q). The type of interarrival-time distribution for arrival (A) and service (B) are denoted by M for Markovian, D for deterministic, and G for general. Number of servers m, system capacity K, and population size N are specified by a number or ∞ if they are unlimited. An abbreviation such as FCFS denotes the queuing discipline Q.

Default values (M for distributions, unlimited system capacity, one server, and FCFS discipline) may be omitted, and even the separating slashes are only necessary if the notation would otherwise be misunderstood. However, different versions exist, which do not mention all of the attributes. The notation for an open QS does not incorporate a population size.

Examples include the M/M/1 queue: a QS with exponentially distributed arrival and service times, single server, unlimited waiting room, and FCFS discipline. A more complex QS is, e.g., denoted by M/D/m/K, which means an open m-server queue with K waiting slots, exponential customer interarrival times, and deterministic service time.

Queuing Networks

QS models are sufficient for single queuing situations, and an obvious advantage of their simplicity is the availability of efficient solution methods. There are, however, many technical systems comprising numerous waiting lines and service requests. Customers may arrive from outside the system, leave it after service, or are probabilistically routed to another queue after service. Models with such an extended combination of several QS nodes are called **queuing networks** (QN). Any QS is thus a special case of a simple QN.

Closed QS can for instance be interpreted as a circular QN with two nodes: one for customers that are actually waiting and receiving service, and another for customers that are currently not competing for service.

Figure 4.2 depicts an example of a QN. Parts to be processed by the modeled production cell arrive from the left; a *source* of new customers is thus connected to node `Machine1`. After the processing step in that machine, there is a probability for each part to be sent to `Machine2` or `Machine3`, in which the second production step is carried out. Every part undergoes a quality check in the `Inspection` station. A certain percentage of parts is finished as a result and leaves the model. These parts are routed from node `Inspection` to a *sink* of the model. The remaining ones enqueue at the `Repair` station and are inspected again.

The example shown is an *open network* because customers (parts in that case) cross the borders of the model when arriving or leaving. Networks without sources and sinks are called *closed*, and the number of customers in them obviously remains constant.

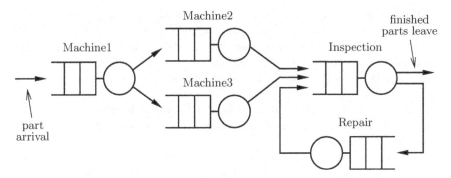

Fig. 4.2. A queuing network example

The following issues thus need to be specified in a QN:

- All individual details described above for every QS node that is a part of the QN. This includes the *arrival patterns* if the network is open. For the later modeling of departures from the model, an imaginary *sink node* without waiting and service is assumed, where customers vanish.
- The *routing behavior* of a customer describes to which subsequent queue each customer is sent after service. This must be set for each pair of queues individually. The sink node may be chosen as the destination to model leaving customers.
- Finally, the *initial customer population* must be given for every node in the QN.

The special case of a *Jackson network* requires the interarrival-times from the outside as well as the service time at each node of the QN to be independent and exponentially distributed, and a fixed probability r_{ij} for a customer to be routed to node q_j after service at node i. An efficient method for their steady-state analysis is based on a *product-form solution* [177], which does not require the construction of the reachability graph.

Performance Measures of QN

As for the other SDES models in this text, queuing systems and networks are meant to obtain quantitative measures from them. An actual QN model includes reward variables, for which the values should be computed by an evaluation algorithm.

The following list describes the most important measures for QN.

- The probability that the number of customers in queue q_i equals j is denoted by $\pi_j(q_i)$. It represents a basic measure from which several others can be derived. The *loss probability* or *blocking probability*, i.e., the probability that an arriving customer does not find an empty waiting slot, is obviously a special case. This probability equals π_n for queues with a finite system capacity of n. Another special case is the *queuing probability*, i.e., the probability that a customer arriving at a queue does not find an idle server and is thus required to wait. If n denotes again the system capacity (n might be infinity for a QS without waiting room restriction) and m the number of servers at the queue of interest, the queuing probability is $\sum_{i=m+1}^{n} \pi_i$.

– The number of customers at a queue q_i (the **queue length**) is denoted by $X(q_i)$. It is an important measure to assess the utilization of waiting room, work in progress, etc.

– An owner of a service facility will be interested in the **server utilization**, which gives the amount of time that the server is actively serving a customer. This amount is denoted by $U(q_i)$ for a queue q_i. In the case of a node with more than one server, this value equals the number of busy servers.

– Another important issue is the number of customers served per time unit, i.e., the **throughput** $T(q_i)$ of a queue q_i.

– From the customer's point of view, the **customer waiting time** at a queue q_i is interesting, which is denoted by $W(q_i)$. This time does only take into account the "unproductive" waiting time before service starts.

– When the service time is added to the waiting time W, the **system time** or **response time** is obtained. This measure is denoted by $S(q_i)$ for queue q_i.

One of the most important results in queuing theory relates three measures: The mean number of customers in a queue is equal to the product of mean arrival rate λ and the mean system time at that queue.

$$E[X] = \lambda \, E[S]$$

This is known as **Little's Law** and proved in Ref. [231]. Its power lies in its independence of the type of arrival and service pattern. Moreover, it can be applied to any subsystem with a defined boundary, and not only to a single queue.

Per-customer delays would require a reward variable that accumulates individual waiting times of all customers, and a division by the overall number of customers that were in the queue. This cannot be described directly as one reward variable in our framework. However, the values can be easily derived in steady-state using Little's law. The mean system time $S(q_i)$ per customer at queue q_i is thus given by

$$E[S(q_i)] = \frac{E[X(q_i)]}{\lambda}$$

The pure waiting time per customer excludes the service time, and thus the mean number of customers in the inspected subsystem is smaller by the utilization value.

$$E[W(q_i)] = \frac{E[X(q_i) - U(q_i)]}{\lambda} = \frac{E[X(q_i)] - E[U(q_i)]}{\lambda}$$

The necessary values of $E[X(q_i)]$ and $E[U(q_i)]$ can be derived by reward variable definitions as stated below. From these results the mean waiting and system times can be obtained.

4.2 Model Class Definition

A **queuing network** QN is defined as

$$QN = (Q, \Lambda, K, \mu, M, R, I)$$

Q is the set of simple queuing systems or **nodes** with queuing space and connected server(s) in the network. To simplify the following definitions of node attributes, we number the finite number nodes such that

$$Q = \{q_1, q_2, \ldots, q_{|Q|}\}$$

and attributes of a single node can be written with the index number of the node. The **state** X of one node is given by the number of customers currently in the queue,[2] incorporating both waiting and those being served.

$$X : Q \to \mathbb{N}$$

A complete state of the queuing network can also be written as a vector of natural numbers

$$X \in \mathbb{N}^{|Q|} = \big(X(q_1), X(q_2), \ldots, X(q_{|Q|})\big)$$

and the set of all theoretically possible states of a queuing network model is denoted by X^*.

$$X^* = \{X \mid \forall q \in Q : X(q) \in \mathbb{N}\}$$

Arrival of customers from outside of the model can happen in an open network. The interarrival time between two consecutive customer arrivals at node $q_i \in Q$ is a random variable. The probability distribution for it is denoted by Λ.[3]

$$\Lambda : Q \to \mathcal{F}^+$$

The **capacity** or length of the queuing space of one node $q_i \in Q$ specifies the number of waiting slots for customers. It is denoted by K and either a positive natural number or infinity for unrestricted waiting space.

$$K : Q \to \big(\mathbb{N}^+ \cup \{\infty\}\big)$$

---

[2] Please note that this is intentionally denoted by the same symbol as the performance measure *queue length*, because both have identical values.

[3] See Sect. 1.4 for the definition of $\mathcal{F}^+$.

Customers in a queue are being served if there is a server available for them. The **service time** of a node is a random variable for which the probability distribution is denoted by μ.

$$\mu : Q \to \mathcal{F}^+$$

There might be several identical servers available at one node, all serving one queue. The maximum reachable degree of concurrency at a node, the **number of servers**, is denoted by M and a natural number.

$$M : Q \to \mathbb{N}^+$$

If $M_i > 1$ we say that node q_i is a **multiserver node**.

After being served at a node q_i, a customer in a queuing network either leaves the model or queues at another node q_j. Leaving the system is captured as being routed to a virtual sink node **sink** with index zero.[4] The probability for a customer to go from node q_i to q_j is denoted by the **routing probability** r_{ij}. All routing probabilities are collected in the **routing matrix** R. If a customer is routed to a queue that has already reached its maximum capacity, the customer is lost.

$$r : Q \times (Q \cup \{\mathbf{sink}\}) \to \mathbb{R} \quad \text{or simply} \quad R \in \mathbb{R}^{0\ldots|Q| \times 0 \ldots |Q|}$$

For an evaluation of the behavior we need to know the queue contents at the beginning. The **initial number of customers** defines for every node the number of customers stored at the start. This number must of course not be greater than the queue capacity.

$$I : Q \to \mathbb{N} \quad \text{with} \quad \forall q \in Q : I(q) \le K(q)$$

RV specifies the set of **reward variables** of a queuing model. Four basic types have been informally introduced above. Throughput of servers corresponds to an impulse reward, while the other three measure types can be expressed with a rate reward. We differ between the two types by using a boolean variable $rtype$ that denotes the type of reward variable. An additional parameter $rexpr$ either specifies a state-dependent numerical expression for rate-related measures or the queue of interest in the case of a throughput computation.

$$\forall rvar \in RV : rvar = (rtype, rexpr)$$
$$\text{with} \quad rtype \in \mathbb{B}$$
$$\text{and} \quad \begin{cases} rexpr \in Q & \text{if } rtype = \text{False (throughput)} \\ rexpr : X^* \to \mathbb{N} & \text{if } rtype = \text{True (otherwise)} \end{cases}$$

The four basic measures are then expressed by setting the variables as shown in Table 4.1.

---

[4] The "normal" nodes are numbered from one upwards.

Table 4.1. Settings for queuing model reward variables

	Symbol	*rtype*	*rexpr*
Customer probability	$\pi_j(q_i)$	True	$\begin{cases} 1 & \text{if } X(q_i) = j \\ 0 & \text{otherwise} \end{cases}$
Queue length	$X(q_i)$	True	$X(q_i)$
Server utilization	$U(q_i)$	True	$\min\big(X(q_i), M(q_i)\big)$
Queue throughput	$T(q_i)$	False	q_i

4.3 Representing Queuing Networks as SDES

Before we start with the actual SDES definition of a QN, a remark is necessary. The node servers of the QN are obvious candidates for actions of the corresponding SDES. However, there are other classes of actions, the first one being indirectly introduced by the routing decisions. Remember that after being served, a customer leaves its queue and is transferred either to another queue or deleted from the system. The latter is described by a movement to the **sink** queue. The transfer of a customer is an SDES action as well, and is named **transfer action** in the following. It has priority over normal service actions because it must be executed first and happens in zero time. Their set is denoted by *Trans*.

The transfer actions could also be modeled locally within the actions by using different action modes. However, this would imply a decision about the later routing at the time when the service starts. This would not be a problem as long as these decisions are fixed priorities; however, the actual decision is made after the service in reality, and thus the QN model class could not be extended for more complex cases.

The second set of additional actions are the ones related to customer arrivals from outside of the model. These **arrival actions** need to be modeled by actions *Arr* with their interarrival time distributions. As they are always enabled, no additional state variables are necessary for them.

An SDES is defined as a tuple of state variables, actions, sorts, and reward variables

$$\text{SDES} = (SV^\star, A^\star, S^\star, RV^\star)$$

to which a mapping of the queuing network definition is given in the following.

There is one state variable for each node of the QN storing the current number of customers in the queue of the node. In addition to that we need to capture the enabling of transfer actions. As there is one routing decision to be made after a customer has received service at a node, there is one additional boolean state variable per node that signals a finished service and thus the enabling of the corresponding transfer actions. We denote the set of these boolean state variables by E.

$$SV^* = Q \cup E \quad \text{with} \quad |E| = |Q|$$

The sort of the SDES state variables is thus either a natural number as in the QN definition or Boolean for the additional state variables.

$$\forall sv \in SV^* : S^*(sv) = \begin{cases} \mathbb{N} & \text{if } sv \in Q \\ \mathbb{B} & \text{if } sv \in E \end{cases}$$

A normal state of the QN refers to one state of the corresponding SDES, where all additional state variables E are False. The value of a state variable from Q in a state refers to the number of tokens in the node of the QN.

The set of SDES actions A^* equals the set of servers of the queuing network plus the transfer actions $Trans$ and arrival actions Arr as mentioned above. As we have one normal server (which may be multiple) and one arrival action per node, and one transfer action per node pair plus the sink destination, we define

$$A^* = Q \cup Arr \cup Trans \quad \text{with} \quad Trans = Q \times (Q \cup \mathbf{sink}) \text{ and } Arr = Q$$

There are no action variables necessary to describe a QN, thus no sort definition is given for them.

The set of all theoretically possible states Σ is a combination of the number of customers X in the queues plus the enabling state of transfer actions E. There is thus a natural number and a boolean value per node.

$$\Sigma = (\mathbb{N} \times \mathbb{B})^{|Q|}$$

The condition function is used to capture the capacities of queues.

$$Cond^*(a, \sigma) = \begin{cases} \text{False} & \text{if } a \in Q \wedge \sigma(a) = (k, \cdot) \text{ with } k > K(a) \\ \text{True} & \text{otherwise} \end{cases}$$

The initial value of a state variable is given by the initial number of customers of the corresponding node. Transfer actions are disabled initially.

$$\forall sv \in SV^* : Val_0{}^*(sv) = \begin{cases} I(sv) & \text{if } sv \in Q \\ \text{False} & \text{if } sv \in Trans \end{cases}$$

There are no priorities defined among different servers of the QN. Transfer actions have a higher priority than the service-related actions. Because such a transfer action becomes enabled only after a service has finished, and service takes some time, transfer actions are never in actual conflict with each other, and their priorities are defined to be the same. Arrival actions are assumed to have the same priority as normal service actions.

$$\forall a \in A^\star : Pri^\star(a) = \begin{cases} 1 & \text{if } a \in Q \cup Arr \\ 2 & \text{if } a \in Trans \end{cases}$$

The enabling degree $Deg^\star$ of a server is directly given by the number of servers at that node. The actual enabling degree in a state depends on the number of customers that are waiting in the queue to be serviced as well. It is thus the smaller number of the QN enabling degree and the number of customers. Transfer and arrival actions may not be concurrently enabled with themselves.

$$\forall a \in A^\star : \qquad Deg^\star(a) = \begin{cases} M(a) & \text{if } a \in Q \\ 1 & \text{if } a \in Trans \cup Arr \end{cases}$$

$$VDeg^\star(a, \cdot, \sigma) = \begin{cases} \min(n, M(a)) & \text{if } a \in Q, \sigma(a) = (n, \cdot) \\ 1 & \text{if } a \in Trans \cup Arr \end{cases}$$

An action of an SDES for a QN is completely described by the attributes of a node server. No action variables are necessary because there are no different variants or modes of servers.

$$\forall a \in A^\star : Vars^\star(a) = \emptyset$$

Thus there is exactly one action mode for each node, $|Modes^\star(a)| = 1$, and we omit the action mode in the following mappings of QN attributes to SDES elements.

A QN server is active iff there is at least one customer in its queue, while a transfer action is enabled if its corresponding state variable is True. Arrival actions are always enabled.

$$\forall q \in Q, \sigma \in \Sigma :$$

$$Ena^\star(q, \sigma) = \begin{cases} \text{True} & \text{if } a \in Q \wedge \sigma(q) = (n, \cdot) \text{ with } n > 0 \\ \text{True} & \text{if } a \in Trans, a = (q_i, q_j) \wedge \sigma(q_i) = (\cdot, \text{True}) \\ \text{True} & \text{if } a \in Arr \\ \text{False} & \text{otherwise} \end{cases}$$

The delay of an action is distributed as specified by the service time function and zero for transfer actions. The interarrival time distribution of arrival actions is defined in the QN net class.[5]

$$\forall a \in A^\star : Delay^\star(a) = \begin{cases} 1 - e^{-\mu(a)\,x} & \text{if } a \in Q \\ 1 - e^{-\Lambda(a)\,x} & \text{if } a \in Arr \\ s(x) & \text{if } a \in Trans \end{cases}$$

---

[5] $s(x)$ denotes the step function, see Sect. 1.4.

Weights of actions are used to decide which one is executed first if there are more than one scheduled to do so at the same time. There is no corresponding information in a QN, and thus the weights are just set to one. For the considered class of queuing networks, the selection of differing weights for service and arrival actions would not change the behavior, because the sequence of service completions at the same time does not matter. The same applies to transfer actions, because neither can be in conflict with different solution results. The weights of transfer actions are given by the routing probabilities.

$$\forall a \in A^{\star} : Weight^{\star}(a) = \begin{cases} 1 & \text{if } a \in Q \cup Arr \\ r(i,j) & \text{if } a \in Trans \wedge a = (i,j) \end{cases}$$

If a service is executed, i.e., an active server has finished to serve a customer, this customer is taken from the queue. Depending on the routing probabilities the customer is transferred to another queue or deleted from the model. However, if the randomly chosen destination queue of the customer is full or the sink node **sink**, the customer is lost. The execution of an arrival action leads to an additional customer in the associated queue, if the capacity is not exceeded. The change of the state $\sigma = (a,b)^{|Q|}$ to a subsequent state $(a',b')^{|Q|}$ by the execution of an enabled action a is defined as follows.

$$\forall a \in A^{\star}, \sigma = (n,b)^{|Q|} \in \Sigma, (n,b)_i \in \sigma :$$

$$Exec^{\star}(a,\sigma)_i = \begin{cases} (\max(n-1,0), \text{True}) & \text{if } a \in Q \wedge a = q_i \\ (\min(n+1, K(q_i)), \text{False}) & \text{if } a \in Trans \wedge a = (\cdot, q_i) \\ (\min(n+1, K(q_i)), \text{False}) & \text{if } a \in Arr \wedge a = q_i \\ (n, \text{False}) & \text{otherwise} \end{cases}$$

The set RV of the QN reward variables is mapped to the set of SDES reward variables as follows. The last two elements $rint^{\star}$ and $ravg^{\star}$ specify the interval of interest and whether the result should be averaged. Both are not related to the model or reward variable; they correspond to the type of results that the modeler is interested in. They are therefore set according to the type of analysis and not considered here further.

There is one SDES reward variable for each QN reward variable such that

$$RV^{\star} = RV$$

The parameters of each reward variable are set as follows.

$$\forall rvar \in RV^{\star} : \begin{cases} rrate^{\star} = rexpr, rimp^{\star} = 0 & \text{if } rvar = (\text{True}, rexpr) \\ \\ rrate^{\star} = 0, \forall q_i \in Q : \\ rimp^{\star}(q_i) = \begin{cases} 1 & \text{if } q_i = q \\ 0 & \text{otherwise} \end{cases} & \text{if } rvar = (\text{False}, q) \end{cases}$$

Thus a rate reward given by the state-dependent numerical value *rexpr* is earned in the first case. The second case covers throughput results, and an impulse reward of one is collected if a server of the measured queue finishes its service.

Notes

Queuing theory has its origins in the work of Erlang [99] in the area of telephone system planning. Questions like how many telephone lines are necessary to provide a certain quality of service (probability of having a free line) for random phone calls lead to this development. Later work extended the initial formulae to cover more complex queues, e.g., with non-Markovian times and queuing networks [177]. The abbreviated notation has its roots in Kendall [202].

Selected textbooks on the subject include [31, 149, 207, 305]. Queuing systems in perspective with other discrete event models as well as some software tools for QN are described in Cassandras and Lafortune [42].

Important application areas include telecommunication and networks [25], manufacturing [32], computer systems [31, 220], traffic and transport systems, and business management.

More complex queuing models than covered in this chapter have been considered in the literature. Routing probabilities may for instance be state-dependent, which can easily be captured by adding X^* to the arguments of r. The same applies to the distributions of the service and arrival times. The model presented in this chapter assumed the loss of a customer that tries to enter a full queue. Another possibility is *blocking*, in which case the customer stays in the source queue until an empty slot in the destination queue is available. This could be included in the definition by adapting the enabling and execution functions.

Another extension of the QN model is to allow customer arrival and service to happen in **batches** (also called bulk input or service). Batch sizes may be fixed or random, and even depend on the model state. To include this in the model of this chapter, a discrete probability distribution function describing the batch sizes for all arrivals and services would need to be specified first. Execution of arrival, service, and transfer actions would have to be adapted to cover multiple customers.

The extensions listed so far are quite simple to adopt in the model of this chapter. Things become a bit more complex if *impatient customers* have to be taken into account. One possibility is *balking*: a customer may decide to not enter a queue based on its contents. This can be modeled with a state-dependent routing probability. *Reneging* or *defection from queue* happens when a customer leaves the queue after having waited for some time. Additional actions would be required in the SDES model presented in this chapter, which correspond to the customer actions in the queue.

A principal extension of the model class is achieved when different customer types are considered. Service times, arrival rates, and routing probabilities can then be defined individually for each customer class. An example are *multiclass Jackson networks*, for which still a product-form solution exists [18] under certain restrictions. The state of each queue then obviously must be extended to a list of all customers, described by their class. SDES action enabling and execution would need to be redefined to work with lists instead of just the numbers of customers.

When different customer classes are adopted, queuing disciplines other than FCFS change the performance results. The options considered in the literature include last-come-first-serve, random selection, as well as priorities of customer classes. Preemption of the running service can be considered in the case of priorities.

5

Simple Petri Nets

This chapter covers the use of classic **stochastic Petri nets** for the modeling and evaluation of stochastic discrete event systems. They represent a graphical and mathematical method for their convenient specification. Petri nets are especially useful for systems with concurrent, synchronized, and conflicting or nondeterministic activities. The graphical representation of Petri nets comprises only a few basic elements. They are, therefore, useful for documentation and a figurative aid for communication between system designers. Complex systems can be described in a modular way, where only local states and state changes need to be considered. The mathematical foundation of Petri nets allows their qualitative analysis based on state equations or reachability graph, and their quantitative evaluation based on the reachability graph or by simulation.

Petri nets come in many different flavors. Common to all of them is that they contain **places** (depicted by circles), **transitions** (depicted by boxes or bars), and directed **arcs** connecting them. A Petri net can thus be mathematically classified as a directed, bipartite graph. Places may hold **tokens**, and a certain assignment of tokens to the places of a model corresponds to its model state (called **marking** in Petri net terms). Transitions model activities (state changes, events). Just like in other discrete event system descriptions, events may be possible in a state – the transition is said to be **enabled** in the marking. If so, they may happen atomically (the transition **fires**) and change the system state. Transition enabling and firing as well as the consequential marking change are defined by the **enabling rule** and **firing rule** of the actual Petri net class. In our understanding of SDES, activities may take some time, thus allowing the description and evaluation of performance-related issues. Basic quantitative measures like the throughput, loss probabilities, utilization, and others can be computed. In the Petri net environment, a **firing delay** is associated to each transition, which may be stochastic (a random variable) and thus described by a probability distribution. It is interpreted as the time that needs to pass between the enabling and subsequent firing of a transition.

This text presents different Petri net model classes. Depending on the type of tokens, we distinguish between standard Petri nets with identical tokens (referred to as **simple Petri nets** here) and **colored Petri nets** . Tokens can not be distinguished in the first model type and are therefore depicted as black dots or just their number in a place. For many complex applications, a more natural and compact description is possible if tokens carry information. This lead to the development of colored Petri nets, of which two variants are covered in Chap. 6.

The remainder of this chapter deals with stochastic Petri nets with identical tokens. A continuous underlying time scale of the stochastic Petri net is adopted. An informal introduction into the class of stochastic Petri nets (SPNs) with a small toy example is given in Sect. 5.1. The dynamic behavior of an SPN is described informally in Sect. 5.2, followed by a formal definition of SPNs in Sect. 5.3. The interpretation of an SPN model in terms of an SDES covers Sect. 5.4. The chapter closes with some historical and bibliographical remarks about Petri nets with identical tokens.

Relevant analysis techniques are presented in Part II. More complex application examples from the fields of manufacturing and communication are presented in Chaps. 13 and 14.

5.1 Introduction to Stochastic Petri Nets

Throughout this section the model class of generalized stochastic Petri nets (GSPN) is introduced informally. This is done along the way of the stepwise construction of a flexible manufacturing system (FMS) modeling example. The model used here is a slightly changed version of a GSPN model presented in p. 208 of [4]. Petri net model construction in a modular way and by refining coarse-grain structures is demonstrated as a byproduct. New model elements are explained when they are necessary. Other variants of stochastic Petri net classes with more transition types than GSPNs are described in the final notes of the chapter.

Tokens in a Petri net model the changing states and locations of objects, while the places they are located in model buffers or attributes of the system. The number of tokens in every place of a model corresponds to the overall system state and is called Petri net **marking**. The natural choice for our example is thus to model parts as tokens, and locations in which they might reside in as places. Transitions model activities that change the system state or object location. When an activity is executed, the transition **fires**. It changes the Petri net marking by removing tokens from input places (the ones that are connected to the transition by a directed arc), and adding tokens to the output places. The number of tokens that are removed and added via one arc is called the **arc cardinality** (or sometimes **multiplicity**). It is written besides the arc in the graphical representation, but the default value of one is omitted. Before it can fire, a transition must be **enabled**, requiring that enough tokens

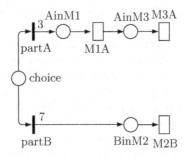

Fig. 5.1. First modeling step for FMS example

exist in its input places. Typical flows of objects like the path of parts along their work plans can be modeled by sequences of transitions and places that correspond to the subsequent production stages.

Figure 5.1 shows a first partial model for the FMS. The flexible manufacturing system has three machines, which are called M1 through M3. It handles two types of products named A and B. Parts of type A go through machine 1 and machine 3, while parts of type B are processed by machine 2. The upper part of the model corresponds to part A in its different stages: a token in AinM1 means the part is processed by machine 1. The activity of manufacturing part A in machine 1 is modeled by transition M1A. Part A in machine 3 and part B in machine 2 are modeled in a similar way by AinM3, M3A, BinM2, and M2B.

All manufacturing steps take a certain amount of time. The associated transitions are hence **timed transitions**, which are drawn as rectangles in the figure. The **firing delay** of a transition is the amount of time that needs to elapse between its enabling and firing. In a GSPN, the firing delays of all timed transitions are individual random variables with an exponential probability distribution function (c.f. Sect. 1.4). Each timed transition, therefore, has one parameter that specifies the **rate** of the exponential distribution. Letters λ or μ are normally used to denote these **firing rates**.

New raw material arrives in place choice. A decision about what type of part should be produced from one part of raw material needs to be made at that point, before the resulting work plans are started. Activities like this, which typically take no time from the modeling point of view, are modeled by **immediate transitions**. They might take some time in reality, but we decide to abstract from that in the model. We will see later that this distinction between timed and immediate activities can make the evaluation process more efficient.

The firing of transitions partA and partB decides what kind of part will be produced. Only one outcome of the decision is possible for a token in place choice, which corresponds to a **conflict** between transitions partA and partB. The decision should be made in the model such that a fixed percentage of parts of each type are manufactured. The conflict resolution can be influenced accordingly by associating a **firing weight** to each of the concerned transitions. In fact, every immediate transition has an associated firing weight, but usually the default value of 1.0 is unchanged. Firing weights of 3 and 7 are set for the two transitions as shown in the figure to achieve the desired behavior. Please note that these values are usually not shown in the graphical model. Later on the detailed semantics of firing weights is defined. Informally, they specify the probability of firing a transition relative to the weights of all other conflicting transitions. Transition partA thus fires in $3/(3+7) = 30\%$ of all cases.

The first model reflected only a part of the original work plans for the two parts. In fact, the manufacturing steps carried out in machines 2 and 3 can be done on either machine for the two part types. We need to incorporate the variants in the two work plans as well as the decision, on which machine the actual processing is done. Figure 5.2 shows the resulting refined model.

Parts of both types may now be in machines 2 and 3. However, the model needs to distinguish between them. As tokens in simple Petri nets are identical, the distinction must be made by a duplication of the corresponding net portion. The manufacturing place inside machine 2 is now modeled by places AinM2 and BinM2 (and similarly for machine 3). This is also necessary to specify the possibly differing properties of both manufacturing steps in one machine using one transition for each operation. For example, processing of part A could take more time than for part B in machine 3. With the duplication it is easy to specify the corresponding firing delays of transitions M3A and M3B.

Another difference to the previous model is the insertion of four immediate transitions similar to AsM3. Its firing models the start of the processing of part A in machine 3. The other three immediate transitions below AsM3 serve the same purpose for parts A and B in machines 2 and 3. It is often a good idea to separate the start from the ongoing activity in a Petri net model. Without

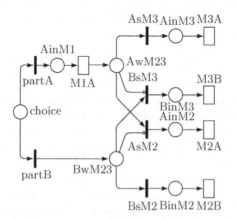

Fig. 5.2. First refinement of the FMS model

immediate transitions, the variants of the work plans would lead to conflicting timed transitions. This would be a modeling error, because then the transition who fires first would disable the other one, resulting in a conflict resolution probability that would depend only on the firing speeds of the transitions. In most models this needs to be treated separately.

An additional advantage of transitions like AsM3, which model the beginning of an activity, is that the desired conflict resolution can be specified. This might include a probability of a decision as it has been shown for transitions partA and partB. In the case of the decision between the two machines in the work plans, the desired behavior is that machine 3 should be used for parts A, and machine 2 for parts B as long as there are any of them waiting. If not, the machine may process parts of the other type. This machine allocation strategy is assumed to be useful in the example because the machines need less time for the mentioned steps. The outcomes of strategies like this one can be evaluated using performance evaluation, to select the best one out of several alternatives (c.f. Part II).

The described machine allocation is achieved in the model using **inhibitor arcs** that connect place AwM23 with transition AsM2 and place BwM23 with transition BsM3. Inhibitor arcs always go from a place to a transition, and have a small circle at the transition in the graphical representation. A transition is not enabled in a marking if the number of tokens in the place that is connected by an inhibitor arc is at least as high as the cardinality of the inhibitor arc. In the default case (as in the model), the transition may not fire if there is any token in the connected place.

It should be noted that the same machine allocation behavior could also be reached by associating a higher **priority** to transitions AsM3 and BsM2. Immediate transitions have a priority that specifies which one of them should be fired first in a marking. There is no specific firing priority associated to timed transitions; they all share one priority level, which is less than that of all immediate transitions.

To use firing weights and priorities to specify the later resolution of conflicts, the modeler must be sure to influence the reachability graph generation in the desired way. Both are ways to tell what should be done in cases where different activities are scheduled to be executed at the same time instant. There is thus no timing constraint specifying which one should be executed first. The notion of "first" is a bit strange in that context, because both happen at the same point in time. However, because of the general requirement of atomic action execution, we need to select the one to be executed first. Although the model definition allows for an unambiguous derivation of an underlying semantics, i.e., the stochastic process, the modeler only works at the model level. Firing weights are meaningful only locally inside the sets of transitions (and relative to the other ones) that are in conflict in a marking. Those sets are called **extended conflict sets** (ECS) in the literature. The problem during modeling is now that there is no direct way to specify which transitions should belong to one ECS, except maybe to assign a different priority to transitions that should go into another one. If the modeler makes a mistake due to an overlooked conflict, the model might not be correct. This is particularly hard to detect manually if a model is structurally constructed in a way that there are transitions which are not in direct conflict, but for which the order of firing leads to different model evolutions. This is mostly due to extended conflict sets in which transitions have different input places, or models with subgraphs of immediate transitions with different priorities. The general problem is known under the term of **confusion** and generally seen as a modeling error. There are approaches that ensure a confusion-free net by analyzing the structure ("at the net level"), or by checking on-the-fly during the reachability graph generation. We will not go into the details of this issue further here; the problem has been extensively studied in the literature [4,53,70,303], to which the interested reader is deferred. This topic is discussed in other settings in this text as well, for instance in Sect. 7.1.

Until now, only the processing of parts has been considered. The model shown in Fig. 5.2 reflects the work plans of the two parts. In addition to that there are transport operations needed in the FMS. Likewise, in communication systems we have packet transfer operations in addition to information processing stages. In our example FMS, an automated guided vehicle system (AGV) and a conveyor move parts between processing stations. Moreover, parts are transported on pallets throughout the system. The according model refinements are included in Fig. 5.3.

Empty pallets that are waiting for raw parts to be mounted are modeled by tokens in place emptyP on the left. The loading of a raw part corresponds to transition loadP, which takes one pallet token and creates a token in place loadedP. Usually such an assembly operation (between pallet and part) would result in two input places of the corresponding transition. However, in our case we assume that no shortage of raw material takes place, thus the always

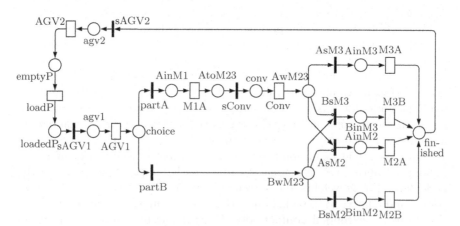

Fig. 5.3. Second refinement of the FMS example

available raw parts do not need to be modeled. This is an implicit input over the model boundaries.

During the first transport operation, a loaded pallet is transported by an AGV from the loading station to the first processing operation of the part. The start of the transport is modeled by sAGV1, in-transport state by agv1, and the completion and delay by AGV1. The second transport moves parts from machine 2 to 3, which is done by a conveyor belt. The corresponding model parts are sConv, conv, and Conv.

Another extension concerns transporting the pallets finally back to the initial buffer. When any one of the production steps carried out by machines 2 and 3 is finished, a part is completed and leaves the scope of the model. The resulting empty pallet in place finished is transported back to emptyP by the second AGV operation, which is modeled by sAGV2, agv2, and AGV2. Patterns like this circle of moving pallets are often found in Petri net models where objects are transported or change their status, while their total number remains constant.

The model refinement explained so far considers all processing and transport operations. What is still missing are resource constraints. Every operation has been modeled by a starting immediate transition, a place, and a finishing timed transition. We now add one place for every resource type we find in the model, namely each machine, the conveyor, and the AGVs. The start of an operation requires and blocks one resource, while the completion releases it. Therefore, an input arc is added from the required resource of an operation to its start transition, and an output arc from the timed transition back to the resource place. Please note that the model contains only this simple type of resource usage. It is obvious that multiple necessary resources of different types could be modeled in the same way by adding more arcs.

In the case of the two AGV operations, a conflict may arise between the two possible transports, corresponding to a conflict between immediate transitions sAGV1 and sAGV2. The chosen Petri net model pattern assures a mutual exclusion between operations that require exclusive use of a unique resource.

Related to the issue of resource constraints is the specification of the number of resources that are available initially. One token in each of the added places models the availability or idle state of one resource of the corresponding type. The initial marking of the model thus contains as many tokens in every resource place as there are resources available. In our example, every machine as well as the AGV is available only once. The conveyor place models available pallet spaces on the conveyor belt, which is assumed to be three here. The initial marking is depicted by black dots inside the place.

Transition Conv, the conveyor transport operation, has a special behavior: the associated operation (and required delay) applies to every pallet that is located on it. This is due to the fact that the transport of one part from the start to the end of the belt needs the same time, independent of how many pallets are otherwise located on it. To simplify the model, we still want to specify the time of one transport as the firing delay of transition Conv.

During the model evolution, the conveyor behaves as if there is one individual transport operation for every single part on it evolving in parallel. This type of behavior is often found in modeled systems, and is called **infinite server** (in contrast to the standard **single server**) firing semantics like in queuing models. It is depicted in the figure by IS near the transition.

All parts are transported on pallets. The number of available pallets thus limits the overall number of parts in the system. More pallets may lead to a better machine utilization and throughput, but they cost money and increase the amount of work in progress. Place emptyP contains empty pallets, for which the initial number is given by P. The marking of a Petri net model may be depicted by black dots, a natural number, or an identifier like P for which the value needs to be specified for an analysis.

More sophisticated features of GSPNs that have not been used in the example include **marking-dependent** model elements. A **marking-dependent arc cardinality** may be specified, which means that the actual number of tokens to be removed or created in a place is set according to the current model state during the model evolution. The use of marking-dependent arc cardinalities can make a model elegant and readable, but might also lead to a confusing behavior. One should restrict the use to cases in which they are easy to understand, like in the case where all tokens are removed from one place.

Firing delay parameters of timed transitions as well as firing weights of immediate transitions might also be defined depending on the marking. The speed of a transition can, therefore, be changed according to the number of tokens in a place. Timed transitions with infinite server semantics are a special case, because in the simple case they can be treated as a transition where the firing delay parameter is divided by the number of tokens in its input place.

It should be noted that the values of all marking-dependent attributes are computed before the transition fires. This is important because the transition firing might affect the marking from which its own attributes are depending.

Performance Measures

Performance measures need to be added to the structural Petri net model before any meaningful quantitative evaluation. They define what is computed during an analysis. A typical value would be the mean number of tokens in a place. Depending on the model, this measure may correspond to the mean queue length of customers waiting for service or to the expected level of work pieces in a buffer. A set of elementary measures is given below, followed by an explanation of how they are typically used.

For the definition of measures a special grammar is used, which follows the one used in the software tool TimeNET (see Sect. 12.1). A performance measure from the user's perspective is an expression that can contain numbers, marking and delay parameters, algebraic operators, and the following *basic measures*:

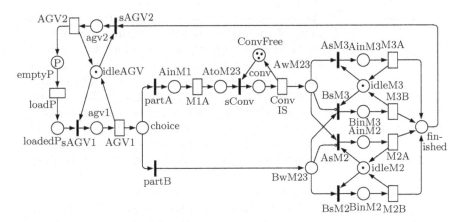

Fig. 5.4. Complete GSPN model of the FMS example

1. P{ <logic_cond> } corresponds to the probability of a logic condition, usually containing a comparisons of token numbers in places and numbers.[1]
   Example: P{#emptyP<3}.
2. E{ #<place> } refers to the expected number of tokens in a place. Example: E{#conv}
3. TP{ #<transition> } computes the number of transition firings (interpretable as the transition throughput). Example: TP{#M1A}

Typical ways of using performance measures are now informally explained
using the example in Fig. 5.4.

Buffer utilization   can be measured with the number of tokens in the place
that models the buffer. The number of available empty
pallets before loading can be computed using a performance measure E{#emptyP}. The same type of measures
are used for similar questions in different application areas, like communication packets in a buffer, vehicles in a
parking lot, and so forth. By adding the values for several
places, we can easily calculate the number of objects in
a subnet.

Machine utilization corresponds to a resource's probability of being in one
of its states, namely to being working. The utilization
of machine M2 can thus be computed with a measure
P{#idleM2} > 0, because if there is a token in place
idleM2, the machine is not working. Similar measures
are used if we are asking for resource state probabilities
in other applications, like the utilization of a server, a
telephone line, or a worker.

---

[1] Or any marking-dependent boolean function.

Throughput   is also an important issue for quantitative evaluations, thinking of production capacity, communication link load, or traffic intensity. The throughput of machine M1 in our model equals the production of parts of type A (at least in the long run, where it is equal to the added throughputs of transitions M2A and M3A). Both values can thus be computed with a performance measure TP{#M1A}.

Processing times   and other delays can be computed indirectly using Little's Law (see p. 70), which states that the number of customers (tokens) in a subnet is equal to the throughput into the subnet times the mean time to traverse it. The overall AGV transportation time for loaded pallets (including waiting for an available AGV and the actual transportation time) can thus be computed as the number of tokens in places loadedP plus agv1 divided by the throughput of transition loadP:
(E{#loadedP} + E{#agv1}) / TP{#loadP}.

5.2 The Dynamic Behavior of a SPN

A part of the FMS example is used in the following to informally explain the semantics of a generalized stochastic Petri net, which define the dynamic behavior of the model. Figure 5.5 shows the selected part and its first considered marking at the left.

It has already been stressed that places with tokens model states of a system, while transitions stand for operations or actions. The dynamic behavior is determined by the activation and execution of actions, which in turn change the model state. In a Petri net, transition enabling models activation, while transition firing corresponds to execution. The model state is given by the Petri net marking.

Transitions in a Petri net are activated from a structural point of view if there are enough tokens available in their input places. This means that for every input arc directed to that transition, at least as many tokens as the arc cardinality must be contained in the connected place. If there are inhibitor arcs present at the transition, there must be fewer tokens in the connected place than the inhibitor arc cardinality. If all these conditions are true, the transition is said to **have concession**.

A transition with concession is not always allowed to fire in a marking, because there might be other transitions with a higher priority having concession at the same time. Remember that immediate transitions have arbitrary priorities, and are always prioritized over timed transitions. A transition is thus said to be **enabled** in a marking if it has concession and there is no other transition with a higher priority that also has concession. In the left part of Fig. 5.5, only transition Conv has concession, and is thus enabled. It is

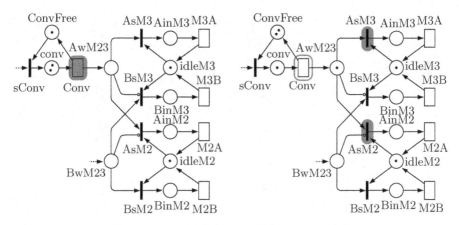

Fig. 5.5. Model behavior: Markings 1 and 2

marked with a grey shadow in the picture to point this out. The transition has no inhibitor arc, and there is a token available in the only input place `conv`.

The execution or completion of an activity corresponds to the **firing** of a transition in a Petri net. A transition fires when it had concession over a period of time given by its delay, provided that there are no other transitions firing first due to their higher priority. During the atomic firing, the number of tokens specified by the input arc cardinalities is taken from each corresponding input place, and the number of tokens given by the output arc cardinalities is added to every output place. Firing transition `Conv` removes one token from `conv`, and adds one token to `ConvFree` and `AwM23`.

The subsequent marking after firing `Conv` is shown on the right of Fig. 5.5. After the firing of a transition, a new marking is reached, and the enabling of transitions must be checked again. In our example, transition `Conv` still has concession, because there is one token left in place `conv`. It is, however, not enabled, because the two immediate transitions `AsM3` and `AsM2` have concession too. The latter are both enabled, because they share the default priority for immediate transitions. Transitions with concession are depicted with a thin grey shadow. Transition `AsM2` is enabled because there is one token in places `idleM2` and `AwM23`, while there is no token in place `BwM23`, which is connected to it by an inhibitor arc. Markings in which at least one immediate transition is enabled are somehow special, because they are left without spending any time in them due to the instantaneous transition firing. Those markings are thus called **vanishing markings** as opposed to tangible markings in which only timed transitions are enabled.

The two enabled transitions both need the token in place of `AwM23` to fire. Therefore, they can not fire both. This situation is called a conflict, and the evolution of the model behavior requires a decision, which one can fire. It has been explained already that conflicting immediate transitions fire with relative probabilities that are derived from their weights. For the two

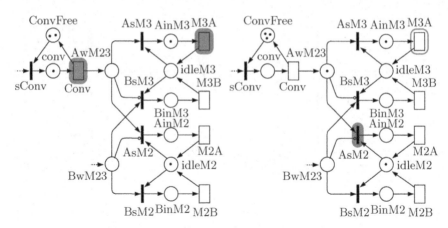

Fig. 5.6. Model behavior: Markings 3 and 4

transitions enabled now, no weight was specified, which means they both have the default weight of one. The probability of firing either one of them is thus 50%.

Let us assume that transition AsM3 fires, which leads to the marking shown in the left part of Fig. 5.6. Nothing has been changed for transition Conv, which therefore still has concession. Because of the token added to place AinM3, transition M3A gets concession as well. There are no immediate transitions with concession, thus the mentioned two timed transitions are enabled in the shown marking.

For the further model evolution, we need to know the probability of each transition to fire first. Moreover, for the later analysis, it is necessary to derive the time that the model spends in this marking. This is specified by the timing behavior of a timed Petri net. Let us first consider the simplest case in which only one transition is enabled, just like transition Conv in the first marking shown in Fig. 5.5. It is clear that the probability of firing this transition first is equal to one, while the time that the model spends in the marking is directly given by the firing time distribution of the transition.

In the general case, a **remaining firing time** (RFT) is associated to every timed transition, which counts the time that still has to elapse before the transition fires. In a GSPN this time is always exponentially distributed (c.f. Sect. 1.4), which makes the RFT handling much easier because of the memoryless property of the exponential distribution. When a transition becomes newly enabled, this RFT is set to a value that is sampled from the firing time distribution of the transition. The RFT decreases with model time as long as the transition has concession. When it reaches zero, the transition fires and the RFT becomes undefined. If the transition loses concession, the RFT is also set to undefined.[2] If a transition is of type infinite server, one

---

[2] If it is disabled temporarily while still having concession, the time does not change.

RFT is maintained as explained for every set of input tokens that lead to an independent transition firing. The resulting **enabling degree** of transition Conv in the initial marking was 2, because there are two tokens on its only input place. Hence two independent RFT were sampled.

In the subsequent second marking of our example Conv still had concession, but was not enabled because of immediate transitions. The enabling degree was one, because of the only token remaining in place conv. The RFT for the transition was already set in the first marking, and is still running. In the third marking (Fig. 5.6, left) transition M3A becomes newly enabled, and a RFT is sampled for it therefore. The decision which one of the transitions Conv or M3A fires first depends on the values of the corresponding RFT – the one that reaches zero first (the smaller one) fires. We assume that Conv fires in the third marking, leading to the marking shown right in Fig. 5.6. In this final considered marking, transition M3A still has concession, but may not fire because of the enabled immediate transition AsM2.

5.3 A Formal Definition

A stochastic Petri net can be formally defined as a tuple

$$\mathrm{SPN} = (P, T, \Pi, \mathbf{Pre}, \mathbf{Post}, \mathbf{Inh}, \Lambda, W, \mathbf{m}_0, RV)$$

with the elements described in the following.

P is the set of **places**, which may contain tokens. The **marking m** of the Petri net associates a (nonnegative integer) number of tokens to each place.

$$\mathbf{m} : P \to \mathbb{N}$$

The marking can also be viewed as a vector of natural numbers with the size of the number of places.

$$\mathbf{m} \in \mathbb{N}^{|P|} = \big(\mathbf{m}(p_1), \mathbf{m}(p_2), \dots, \mathbf{m}(p_{|P|})\big)$$

We denote by M the set of all theoretically possible markings of a Petri net.

$$M = \{\mathbf{m} \mid \forall p \in P : \mathbf{m}(p) \in \mathbb{N}\}$$

T denotes the set of **transitions**, which contains the set of timed T^{tim} and immediate transitions[3] T^{im}. It is quite obvious that a node of a Petri net may either be a place or transition, and that a net should not be empty.

$$T \cap P = \emptyset, T \cup P \neq \emptyset$$

---

[3] The partition depends on the firing delay distributions of the transitions and is thus given below.

The priority Π is a function that maps every transition to a natural number.

$$\Pi : T \to \mathbb{N}$$

Higher numbers mean a higher priority, and only immediate transitions may have a priority greater than zero. The priority thus implicitly defines a mapping to the transition types.

$$T^{tim} = \{t \in T \mid \Pi(t) = 0\} \quad \text{and} \quad T^{im} = \{t \in T \mid \Pi(t) > 0\}$$

In the graphical representation, transitions may be labeled with their priority. This is usually left out for transitions having default priority, namely all timed transitions and immediate transitions with priority equal to one.

Pre describes the multiplicities of the **input arcs** that connect places to transitions. The most general case is a marking-dependent multiplicity of an input arc. Thus **Pre** is defined as a function that maps each place-transition pair together with a marking vector to a natural number (the arc cardinality).

$$\mathbf{Pre} : P \times T \times \mathbb{N}^{|P|} \to \mathbb{N}$$

For the simple case of a cardinality independent of the marking we write

$$\mathbf{Pre}(p_i, t_j, \cdot) \in \mathbb{N}$$

and if there is no input arc connecting place p_i to transition t_j, $\mathbf{Pre}(p_i, t_j, \cdot) = 0$.

Post denotes the multiplicities of **output arcs** connecting transitions to places. The definition is similar to input arcs.

$$\mathbf{Post} : P \times T \times \mathbb{N}^{|P|} \to \mathbb{N}$$

Inh specifies the multiplicities of **inhibitor arcs**, similar to the definition of input arcs. A zero value means that there is no arc for a place-transition pair.

$$\mathbf{Inh} : P \times T \times \mathbb{N}^{|P|} \to \mathbb{N}$$

The **delay** Λ of a transition specifies the time that a transition needs to be enabled before it fires. The delay is defined by a probability distribution function (compare Sect. 1.4) that describes the possibly random delay time.

$$\Lambda : T \to \mathcal{F}^+$$

W maps each immediate transition to a real number. This value is interpreted as the **firing weight** for immediate transitions.

$$W : T^{im} \to \mathbb{R}$$

Firing weights for immediate transitions may be written near the transition in the graphical representation, if the value differs from the default 1.

Deg describes the degree of concurrency for each transition.

$$Deg : T \rightarrow \{SS, IS\}$$

SS means **single server** and *IS* **infinite server**.

$\mathbf{m}_0$ denotes the **initial marking** of the model. Because $\mathbf{m}_0$ is a marking, it is of the form

$$\mathbf{m}_0 : P \rightarrow \mathbb{N}$$

RV specifies the set of **reward variables** of the stochastic Petri net model. Three basic types[4] have been informally introduced in Sect. 5.1. From the mathematical standpoint, there is no big difference between the first two cases, because both are related to rate rewards. If we extend the notion of $\mathbf{E}\{\cdot\}$ from places to any marking-dependent expression, we can express probability-type measures simply by assuming a numerical result of one if the logic condition is true and zero otherwise (as it is done in some programming languages as well, or understood as an indicator variable):

$$\mathrm{P}\{ \text{ log\_cond } \} = \mathrm{E}\{ \begin{cases} 1 & \text{if } \texttt{log\_cond} = \text{True} \\ 0 & \text{otherwise} \end{cases} \}$$

Only two types of reward variables are left to be specified after this simplification. This is done by a flag *rtype* that denotes the type of reward variable, and a parameter *rexpr* that either specifies the marking-dependent expression ($\mathrm{E}\{rexpr\}$) or a transition for which the throughput should be computed in the second case:

$$\forall rvar \in RV : rvar = (rtype, rexpr)$$
$$\text{with} \quad rtype \in \mathbb{B}$$
$$\text{and} \quad \begin{cases} rexpr : M \rightarrow \mathbb{N} & \text{if } rtype = \text{True} \ \text{(E-case)} \\ rexpr \in T & \text{if } rtype = \text{False} \ \text{(TP-case)} \end{cases}$$

5.4 An **SDES** description of SPNs

Remember that a stochastic discrete event system was defined as a tuple

$$\mathsf{SDES} = (SV^\star, A^\star, S^\star, RV^\star)$$

The set of state variables $SV^\star$ equals the set of Petri net places.

$$SV^\star = P$$

---

[4] Recall the syntax: P for the probability of a condition, E for the expected number of tokens in a place, and TP for the throughput of a transition.

A marking of the Petri net is equivalent to a state of the SDES. The value of a state variable in a state is given by the number of tokens in the corresponding place in the marking.

The set of SDES actions $A^\star$ is given by the set of transitions of the Petri net.

$$A^\star = T$$

The sort function of the SDES maps Petri net places to natural numbers.

$$\forall p \in P : S^\star(p) = \mathbb{N}$$

The set of all possible states Σ is defined by

$$\Sigma = \mathbb{N}^{|P|}$$

The condition function is always true, because there are no restrictions on the place markings.

$$Cond^\star(\cdot, \cdot) = \text{True}$$

The initial value of a state variable is given by the initial marking (number of tokens) of the corresponding place.

$$\forall p \in P : Val_0{}^\star(p) = \mathbf{m}_0(p)$$

Transitions of the Petri net are the actions of the corresponding SDES. The priority $Pri^\star$ is the same as the one defined for the Petri net.

$$\forall t \in T : Pri^\star(t) = \Pi(t)$$

The enabling degree of transitions is either 1 for infinite server transitions, or ∞ for those of type infinite server. The actual enabling degree $VDeg^\star$ in a state returns the actual number of concurrent enablings in one state, and equals thus the maximum number of possible transition firings in a state for infinite server transitions. This number can be computed by dividing the number of tokens in an input place of the transition by the arc cardinality of the connecting arc. In the general case of several input places, the minimum over the computed numbers is used. In any case the result needs to be rounded downwards. In the presence of inhibitor arcs this might be a simplification.

$$\forall t \in T, \mathbf{m} \in M :$$

$$Deg^\star(t) = \begin{cases} 1 & \text{if } Deg(t) = SS \\ \infty & \text{if } Deg(t) = IS \end{cases}$$

$$VDeg^\star(t, \cdot, \mathbf{m}) = \begin{cases} 1 & \text{if } Deg(t) = SS \\ \min_{p \in P,\, \mathbf{Pre}(p,t,\mathbf{m})>0} \left\lfloor \frac{\mathbf{m}(p)}{\mathbf{Pre}(p,t,\mathbf{m})} \right\rfloor & \text{if } Deg(t) = IS \end{cases}$$

An action of a SDES for a SPN is completely described by the attributes of a single transition of the Petri net. There are no different variants or modes of transitions, there is no need for action variables.

$$\forall t \in T : Vars^\star(t) = \emptyset$$

Thus, there is exactly one action mode for each transition, $|Modes^\star(t)| = 1$, and we omit to mention the action mode in the following mappings of Petri net attributes to SDES elements.

A transition of a SPN is enabled if and only if (1) there are enough tokens in the input places of the transition, and (2) the number of tokens in places that are connected to the transition by an inhibitor arc does not exceed the arc multiplicity.

$$\forall t \in T, \forall \mathbf{m} \in M : Ena^\star(t, \mathbf{m}) =$$
$$\quad \forall p \in P : \mathbf{Pre}(p, t, \mathbf{m}) \le \mathbf{m}(p) \quad \text{(enough input tokens)}$$
$$\wedge \ \forall p \in P : \big(\mathbf{Inh}(p, t, \mathbf{m}) > 0\big) \longrightarrow \big(\mathbf{Inh}(p, t, \mathbf{m}) > \mathbf{m}(p)\big) \quad \text{(inhibitor arcs)}$$

The delay of an action is distributed as specified by the Petri net delay function.

$$\forall t \in T : Delay^\star(t) = \Lambda(t)$$

With the delay of a transition defined, we can now formally decide whether it belongs to the set of timed or immediate transitions. The latter type fires immediately without a delay after becoming enabled, the firing time is thus deterministically "distributed" with the result always being equal to zero.

$$T^{im} = \{t \in T \mid \Lambda(t) = \mathcal{F}^{im}\}$$

The set of timed transition is defined as the ones not being immediate. However, we require the firing time distribution of timed transitions to not to have a discrete probability mass at point zero, and to have an expectation greater than zero.

$$T^{tim} = T \setminus T^{im} \quad \text{with} \quad \forall t \in T^{tim} : \Lambda(t)(0) = 0$$

The weight of an action is determined by the transition weight in the case of immediate transitions. For timed transitions, SPN do not define an explicit weight, because the probability of firing two timed transitions at the same time is zero. This is due to the fact that the continuous exponential distributions have a zero probability of firing at a given point in time. However, during a simulation, it is not impossible to have two timed transitions who are randomly scheduled to fire at the same instant of time because of the finite representation of real numbers in computers. We assume an equal weight of 1 for timed transitions to resolve ambiguities in these rare cases.

$$\forall t \in T : Weight^\star(t) = \begin{cases} 1 & \text{if } t \in T^{tim} \\ W(t) & \text{if } t \in T^{im} \end{cases}$$

If an enabled action is executed, i.e., an enabled SPN transition t fires, tokens are removed from the input places and added to the output places as determined by **Pre** and **Post**. The change of the marking **m** to a subsequent marking **m**$'$ is defined as follows.

$$\forall t \in T, \mathbf{m} \in M, p \in P : Exec^\star(t, \mathbf{m})(p) = \mathbf{m}(p) - \mathbf{Pre}(t, p, \mathbf{m}) + \mathbf{Post}(t, p, \mathbf{m})$$

The set of reward variables of the Petri net RV is converted into the set of SDES reward variables as follows. The last two elements $rint^\star$ and $ravg^\star$ specify the interval of interest and whether the result should be averaged. Both are not related to the model or the reward variable definition of the Petri net, but correspond to the type of results one is interested in (and the related analysis algorithm). Hence they are set according to the type of analysis and not considered here. There is one SDES reward variable for each Petri net reward variable, such that

$$RV^\star = RV$$

and the parameters are set as follows:

$$\forall rvar \in RV^\star : \begin{cases} rrate^\star = rexpr, rimp^\star = 0 & \text{if } rvar = (\text{True}, rexpr) \\ \\ \left. \begin{array}{l} rrate^\star = 0, \forall t_i \in T : \\ \\ rimp^\star(t_i) = \begin{cases} 1 & \text{if } t_i = t \\ 0 & \text{otherwise} \end{cases} \end{array} \right\} & \text{if } rvar = (\text{False}, t) \end{cases}$$

This ensures that in the first case the rate rewards are earned according to the value of the marking-dependent expression of the Petri net, while in the second case an impulse reward of one is collected if the measured transition fires.

Notes

The foundation of Petri nets has been laid in Carl Adam Petri's Ph.D. thesis [263] on communication with automata in the early 1960s. One of the main issues was the description of causal relationships between events. Later work showed how the resulting models can be used to describe and analyze concurrent systems with more descriptional power than automata. The earliest results as well as the first books [262, 279] focus on qualitative properties and structural analysis. Petri nets were later used in engineering applications and for general discrete event systems, some selected books include [80, 88, 150, 290]. Petri nets together with their related way of interpreting, modeling, and analyzing systems characterizes them as a *conceptual framework* or *paradigm* [291].

As with SDES model classes in general, there are different subclasses of
Petri nets that are easier to analyze in exchange for certain restrictions in
their modeling power. Petri nets in which every transition has exactly one
input and one output arc are called **state machines**. The dual in which the
number of input and output arcs connected to every place equals one is coined
a **marked graph**. They only allow either conflicts or synchronization, respectively. **Free-choice nets** allow conflicts only between transitions with the same
input places. **Place/Transition nets** are not structurally restricted, but do not
contain extensions such like priorities, inhibitor, or flush arcs. The mentioned
additions extend the expressive power of Petri nets to the one of Turing machines, but hinder some structural analysis techniques. This is, however, not
an issue for a quantitative evaluation as intended in this text.

Time has been added to Petri net models starting in the 1970s [238, 275,
289]. The use of random firing delays and the relation between reachability
graphs and Markov chains in the case of a memoryless distribution was subsequently proposed by different authors [246, 248, 251, 300]. A literature overview
of timed Petri net extensions is given in [311].

Many classes of stochastic Petri nets (SPNs) with different modeling power
were developed since their first proposal. Firing delays of transitions are often exponentially distributed because of their analytical simplicity. The class
of generalized stochastic Petri nets (GSPNs, [4, 53]) adds immediate transitions which fire without delay. The underlying stochastic process of a GSPN
and other variants of stochastic Petri nets is a continuous-time Markov chain.
Those *Markovian SPNs* are well-accepted because of the availability of software tools for their automated evaluation. Chapter 12 lists some of them.
However, the assumption of a memoryless firing delay distribution is not realistic in many cases and can lead to significant differences for the computed
measures. Transitions with deterministic or more generally distributed firing
delays are needed for the modeling of systems with clocking or fixed operation
times. Many technical systems of great interest from the fields of communication, manufacturing, and computing belong to this class.

Examples of *Non-Markovian SPNs* are deterministic and stochastic Petri
nets (DSPNs, [6, 229]) and extended deterministic and stochastic Petri nets
(eDSPNs, [67,130]), also referred to as *Markov regenerative SPNs* [57]. DSPNs
add transitions with fixed firing delay to the class of GSPNs, and eDSPNs increase the modeling power further by allowing transitions with expolynomially
distributed firing delay. Expolynomial distributions can be piecewise defined
by exponential polynomials (cf. Sect. 1.4).

Stochastic Petri nets with a discrete time scale have been proposed
in [58, 247] among others, reference [311] contains a more thorough bibli-
ography on discrete-time Petri nets. This model type was later extended to
discrete (time) deterministic and stochastic Petri nets in [337, 338]. The latter are comparable with DSPNs w.r.t. their modeling power: immediate and
deterministic transitions are allowed, as well as geometrically distributed delays. The notes on p. 154 briefly comment on the analysis of models with

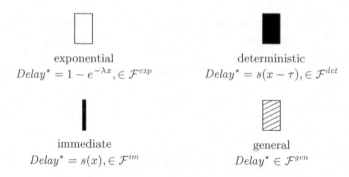

$$Delay^{\star} = 1 - e^{-\lambda x}, \in \mathcal{F}^{exp}$$

deterministic
$$Delay^{\star} = s(x - \tau), \in \mathcal{F}^{det}$$

immediate
$$Delay^{\star} = s(x), \in \mathcal{F}^{im}$$

general
$$Delay^{\star} \in \mathcal{F}^{gen}$$

Fig. 5.7. Types of transitions in a stochastic Petri net

an underlying discrete time scale. A good overview of Petri net performance models and analysis methods is contained in [15]. Please refer to e.g., [130] for a more thorough annotated list of relevant references.

It has been mentioned above that the class of a stochastic Petri net (and thus the available analytical methodology) depends on the used types of transitions. Figure 5.7 depicts the most important ones found in the literature and used in this text. It should be noted that despite the various graphical appearances the *structural* behavior of all these transitions is the same; they only differ in the associated firing delay distribution.

Immediate transitions are drawn as thin black rectangles, and their firing delay is zero. The other ones are called **timed transitions**, and contain types where the firing delay is exponentially distributed, deterministic, or general. Please refer to Sect. 1.4 for more details and the set of allowed general delays.

The presentation in this chapter has especially been influenced by the work on modeling and analysis of GSPNs [4] and eDSPN models [130].

Evaluation techniques for stochastic Petri nets are explained in Part II of this text. GSPN models of manufacturing systems are used in Chap. 13 for an automated derivation of heuristically optimal parameter sets. Chapter 14 models and analyzes a train control application example with eDSPNs. Creation and analysis of Petri net models requires a software tool for nontrivial examples. Chapter 12 contains more information on that issue, including the software tool TimeNET [359], which has been used throughout this text.

6

Colored Petri Nets

This chapter describes **stochastic colored Petri nets**, which are especially useful to describe complex stochastic discrete event systems. They can be seen as an extension of simple Petri nets as covered in Chap. 5. Places and transitions of a Petri net naturally map to buffers and activities or similar entities. Objects that are created, changed, and moved through a system are usually described by tokens in places. The application of classic Petri nets to examples in which these objects carry some significant attributes leads to cluttered models in which places and transitions need to be unfolded to keep track of the individual attributes. These problems motivated the development of **high-level Petri nets**, a set of net classes with distinguishable tokens.

The introduction of individual tokens leads to some questions with respect to the Petri net syntax and semantics. The attributes of tokens need to be structured and specified, resulting in **colors** (or **types**). Numbers as arc information are no longer sufficient as in simple Petri nets. Transition firings may depend on token attribute values and change them at firing time. A transition might have different modes of enabling and firing depending on its input tokens. The class of **stochastic colored Petri nets** that is presented in the following sections uses arc variables to describe these alternatives and is motivated by colored Petri nets as defined in [188].

Stochastic colored Petri nets are informally introduced in the first section. The dynamic behavior of SCPNs is covered in the Sect. 6.2. A formal definition of stochastic colored Petri nets as well as their interpretation in the SDES framework is given in Sects. 6.3 and 6.4. The special case of colored Petri nets with constant arc inscriptions is covered afterwards. The chapter closes with some remarks, clarifying the differences of the used model class with respect to other colored Petri net definitions. Application examples are presented in Chaps. 15 and 16.

6.1 Informal Introduction

This section informally introduces stochastic colored Petri nets in the sense that is used in this text. It mostly points out differences to simple Petri nets, which have been introduced in Chap. 5 already. The syntax of textual model inscriptions is chosen similar to programming languages like C++ or Java; it is completely specified in Backus Naur form in [327].

The main difference between simple Petri nets and colored models is that tokens may have arbitrarily defined attributes. It is thus possible to identify different tokens in contrast to the identical tokens of simple Petri nets. Most of the added complexity of transitions, places, and arcs comes from this simple extension.

6.1.1 Token Types or Colors

Tokens belong to a specific **type** or **color**, which specifies their range of attribute values as well as the applicable operations just like a type of a variable does in a programming language. It will be shown in the following that types are important in the context of tokens, places, arc variables, and thus transition behavior. The terms *color* and *type* are used synonymously in the following.

Types are either **base types** or **structured types**, the latter being user-defined. Table 6.1 lists the available base types in the software tool TimeNET (see Sect. 12.1) on which this chapter is based. The types are for instance used in the example model of Chap. 15, but could easily be extended if necessary.

The **empty type** is similar to the "type" of tokens in simple Petri nets. Tokens of this type cannot be distinguished, they are depicted as black dots and do not possess any attributes. Integers and real numbers are numerical values, which can be compared and used in arithmetic expressions. They are similar to `int` and `double` types in a programming language. Boolean values can be compared, negated, and used in AND- and OR-expressions. Strings represent character arrays as usual. String constants are enclosed in quotation marks like `"hello"`. Model times can be stored in attributes of `DateTime` type. It includes the time in hours:minutes:seconds format and the date in month/day/year

Table 6.1. List of base types for stochastic colored Petri nets

Type	Name	Default value	Examples
Empty type	—	●	●
Integer	`int`	0	123
Real	`real`	0.0	12.29
Boolean	`bool`	false	true, false
String	`string`	" "	"hello"
Date and time	`DateTime`	0:0:0@1/1/0	NOW, 14:12:15@11/03/2005

Table 6.2. Example token type definitions

Type	Element	Element type	Remarks
Product	name	string	Product name
	step	int	Production step number
Container	id	int	Container identification number
	contents	Product	The contained product

format, separated by a @ symbol. Time values can be subtracted resulting in the number of seconds between the two times. Adding or subtracting integers works in a similar way. Their comparison is defined as one would expect. The current model time during an evaluation is denoted by NOW.

Structured types are user-defined and may contain any number of base types or other structured types just like a Pascal **record** or a C **struct**. An example would be the definition of a product and a container in a manufacturing example as shown in Table 6.2. Circular definitions are obviously not allowed.

Notation of access to structured types as well as the specification of structured objects is done within braces {}. They enclose a list of attribute values together with the attribute names to define the value of a complex token. For the example in Table 6.2, a token of type *Product* could be specified by { name = "lever", step = 2 }, while the notation of a *Container* object might be { id = 42, contents = { name = "fuse", step = 6 }}. Default values in token creations and unchanged attributes in token operations can be omitted. The empty type may not be used as a part of structured types, and there is no textual notation for it.

The only allowed operation on a structured type is a comparison. Two objects of a structured type are equal if all of their element attributes are equal. Although an attribute of a structured type may be internally implemented as a pointer, there are no references accessible at the model level. Token objects may only be copied, it is not possible to have different references to the same token or attribute object.

Types and variables are textually specified in a declarational part of the model in the original definition of colored Petri nets [188]. The same applies to structured types in the described SCPN model class. This is done with type objects in the graphical user interface of TimeNET, but is omitted in the model figures. User-defined types are given in the text whenever necessary. Variable definitions are not necessary in difference to standard colored Petri nets as explained later.

6.1.2 Places

Places are similar to those in simple Petri nets in that they are drawn as circles and serve as containers of tokens. By doing so they represent passive elements of the model and their contents correspond to the local state of the model. As tokens have types in a colored Petri net, it is useful to restrict the type of tokens that may exist in one place to one type, which is then also the type or color of the place. This type may either be a predefined base type or a model-defined structured type. In any case it is shown in italics near the place in figures. The empty type is omitted.

The unique name of a place is written close to it in figures as well as the type. The **initial marking** of a place is a collection of individual tokens of the correct type. It describes the contents of the place at the beginning of an evaluation. There might be different tokens with identical attributes, which makes them alike, but not the same. The place marking is thus a multiset of tokens. Only the number of initial tokens is shown in drawings in this text, while the actual tokens are listed elsewhere when needed. A useful extension that is valuable for many real-life applications is the specification of a place **capacity**. This maximum number of tokens that may exist in the place is shown in square brackets near the place in a figure, but omitted if the capacity is unlimited (the default). An example is given below.

6.1.3 Arcs and Arc Inscriptions

Places and transitions are connected by directed arcs as in any other type of Petri net. An arc going from a place to a transition is called input arc of that transition, and the connected place is also called input place (and vice versa for output places and output arcs). In contrast to simple Petri nets, where a number is the only attribute of an arc, the modeler must be able to specify what kinds of tokens should be affected and what operations on the token attributes are carried out when a transition fires. This is done with **arc inscriptions**. Arc inscriptions are enclosed in angle brackets $<>$ in figures.

Input arcs of transitions and their inscriptions describe how many tokens are removed during a transition firing, and attach a name to these tokens under which they may be referenced in output arc and guard expressions. They carry a variable name in pointed brackets for the latter task, optionally extended by a leading integer specifying the number of tokens to be removed from the place. The default value for omitted multiplicities is one. A token from the input place is given to the variable as its current value, and removed from the place during firing. If a multiplicity greater than one is specified, the corresponding number of tokens are bound to the variable and removed during firing. Each input variable identifier may be used in only one input arc of a transition to avoid ambiguities.

A transition's **output arcs** define what tokens are added to the connected place at the time of the transition firing. There are two general possibilities for this: either existing tokens are transferred[1], or new tokens are created.

---

[1] Note that there is no theoretical difference between thinking of a transferred token and a token from the input place that is deleted with an identical one that is created in the output place.

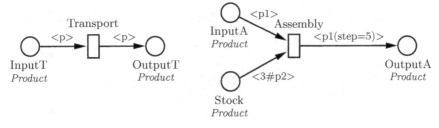

Fig. 6.1. Simple arc inscription example

In the transfer/copy case, the name of the chosen input token is used at the output arc. The token that was bound to this input variable is moved to the corresponding output place. The multiplicity of tokens must be the same to avoid ambiguities, i.e., if three tokens are taken away from a place, there is no possibility of transferring only two of them and arbitrarily removing one. For the same reason, it is not possible to use the same input variable at several output places. Arbitrary numbers of input token copies can be made by creating new tokens and setting the attributes accordingly (see later).

Figure 6.1 shows two examples of transitions with arc inscriptions. The trivial case of a token transfer without any attribute changes is depicted on the left side. Transition `Transport` is connected to places `InputT` and `OutputT`, and both have *Product* as their type. `Transport` is enabled by any token in its input place `InputT`, which is bound to variable `p`. The token is transferred from `InputT` to `OutputT` because the output arc inscription uses the same variable `p`.

The model on the right of Fig. 6.1 depicts a slightly more complex example. Transition `Assembly` has two input places `InputA` and `Stock`, from which one product `p1` and three products `p2` are removed, respectively. The firing of the transition transfers the `p1`-token to the output place `OutputA`, and changes the attribute `step` to the new value 5. Tokens bound to input variables that are not used on output arcs are destroyed at the end of the firing. This applies to the three `p2`-tokens, in which model parts are assembled to `p1`.

New tokens of the output place type are created if no input variable is specified at an output arc. The attributes of a new token are set to their default values initially (cf. Table 6.1). The firing of transition `Arrival` in the example shown in Fig. 6.2 creates a token of type *Product* in place `InBuffer` with the default attribute values {`name` = `""`, `step` = `0`}.

Attributes of new tokens can be set to specific values using the same syntax as described earlier for transferred tokens. Individual attributes of a token (or the value if it is a base type) may be set to a constant value of the type or to a value that depends on other input tokens. Elements of a structured type are set in braces {}. Firing of transition `Pack` in Fig. 6.2 removes a token `p` of color *Product* from place `InBuffer`. It creates a new token of type *int* in `Counter` with the value 3. An additional token with type *Container* is created in place

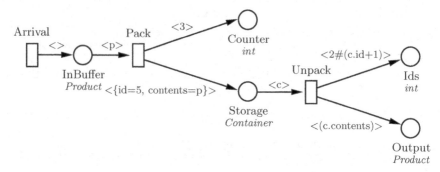

Fig. 6.2. More complex arc inscription examples

Storage. Its attribute `id` is set to 5, while the `contents` attribute is set to the value of the structured token that was removed from `InBuffer`.

Multiple new tokens can be constructed with a leading number and #-symbol. The firing of transition `Unpack` in Fig. 6.2 creates two Integer tokens in `Ids`. Their values are set to the `id` attribute of the consumed container token `c` plus one. Operators are allowed in expressions as long as their resulting type corresponds to the required one; see further for a list.

In the special case where an element of a structured token should be copied to create a new token, it is not necessary to write down all member attributes in assignments. The member attribute in brackets is sufficient. An example is shown in the token creation in place `Output`. The `product` attribute of token `c` is copied to construct a new token of type *Product*.

The type of the variables contained in the input and output arc inscriptions is implicitly given by the type of the connected place and is thus not defined by the modeler. Restrictions on the input tokens are modeled using transition guards as described later. All variables in arc inscriptions of a transition `t` are denoted as the **transition variables** of `t`.

6.1.4 Transitions

Transitions (drawn as rectangles with different shapes) model activities of the system. They can be activated (enabled) when all necessary input tokens are available and an optional **guard function** is true. Their firing models the occurrence of the activity and changes the marking of places with tokens (the state of the system). There are different transition types with their corresponding shapes: **immediate transitions** firing without delay are drawn as thin rectangles, **timed transitions** bigger and empty, while **substitution transitions** (for details see further) have black rectangles at the top and bottom.

Transitions have several attributes. The name is a string that uniquely identifies the transition on the model page. The **firing delay** (timed transitions only) describes the probability distribution of the delay that needs to elapse between the transition enabling and firing. The examples in this text only use deterministic and exponentially distributed delays. The syntax for their specification in the models is Exp(100) or 200 for a transition with exponentially distributed or deterministic delay, respectively.

Immediate transitions have a **firing weight** (a real number) and a **priority** (integer greater than 0) just like in simple Petri nets, with the same function. Their default value is one in both cases.

Transitions in a colored Petri net have different **firing modes** depending on the token attributes that they remove from their input places. In a state of an SCPN all possible assignments of input tokens to their respective arc variables (bindings) may be valid firing modes. A **guard function** can be used to restrict the tokens for which a transition may be enabled. The guard is a boolean function that may depend on the model state and the input arc variables. It is shown in square brackets close to the transition in figures. The transition is only enabled with a certain binding of tokens to variables in a model state if the guard function evaluates to True for this setting.

Figure 6.3 shows examples for the usage of guard functions. Expressions often contain a comparison of input token attributes with constants or other input token attributes. Operators and syntax are explained below. Transitions ToM1 and ToM2 may be in conflict when a token is in place Input. They decide whether *Product* tokens are sent to M1 or M2. With the selected guards, all tokens with attribute name equal to "bolt" are sent to M2. All other tokens are sent to M2 only if M1 is busy, i.e., place AtM1 is not empty. #AtM1 denotes the number of tokens in that place. The two transitions may never be enabled together in a marking because of the capacity restriction of place AtM1 of one. Transition Pack may only fire if there is a *Container* token available in place ContBuf for which the identification number is stored in Ids. The default guard function of a transition is empty, which evaluates to True by definition, and is not shown in figures.

The **server semantics** specifies whether the transition might be concurrently enabled with itself or not, comparable with the *Deg* attribute of SPN

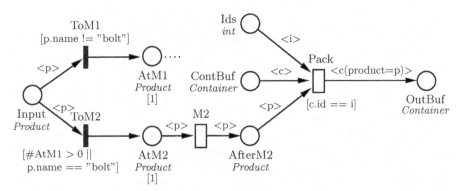

Fig. 6.3. Examples of guard functions

transitions. It may have the (default) value of **single server**, or **infinite server** alternatively. The latter is depicted by a IS at a transition. Single server semantics models the "natural" understanding of a restricted resource that can only perform one action at a time (on one part/token). It thus means that the transition may only be enabled with one binding at a time. Infinite server stands for an arbitrary number of resources, from which one is activated for every token (set) individually. All bindings that may be fired together are enabled concurrently. An incoming token into an input place of an already enabled single server transition is thus ignored, while a new transition enabling is generated for such a token in the case of an infinite server transition. To avoid ambiguities in the model specification, we restrict infinite server transitions to have exactly one input place and one variable in the corresponding arc inscription.

6.1.5 Model Hierarchy

A SCPN model consists of **pages** in a hierarchical tree. There is exactly one **prime page**, which forms the base of the tree structure, and to which other pages are subordinated on different levels of hierarchy via **substitution transitions**. Hierarchical refinement and modular description of complex systems is thus possible.

Substitution transitions act as a placeholder or association to a refining subpage. They have no firing semantic as other normal transitions do. The associated subpage is a place-bordered subnet, i.e., no transitions of the subnet may have direct interactions with elements outside the substitution transition. From the point of view of the upper level model, a substitution transition acts like a generalized transition that may consume and create tokens as well as store them.

Interaction of submodels with the surrounding model only takes place via the places that are connected to the substitution transition. All of these places are known in the submodel and are depicted there as dotted circles. Arcs connecting substitution transitions do not require inscriptions, because they only mark the connected places as known in the subnet.

Figure 6.4 shows an example. Machine is a substitution transition, which is refined by the submodel shown later. The two places InputBuffer and OutputBuffer are visible both in the upper level of hierarchy as well as in the submodel. The submodel describes the behavior of the machine with more detail than one transition would be capable of. In the example, the failure and repair behavior is hidden in the lower level of hierarchy.

Model hierarchy is used in this work as a structuring method for complex models. As such, it is rather a drawing convenience than an additional model class capability. Hierarchy issues can thus be neglected in the formal definition below for a substantial simplification of it. Formal model coverage sees every model as a flat colored Petri net. Such a net would result from simply substituting every substitution transition of the net by its refining subpage,

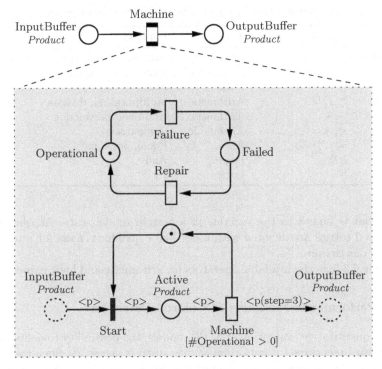

Fig. 6.4. Example of a substitution transition and its refining submodel

and a subsequent merge of referring places (drawn as dotted circles) with their connected place. Objects on each page of the model must be uniquely identifiable by their name. Model elements on nonlocal pages can be simply identified by using a path-like notation as in a file system.

6.1.6 Syntax of Expressions

Wherever variables and constants are allowed in the inscriptions explained earlier, expressions with operators are allowed instead as well (if not stated otherwise). Input arc inscriptions for instance may not carry expressions with operators. The result of an expression must of course match the required type of a place or token color.

The current number of tokens in a place p can be specified by #p. Place identification in expressions happens in the following way: every model element has a unique name on its page. Because of the hierarchical model structure, it is possible to identify any place by its absolute path like Machine/Failed in Fig. 6.4.

The values of input token attributes can be used in output arc expressions as well as guard functions. The name of an input arc variable identifies the

Table 6.3. Operators and their priorities in expressions

Priority	Operator	Type	Remark
8	(...)		Brackets to structure expressions
7	+, −	Arithmetic	Unary plus and minus
	!	Logic	Not
6	*, /, %	Arithmetic	Multiplication, division
5	+, -	Arithmetic	Addition, Subtraction
4	<, <= , >, >=	Arithmetic	Comparison
3	==, ! =	All	Equal, not equal
2	&&	Logic	And
1	||	Logic	Or

token that is bound to the variable in a certain model state. Attributes of structured tokens are denoted with a dot like `c.product.name` for a token c of color *Container*.

Table 6.3 lists the available operators for arithmetic and logic expressions.

6.1.7 Performance Measures of Colored Petri Nets

For the quantitative evaluation of SCPN models, the possibility to define complex reward measures (or performance measures) is necessary. Such measures are defined by expressions containing impulse rewards and rate rewards as explained for general SDES in Sect. 2.4. Impulse rewards may be gained in an SCPN when a transition fires. Rate rewards describe reward that is gained over time, like for instance amortization (constant rate) or inventory costs (marking-dependent rate). They are, therefore, related to the contents of places.

From the perspective of the user of the software tool TimeNET that implements the SCPN model class, a performance measure has a name, an expression, a type, and a computed value if it has been analyzed already. The type may be selected between *instantaneous, cumulative,* and *averaged,* which corresponds to the settings of SDES reward variable elements $rint^\star$ and $ravg^\star$ when the analysis time interval is set. The expression of an SCPN performance measure contains numeric constants and operations as well as the rate and impulse reward elements that are described later.

1. The number of tokens in a place P is measured with the term #P. This follows the syntax in guard functions and arc expressions to simplify model understanding. An example measure for the model in Fig. 6.3 would be `#AfterM2`, resulting in the number of tokens in place `AfterM2`. Whether the mean number of tokens over a time interval, the mean number in steady-state, or the expected number at a certain point in time should be computed is specified by the measure type and analysis time horizon.

More complex measure examples can count the number of tokens in a place that have some property. The number of tokens in place `AfterM2` with attribute value `step` equaling 3 can be measured by `#AfterM2(step == 3)`. Measures of places are rate rewards, which is automatically set by the tool.

2. The number of transition firings can be measured similarly. If we are interested in the number of times that transition M2 in Fig. 6.3 fires, we need to specify `#M2` as the measure expression. Firings can be filtered just like token numbers in places, e.g., by `#M2(p.name != "bolt")` to only count firings where *Product* tokens bound to variable p have their attribute `name` set to something other than `"bolt"`.

Typical examples of measures and how they are specified in a Petri net have already been shown in Sect. 5.1. The examples covered there can be applied to colored Petri nets in a similar way. Chapter 15 contains an application example of SCPN with some performance measures.

Expected numbers of tokens are specified as shown earlier. The throughput of transitions is measured by transition firings and averaging over the time interval of interest. Probabilities of boolean expressions over the model state can be expressed (as explained for simple Petri nets) by a rate reward that gains a reward of one when the expression is true, and zero otherwise. This can be done easily because the numerical results of a true or false boolean expression in a performance measure are defined to be one and zero, respectively.

6.2 On the Dynamic Behavior of Stochastic Colored Petri Nets

After the informal introduction of the static model elements of a stochastic colored Petri nets in the previous section, their dynamic behavior is informally explained in the following. Knowledge of the behavior of simple Petri nets as described in Sect. 5.2 is assumed, and general Petri net issues that do not differ for colored models are not explained again.

Places contain multisets of tokens of the corresponding type – the place marking. All place markings together establish the state of the modeled system (ignoring the transition state for the moment, which is explained later). Transitions can become enabled depending on the current model state and fire, thus changing the state. The initial state of the model is specified by the initial marking m_0.

Transitions in a colored Petri nets have different ways of activation and firing, which correspond to alternative ways of binding tokens in input places to input arc variables. Transitions can, therefore, only be enabled and fire *under a specific binding*. Saying that a transition is enabled is only a shorthand for that at least one of its bindings is enabled.

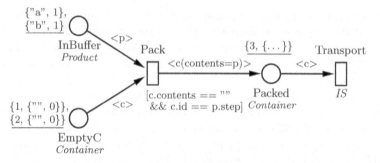

Fig. 6.5. Model behavior example: bindings

The model shown in Fig. 6.5 will serve as an example. The current marking of places is underlined. Place `EmptyC` stores empty container tokens, where initially two tokens are present. One token has an `id` value of one, while the other has an `id` of two. The product `contents` of the containers are empty. All possible bindings have to be evaluated for every transition to test which of them are enabled.

Lets start with the simple case: transition `Transport` has one input place `Packed`, where only one token {3, {...}} is available.[2] Thus, there is exactly one possible binding of the input arc variable c to tokens: c = {3, {...}}. For the moment it is not important that transition `Transport` has *infinite server* semantics. The one found binding can now be checked if it is **enabled**. This requires the guard function of the transition to evaluate to true and enough free capacity in output places of the transition for any tokens created in them. Both requirements are obviously fulfilled, because there is no guard function and no output place. Transition `Transport` and its binding c = {3, {...}} are thus enabled.

The other transition `Pack` has two input places `InBuffer` and `EmptyC`, in which two tokens are stored each. The input arc variables p and c thus may both be bound to two tokens, resulting in four different possible bindings:

1. p = {"a", 1} and c = {1, {"", 0}}
2. p = {"b", 1} and c = {1, {"", 0}}
3. p = {"a", 1} and c = {2, {"", 0}}
4. p = {"b", 1} and c = {2, {"", 0}}

For every one of these bindings, the guard and capacity of output places must be checked. As place `Packed` does not have a restricted capacity, the latter is no issue. The guard function, however, requires that the name of the container contents are empty (which is the case for all two container tokens), and that the container `id` equals the `step` attribute of the

---

[2] The dots denote omitted attributes.

Product token p. This is only true for the first and second binding in the shown list.

Transition `Pack` thus is enabled under the first and second binding. Its firing semantic is however *single server*, modeling that there is only one resource which does the packing of products. Therefore, only one of the two bindings may actually be enabled and there is no concurrency allowed. The selection of one of the bindings is done stochastically with equal probability; we assume that the first one (p = {"a", 1} and c = {1, {"", 0}}) is selected.

For the two enabled transition/binding pairs a **remaining firing delay** is sampled from the transition delay distributions. The one with the smaller delay fires first, while the time of the other one is decreased by the time that has passed. Lets assume that `Pack` fires first. The tokens that were bound to the input variables in the firing binding are removed from the input places now. Output arc inscriptions are evaluated and the resulting tokens are created in the output places. Token c = {1, {"", 0}} is taken and the *Contents* attribute is set to p = {"a", 1}. The resulting token is added to place `Packed`.

Figure 6.6 shows the new marking. Transition `Pack` has fired and thus needs to be checked again if there are enabled bindings. Because of the remaining tokens in `InBuffer` and EmptyC, there is only one possible binding, namely p = {"b", 1} and c = {2, {"", 0}}. `Pack` is not enabled with this binding because the guard function requires `c.id == p.step` (i.e., $2 == 1$). Transition `Pack` is thus not enabled at all now.

Remember that transition `Transport` is still enabled under binding c = {3, {...}}, for which the remaining firing delay is running toward zero. Additional bindings can run concurrently because of the *infinite server* semantics of `Transport`. Thus, the changed marking of place `Packed` is checked if it allows for additional bindings. There is one new token {1, {"a", 1}} after the previous transition firing, which is not yet part of an enabled binding. The corresponding binding c = {1, {"a", 1}} is enabled because there are no restrictions by guard function nor place capacities. Therefore, the two transition bindings `Transport`, c = {3, {...}} and `Transport`, c = {1, {"a", 1}} are enabled in the marking shown in Fig. 6.6, and will eventually fire.

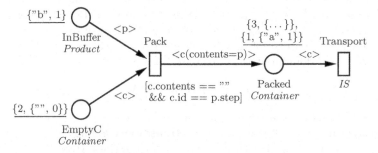

Fig. 6.6. Model behavior example: second marking

6.3 Model Class Definition

An **stochastic colored Petri net** (SCPN) can be formally defined as a tuple

$$\text{SCPN} = (P, T, \mathcal{T}, Cap, \mathbf{Pre}, \mathbf{Post}, G, \Pi, W, \Lambda, Deg, \mathbf{m}_0, RV)$$

with elements that are explained in the following.

P is the set of **places**, which may contain **tokens**, i.e., objects that model some entity with the respective attributes as explained. Places correspond to state variables of the equivalent SDES.

T denotes the set of **transitions**. Just as for simple Petri nets we require places and transitions to be disjoint, and that a net should not be empty:

$$T \cap P = \emptyset, T \cup P \neq \emptyset$$

The set of all **types** (or **colors**) that are allowed inside an SCPN is given by $\mathcal{T} \subset \mathcal{S}^\star$, which is not defined here further. The construction of colors has been covered in the informal introduction. We denote by Var the set of **variables** over the set of types $\mathcal{T}$, Var_T the set of variables over an individual type T, and by $Expr_{Var}$ the set of all expressions built from variables out of Var. The set of variables contained in an expression e is denoted by $Var(e)$, while the individual type of a variable or expression result is denoted by C, with $\forall v \in Var : C(v) \in \mathcal{T}$.

C is the **color domain** function that associates a color (or type) to each place.

$$C : P \to \mathcal{T}$$

Some places have a maximum number of tokens that are allowed to be stored in them. The **capacity** Cap of a place is a function that returns that number. An unrestricted number of tokens is specified by ∞.

$$Cap : P \to \{\mathbb{N}^+ \cup \infty\}$$

The information of input and output arc inscriptions is formally captured in the **backward** and **forward incidence matrices** of an SCPN, **Pre** and **Post**. Several different approaches exist to formally describe the affected tokens. The one used here is similar to the *term form*, described in [139] as probably the most general one. Arcs going from a place p to a transition t correspond to **Pre** and carry a multiset of variables as inscription. The type of the variables equals the color domain of place p, i.e., $C(p)$.

$$\forall p \in P, t \in T : \mathbf{Pre}(p, t) \in \mathcal{M}_{Var_{C(p)}}$$

The firing mode of a transition in a stochastic colored Petri net is represented by a mapping of values to local **transition variables**. These variables are the ones that appear in the input arc inscriptions **Pre**. We define the corresponding set of transition variables $Var(t)$ of a transition t based on the variables that are contained in the **Pre** expressions as

$$\forall t \in T : Var(t) = \bigcup_{p \in P} Var\big(\mathbf{Pre}(p,t)\big)$$

One possible setting of values for all transition variables is called a **binding**[3] of t and denoted by $\beta(t)$:

$$\forall t \in T, v \in Var(t) : \beta(t,v) \in C(v)$$

The set of all theoretically possible bindings of a transition t is denoted by $\beta^*(t)$, which is formed by the cross-product of the sorts (i.e., sets of values) of all individual transition variables of t:

$$\beta^*(t) = C(v_1) \times C(v_2) \times \ldots \times C(v_{|Var(t)|})$$
$$\text{for } Var(t) = \{v_1, v_2, \ldots, v_{|Var(t)|}\}$$

A binding β of a transition t maps the transition variables $Var(t)$ to values and is thus used to derive the actual value of expressions in the model (e.g., of the forward incidence matrix). The value of an expression $Expr_{Var(t)}$ under a binding $\beta \in \beta^*(t)$ is calculated by evaluating the expression after mapping all variables to the values given by the binding. Such an expression result is denoted by $Expr^{\beta}_{Var(t)}$.

Output arcs (going from a transition t to a place p) carry as inscriptions expressions over the transition variables.

$$\forall p \in P, t \in T : \mathbf{Post}(p,t) \in Expr_{Var(t)}$$

On the basis of the bindings, it is now possible to define the result type of the input and output arc expressions. For input arcs, a variable setting obviously leads to a multiset of tokens of the corresponding input place type. Each output arc expression analogously returns a multiset over the color domain of the connected place, when it is evaluated for a mapping of values to the contained variables. We thus define

$$\forall p \in P, t \in T, \beta \in \beta^*(t) : \mathbf{Pre}(p,t)^{\beta} \in \mathcal{M}_{C(p)},$$
$$\mathbf{Post}(p,t)^{\beta} \in \mathcal{M}_{C(p)}$$

In a certain state only selected bindings (or even none of them) might be enabled. The **guard** G of a transition is a boolean function that returns True for a binding β if it is allowed in a model state $\mathbf{m}$. The guard function of a transition is specified by an expression over the transition variables and model state whose syntax was described earlier.

$$\forall t \in T : G(t) : \beta^*(t) \times M \to \mathbb{B}$$

---

[3] Also sometimes referred to as a **transition color** in the literature.

The **priority** of a transition t is a natural number, which is used to decide which transition fires first if there are several scheduled to do so at the same point in time. Higher numbers mean a higher priority.

$$\Pi : T \to \mathbb{N}$$

W maps each transition to a real number, the **firing weight**. This value is interpreted as a relative probability of firing t in a case where there is more than one transition with equal priority, which are scheduled to fire at the same time. The firing weight is used to decide probabilistically which transition fires first. It is usually only applied to immediate transitions, and the default value is 1.

$$W : T \to \mathbb{R}$$

The **delay** Λ of a transition describes the time that must pass while the transition is enabled until it fires (occurs). The delay is defined by a probability distribution function that describes the random delay time.

$$\Lambda : T \to \mathcal{F}^+$$

The Deg specifies the **degree of concurrency** for each transition. Firing modes of transitions (i.e., bindings) may be enabled concurrently to themselves and others of the same transition. This case is called **infinite server** (IS), while the default case with only one concurrent binding is called **single server** (SS).

$$Deg : T \to \{SS, IS\}$$

To avoid an ambiguous semantics and complex preselections, infinite server transitions must have exactly one input place, and only one variable inscribed at the corresponding arc:

$$\forall t \in T : \big(Deg(t) = IS\big) \longrightarrow \sum_{p \in P} |\mathbf{Pre}(p, t)| = 1$$

A state of a stochastic colored Petri net corresponds to a specific association of token multisets to places, and is called **marking**. Each marking $\mathbf{m}$ is thus a vector indexed by the places, whose entries are multisets of colors. A **token** is an object of a color (type).

$$\forall p \in P : \mathbf{m}(p) \in \mathcal{M}_{C(p)}$$

The **initial marking** $\mathbf{m}_0$ denotes the state of the Petri net model from which the dynamic behavior shall start. Because $\mathbf{m}_0$ is a marking, it is also of the form $\mathbf{m}_0 : P \to \mathcal{M}_T$ with the restriction that the tokens in each place must be of the corresponding type. The initial marking must not violate the restriction of the place capacities.

$$\forall p \in P : |\mathbf{m}_0(p)| \leq Cap(p)$$

The set of all theoretically possible markings is denoted by M.

The performance measures or **reward variables** of a stochastic colored Petri net are denoted by RV. Two basic types have been informally introduced earlier, which either correspond to token numbers in a place or to the number of times that a transition fires. The user-level specification of a performance measure allows to arbitrarily mix these basic measures and to use them as terms in a numerical expression with operators. This technically exceeds the formal expressiveness of SDES reward variables. Therefore, only basic reward measures of an SCPN that correspond to places or transitions are considered in the following; it is obvious how a tool implementation can cut a more complex expression into allowed terms, compute them, and derive the final result by applying the original expression.

With this restriction the definition of SCPN reward variables is done as follows. The reward variable applies to either a place or a transition of the model. This **reward variable object** is stored in $robj$. The second parameter $rexpr$ specifies the optional filter expression as a boolean function either on tokens in a place or on transition bindings. If no filter expression has been used in the reward variable specification (the default), $rexpr$ is always true.

$$\forall rvar \in RV : rvar = (robj, rexpr)$$
$$\text{with} \quad robj \in P \cup T$$
$$\text{and} \quad \begin{cases} rexpr : C(robj) \to \mathbb{B} & \text{if } robj \in P \\ rexpr : \beta^*(robj) \to \mathbb{B} & \text{if } robj \in T \end{cases}$$

6.4 A **SDES** Description of Colored Petri Nets

Remember that a stochastic discrete event system has been defined as

$$\text{SDES} = (SV^*, A^*, S^*, RV^*)$$

The specific settings of an SDES representing a CPN model are defined in the following.

The set of state variables SV^* is given by the set of Petri net places

$$SV^* = P$$

and each state of the SDES is equal to a marking of the colored Petri net. The value of a state variable in a state is given by the multiset of colors (tokens) that is contained in the corresponding place in the marking.

The set of SDES actions A^* equals the set of transitions of the Petri net.

$$A^* = T$$

The sort function of the SDES maps to the type function (color) of the Petri net for places and variables of transitions.[4]

$$S^\star(x) = \begin{cases} C(x) & \text{if } x \in P \\ C(x) & \text{if } x \in \bigcup_{t \in T} Var(t) \end{cases}$$

Attributes of the SDES state variables $SV^\star$ correspond to the details of the Petri net places. Namely the condition function $Cond^\star$ and initial value $Val_0{}^\star$ are set for SCPN models as follows. The condition function is true if the capacity of a place is not exceeded.

$$\forall p \in P : Cond^\star(p, \mathbf{m}) = \big(|\mathbf{m}(p)| \leq Cap(p)\big)$$

The initial value is given by the initial marking.

$$\forall p \in P : Val_0{}^\star(p) = \mathbf{m}_0(p)$$

Because the capacities are respected in the initial marking of the Petri net by definition, the condition functions of the SDES are not violated.

Actions $A^\star$ of the SDES are given by transitions as stated above. Their associated attributes are specified using the SCPN elements as follows.

The priority $Pri^\star$ is copied from the one defined in the Petri net.

$$\forall t \in T : Pri^\star(t) = \Pi(t)$$

The set of action variables of an action that corresponds to a transition t is equal to the set of transition variables $Var(t)$ of t. The sort is given by the type in the Petri net, as it has already been specified above.

$$\forall t \in T : Vars^\star(t) = Var(t)$$

As one action mode of an SDES action is equivalent to an individual setting of the action variables, action modes correspond directly to bindings of the SCPN. The set of all action modes $Modes^\star(t)$ equals the set of all possible bindings for a transition t.

The execution of an enabled action mode is done by firing the associated transition t under its binding β, leading from a marking $\mathbf{m}$ to $\mathbf{m}'$. This subsequent marking $\mathbf{m}'$ is calculated by subtracting from $\mathbf{m}$ the multisets of tokens from places given by the backward incidence matrix $\mathbf{Pre}$ and by adding tokens as specified by the expressions in the forward incidence matrix $\mathbf{Post}$. Only the part of both matrices that corresponds to the firing transition t is used, and the expressions are evaluated under the binding β.

$$\forall t \in T, \forall \beta \in \beta^\star(t), \forall \mathbf{m} \in M, \forall p \in P :$$

$$Exec^\star(t, \beta, \mathbf{m})(p) = \mathbf{m}(p) - \mathbf{Pre}(t, p)^\beta + \mathbf{Post}(t, p)^\beta$$

---

[4] More formally, only the *sort* of the type has to be taken in the formula. The term *type* is used in the context of colored Petri nets as it is done in the literature.

The enabling of an action mode now needs to be defined depending on the enabling of the corresponding Petri net binding. More correctly we say that a transition t of an SCPN is enabled under a binding $\beta \in \beta^*(t)$ in a marking $\mathbf{m}$. This is the case if and only if (1) the enabling function of the transition is true for the binding, (2) there are enough tokens in the input places of the transition that match the values, which are bound to the variables of the input arcs, and (3) the execution (defined below) of the transition under the binding would not violate the capacity restrictions. Because the input arc inscriptions are captured as expressions in $\mathbf{Pre}(\cdot, \cdot)$, their actual value under a binding β is denoted by $\mathbf{Pre}(\cdot, \cdot)^\beta$ as introduced earlier.

$\forall t \in T, \beta \in \beta^*(t), \mathbf{m} \in M :$

$$
\begin{aligned}
Ena^\star(t, \beta, \mathbf{m}) &= G(t, \beta, \mathbf{m}) && \text{guard} \\
&\wedge \forall p \in P : \mathbf{Pre}(p, t)^\beta \subseteq \mathbf{m}(p) && \text{input tokens} \\
&\wedge \forall p \in P : |Exec^\star(t, \beta, \mathbf{m})(p)| \leq Cap(p) && \text{capacity}
\end{aligned}
$$

The enabling degree $Deg^\star$ is either 1 for single server transitions or infinity for infinite server transitions. For a specific binding and thus action variant, the actual enabling degree in a state $VDeg^\star$ equals the number of times that the enabled binding could fire in the current marking for those of type infinite server. Because of the restriction of only one input place p_i with one variable v_i in the inscription of an infinite server transition t ($\mathbf{Pre}(p_i, t) = \{v_i\}$), it is not necessary to fire the transition for a derivation of the degree. The number of tokens in p_i that are mapped to v_i by a binding β can be used instead. It appears natural to interpret the enabling degree that way without checking output place capacities after an eventual firing.[5]

$\forall t \in T, \beta \in \beta^*(t), \mathbf{m} \in M :$

$$
Deg^\star(t) = \begin{cases} 1 & \text{if } Deg(t) = SS \\ \infty & \text{if } Deg(t) = IS \end{cases}
$$

$$
VDeg^\star(t, \beta, \mathbf{m}) = \begin{cases} 1 & \text{if } Deg(t) = SS \\ |\{\beta(t, v_i) \in \mathbf{m}(p_i)\}| & \text{if } Deg(t) = IS \end{cases}
$$

The delay of an action mode equals the delay of the transition of the corresponding binding. In an SCPN the delays of all bindings of one transition are always the same by definition.

$$
\forall t \in T : Delay^\star(t, \cdot) = \Lambda(t)
$$

The weight of an action mode is given by the weight of the transition that the associated binding belongs to.

$$
\forall t \in T : Weight^\star(t, \cdot) = W(t)
$$

---

[5] $\beta(t, v_i)$ equals the token value that is bound to variable v_i by binding β.

The set of reward variables of the colored Petri net RV is converted into the set of SDES reward variables as follows. Parameters $rint^\star$ and $ravg^\star$ specify the interval of interest and whether the result should be averaged over $rint^\star$. Both correspond to the type of results the modeler is interested in and the related analysis algorithm. They are not related to the model or the reward variable definition of the model, and thus set according to the type of analysis and not considered here.

There is one SDES reward variable for each SCPN reward variable, such that

$$RV^\star = RV$$

with the following parameter setting

$$\forall rvar \in RV^\star : \begin{cases} \left.\begin{array}{l} rrate^\star = \big|\{x \in \mathbf{m}(robj) \mid rexpr(x)\}\big|, \\ rimp^\star = 0 \end{array}\right\} \quad \text{if } robj \in P \\[2em] \left.\begin{array}{l} rrate^\star = 0, \\ rimp^\star(robj, \beta) = \begin{cases} 1 & \text{if } rexpr(robj, \beta) \\ 0 & \text{otherwise} \end{cases} \end{array}\right\} \quad \text{if } robj \in T \end{cases}$$

This ensures that in the case of a place-related measure the rate rewards are earned according to the number of (matching) tokens of the place, while in the transition-related case an impulse reward of one is collected if the measured transition fires with an accepted binding.

6.5 Variable-Free Colored Petri Nets

The modeling of complex real-life systems with uncolored Petri nets usually leads to large models that are hard to understand and maintain. Colored Petri nets as described before offer more advanced modeling facilities like distinguishable tokens and hierarchical modeling. The pure graphical description method of Petri nets is, however, hampered by the need to define colors and variables comparable to programming languages. Variable-free colored Petri nets (short: vfSCPN) are a modeling class that may be seen between simple and colored Petri nets. The specification of token variables and the issue of bindings is omitted, while tokens can still be distinguished.

vfSCPNs are informally introduced in the following using a small example. Section 6.5.2 gives a more formal definition, which is restricted to the differences with respect to stochastic colored Petri nets. The interpretation of vfSCPN models in terms of a stochastic discrete event system is shown in Sect. 6.5.3. A more complex application example is covered in Chap. 16. References to relevant literature are given in the notes at the end of this chapter.

6.5.1 An Example

Figure 6.7 shows a vfSCPN model of the synchronization between vehicles and trains at a level crossing. Places and transitions beginning with a T specify train behavior, while those with V describe vehicles. The top left part shows the states and state transitions of the gate. It should be noted that the model is only intended to explain vfSCPN models informally, and is no model of guaranteed safe behavior.

Types of places and thus tokens are formally defined as for colored Petri nets. The type of each place and an optional capacity is written besides it as well. The default empty type is again omitted. However, as there are no variables and no operations on types that change attributes of a token allowed, all elements of every type must be explicitly used somewhere in the model. It would otherwise not be possible to create such a token. One may therefore restrict to strings as color identifiers without loss of generality. Another consequence is that it is technically not necessary to define the types, because all elements (values) of types can be extracted from the model syntactically. In our example model places Up and Down have the default empty color. Places and arcs of that color are drawn thin to improve readability. The remaining places contain tokens of a type Obj, which is defined as having the elements $Obj = \{\texttt{train}, \texttt{car}, \texttt{truck}\}$. Places of that type are obviously used to model locations of trains and vehicles.

Having no variables in arc inscriptions means that every arc inscription is a constant token multiset (i.e., a multiset over the sort of the place). Examples of that case are the input and output arcs of train-related transitions in the figure, where the inscription is always train. An example notation of

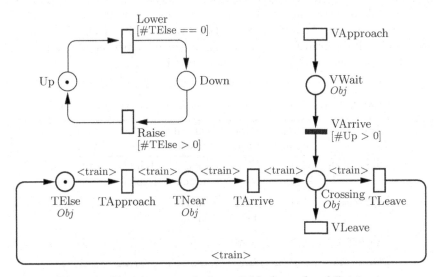

Fig. 6.7. Notation example for variable-free colored Petri nets

a more complex multiset constant would be <3#car + 2#truck>. Arcs may not model behavior that is similar for different token values as in SCPN, but allow for selectivity of token values. An example is transition TLeave, which fires only for train tokens, although there may be vehicle tokens in place Crossing as well. Arcs connecting uncolored places do not need an inscription just like in a simple Petri net (c.f. the upper left part of the example).

vfSCPN models do allow, however, some kind of flexibility in the transition firing with the notion of **transition modes** despite the fixed arc inscriptions. Such a mode describes the input and output tokens, the delay and weight of different transition variants individually. The behavior of a transition mode may thus be described by an uncolored transition, which allows to interprete a vfSCPN transition as a folding of the different modes. Every mode is an explicit description of a SDES action mode in contrast to SCPN bindings, which may be enumerated during the model evolution and for a specific marking only.

The notation of the token value at the arc (like train) is only a visual shorthand for a transition with only one transition mode. In the general case there are different arc inscriptions depending on the mode, which can not be graphically depicted easily. The general way of describing a transition is a table containing all transition modes and their attributes. Table 6.4 shows this table for the transitions in the example that have more than one transition mode.

The table lists the input and output behavior; the places are clear for the example and thus not mentioned. In the general case the token multisets must be specified together with the corresponding input or output place to avoid ambiguities. The interarrival time between two consecutive cars is assumed to be exponentially distributed with a mean of 2 time units (20 for trucks). The time to drive over the crossing is set as 1 and 2 time units. The delay must be consistent with the transition type, i.e., in a timed transition all delays must be greater than zero with probability one and vice versa.

Table 6.4. Transition mode details of the level-crossing example

Transition	Mode number	Input token	Output token	Delay	Firing weight
VApproach	1	–	car	Exp(2)	1
	2	–	truck	Exp(20)	1
VArrive	1	car	car	0	1
	2	truck	truck	0	1
VLeave	1	car	–	1	1
	2	truck	–	2	1

Firing weights may be used to define firing probabilities mainly for immediate transitions, but this was not necessary in the example. Marking-dependent guard functions are available to disable transitions based on the model state. The syntax of expressions is the same as for SCPN. In the example, vehicles may only enter the crossing itself if the gate is up (`[#Up>0]`), and the gate is raised if the train is neither near nor on the crossing. Hierarchical refinement of substitution transitions is possible just as for SCPN models.

The syntax and meaning of performance measures is the same as for stochastic colored Petri nets. However, filter expressions are only available for places, and just specify a token value without an attribute. Examples would be the number of train crossings `#TLeave` and the probability of a forbidden state (`#Crossing(train)>0`) `&&` (`#Crossing(car)+#Crossing(truck)>0`).

The dynamic behavior that is described by a vfSCPN model is defined in a very similar way to that of SCPN. The main difference is that bindings of variables do not have to be computed, because all action modes are already explicitly described by the transition modes.

6.5.2 Model Class Definition

A **stochastic variable-free colored Petri net** (vfSCPN) is similar to a SCPN and defined as follows:

$$\text{vfSCPN} = (P, T, C, \mathit{Cap}, \mathbf{Pre}, \mathbf{Post}, G, \Pi, W, \Lambda, \mathit{Deg}, \mathbf{m}_0)$$

The elements of this definition are detailed below. However, explanations and definitions are only given if they differ from the corresponding SCPN definitions in Sect. 6.3.

P and T denote the transitions and places. The capacity Cap of places, the transition priorities Π, server semantic Deg as well as the guard function G^6 are defined as for SCPN. However, no variables are defined as it is done with Var for colored Petri nets. The guard function is formally defined as for SCPN due to the similarity between transition modes and bindings. It should be noted that the guard represents a marking-dependent boolean function for each transition mode in the vfSCPN context. C associates a color (type) to each place.

States (markings) are defined just like for SCPN. A marking $\mathbf{m}$ thus associates a multiset of tokens of the matching sort to every place. The definitions and explanations of the initial marking $\mathbf{m}_0$ and the set of all theoretically possible markings M are thus also equivalent for vfSCPN.

The **firing modes** of a transition in a stochastic variable-free colored Petri net are directly given by the set of firing possibilities[7]. Most of the other

---

[6] In the original definition on which this section is based [354, 361], there is a distinction between a global guard of the whole transition and local guards defined per mode. This is omitted here for simplification.

[7] Termed **transition table** in [361].

transition properties depend on the mode as explained below. There are no transition variables beside the selection of a certain firing mode. Every mode is thus equivalent to an explicit SCPN binding specification, and the set of all of them corresponds to the set of all theoretically possible bindings of a SCPN transition. In the vfSCPN setting we thus use the bindings symbol $\beta^*(t)$ for the set of firing modes of a transition t. We do not define the element sort of $\beta^*(\cdot)^8$, but use the elements as parameters for the firing mode properties. All mode sets are however required to be finite.

$$\exists k \in \mathbb{N} : \forall t \in T, |\beta^*(t)| < k$$

Backward and forward incidence matrix elements need to be defined depending on the transition mode, as any one of them may remove and add different tokens. Every arc corresponds to a multiset over the color domain of the connected place for a specific transition mode. Thus

$$\forall p \in P, t \in T : \ \mathbf{Pre}(p,t) : \beta^*(t) \to \mathcal{M}_{C(p)}$$
$$\wedge \ \mathbf{Post}(p,t) : \beta^*(t) \to \mathcal{M}_{C(p)}$$

The delay Λ of a vfSCPN transition may depend on the transition mode as well. To simplify understanding, timed and immediate transitions must only possess corresponding delays in their respective mode specifications. In any case the delay is defined as a distribution function like for SCPN.

$$\forall t \in T, \Lambda : \beta^*(t) \to \mathcal{F}^+$$

Firing weights are also be defined depending on the transition mode, while the default value is one.

$$\forall t \in T, \Lambda : \beta^*(t) \to \mathbb{R}^+$$

Reward variables are formally defined exactly like for colored Petri nets. The only difference is in the (simpler) syntax, which has been explained in the informal introduction above. Transition throughput may only be measured for complete transitions, i.e., without filter expressions.

6.5.3 Representing vfSCPNs as SDES

Due to the similarity in the model definitions of SCPN and vfSCPN, their transformation into a SDES model only differs in a few details which are covered below without the repetition of unchanged definitions.

The SDES representation is identical to the one given for SCPN in Sect. 6.4 namely for the set of state variables SV^*, the SDES states and actions, action priorities, guards and enabling degrees as well as the set of all sorts in the

---

[8] It just needs to be a set of identifiers like an enumeration type.

model and the reward variables. The same applies to the condition function (for place capacities) and initial value (for the initial marking).

The set of action variables of a SDES action that corresponds to a vfSCPN transition t contains exactly one element denoted by mv_t. It identifies the different explicitly defined transition modes.

$$\forall t \in T : Vars^\star(t) = \{mv_t\}$$

The SDES sort function equals the sort of the place type for places and the transition mode set for the action variables mv_t (compare the corresponding notes for SCPNs).

$$S^\star(x) = \begin{cases} C(x) & \text{if } x \in P \\ \beta^*(t) & \text{if } x \in Vars^\star(t) \end{cases}$$

One vfSCPN transition mode is equivalent to a SDES action mode just like bindings are for colored Petri nets. The set of all action modes thus equals the set of transition modes $Modes^\star(t) = \beta^*(t)$.

Enabling of transition modes and the corresponding SDES action modes are exactly defined as for SCPN, if the term $\mathbf{Pre}(p,t)^\beta$ is understood as the similar $\mathbf{Pre}(p,t)(\beta)$ in the vfSCPN setting. The execution of transition modes is also identically defined under the mentioned understanding of $\mathbf{Pre}$.

The delay of an action mode and the weights are just copied from the transition attributes. This differs from the colored Petri net definitions where these properties are defined per transition and not per mode.

$$\forall t \in T, \beta \in \beta^*(T) : Delay^\star(t,\beta) = \Lambda(t,\beta)$$
$$\wedge \quad Weight^\star(t,\beta) = W(t,\beta)$$

Notes

There is a large quantity of literature available on Petri nets with individual tokens. Different variants have evolved, namely colored Petri nets [187, 188], algebraic high-level nets [90, 315], and Predicate/Transition nets [120–122]. An overview of the theory and some applications is given in [186]. Theoretical background and qualitative properties of some simple and colored Petri net types are covered in [139].

Extensions of colored Petri nets by stochastic firing times have been introduced in [222, 223, 335]. A proper structuring of the color set leads to the definition of stochastic well-formed nets [11, 55], for which the numerical analysis can be carried out more efficiently [41]. Applications are e.g., reported in [13, 215].

Chapters 15 and 16 give some pointers to application-oriented references using colored Petri nets. The kind of stochastic colored Petri nets presented throughout the chapter was developed within an industrial project [331, 363]. Its creation was mainly influenced by colored Petri nets (CPNs) as defined in [188]. The main differences are easier specification of arc inscriptions, allowing an automated generation of efficient analysis algorithm code, and a true stochastic timing semantics, which is in accordance to the usual understanding of timed Petri nets. Type definitions are simplified, and arc variables do not need to be declared as they have to be in CPNs using the specification language Standard ML.

Remarks on vfSCPN-related work and publications on which the description is based can be found in the Notes of Chap. 6.5 in p. 339. The net class *arc-constant colored Petri nets* defined in [139] is similar to vfSCPN models if we restrict to only one transition mode.

Part II

Evaluation

7

Standard Quantitative Evaluation Methods for **SDES**

Different modeling formalisms and their translation into a general SDES model have been covered in the previous chapters. One of the main reasons for constructing an SDES model is the prediction of properties of the modeled system. This chapter describes a selection of well-known analysis methods that derive quantitative measures from the model, and which can thus be used for a performance and dependability evaluation of a planned system. The starting point is always the dynamic behavior of a model over time, which is given by the stochastic process described by the model as shown in Sect. 2.3.2, and the derivation of the reward measures of interest.

Examples of measures include performance (often in terms of through-put), dependability issues, and combinations of them. They are expressed by reward variables in an SDES model as it has been described in Sect. 2.4.1. The applications described in Part III include different examples. A model-based experiment requires a model, performance measures, and their type (e.g., transient or steady-state). The selection of an evaluation algorithm depends on the type of reward variable as well as the mathematical complexity of the solution, because preferable algorithms might exist for models with certain restrictions.

For simple model classes, it is possible to directly derive mathematical expressions from a specific model, which can then be solved symbolically to compute the values of the measures. This is, however, only possible for very restricted models, and is often hard to implement as an algorithm for a variety of models. This technique is outside the scope of this text, because we are interested in analysis methods that are applicable to certain classes of SDES models in general.

Numerical analysis techniques are based on mathematical formulations of the model dynamics. The stochastic process of a model is, for instance, described by matrix equations or (partial) differential equations. Reward measures can be derived in closed form for special cases, or by numerical algorithms with fewer restrictions. The reachability graph of all possible states and state transitions is the basis for these methods. Numerical techniques such as iterative computations, vector and matrix operations, and numerical integration are examples of the used numeric algorithms. The applicability of this class of techniques depends on the complexity of the stochastic process in terms of the state space size and the combinations of delay distributions.

Delays of action variants may be exponentially distributed or zero, leading to a continuous-time Markov chain (CTMC) as the underlying stochastic process. Those *Markovian* SDES are often used due to the relatively simple analysis. However, the assumption of a memoryless delay distribution is not realistic in many cases and can lead to significant differences for the computed measures. Action variants with deterministic or more generally distributed delays are needed for the modeling of systems with clocking or fixed operation times. Many technical systems of great interest from the fields of communication, manufacturing, and computing belong to this class.

If the restriction of only exponentially distributed and immediate delays is relaxed, the numerical analysis becomes much more complex. The stochastic process is not memoryless any more in states with an executable action variant with nonexponential delay. Examples of *non-Markovian* SDES are given in the final notes of Chap. 5. If, in every reachable state of an SDES model, there is not more than one executable action with a delay that is not zero or exponentially distributed, numerical analysis algorithms have been presented in the literature, which are adopted here.

If the mentioned restriction of delay distributions is violated or the direct numerical analysis is too complex to be handled, **discrete event simulation** can be used to estimate the values of the measures. Large state spaces and non-Markovian models are not a problem for such an algorithm, but a sufficient accuracy of the results and rare events may cause long run times.

Simulation of an SDES model in the sense used throughout this text is always possible. The simulation algorithms implement the dynamic behavior of a system in a way very similar to the stochastic process described in Sect. 2.3. Just like the Latin root of the word *simulare* (to make like, imitate) says, the system behavior is mimicked in the computer model. In doing so, the algorithm follows one of the many possible evolutions of the stochastic process, the state trajectories, or sample paths of the SDES.

Specification and theoretical derivation of performance measures for SDES models have been defined in Sect. 2.4. The reward variable description contained in the model specification includes – besides the most important rate and impulse rewards – the type of the variable derivation. Namely the observation interval $rint^* = [lo, hi]$ and if the result should be averaged over time or not in $ravg^*$ are given. Unfortunately no single evaluation algorithm can be used to compute all possible performance measure variants. Table 7.1 shows which one of the available methods is applied for each case of measure. We differentiate between the cases of performance measures given in Sect. 2.4.1.

The selection of the algorithms presented below is guided by the integration of different model classes into the SDES generalization. Standard

Table 7.1. Types of performance measures and evaluation methods

Type of variable	$[lo, hi]$	$ravg^*$	Appropriate algorithms
Instant-of-time	$[t, t]$	False	TRANSIENTSIMULATION, p. 142
			or transient analysis, p. 148
Interval-of-time	$[0, t]$	$\cdot$	TRANSIENTSIMULATION, p. 142
			or transient analysis, p. 148
Steady-state	$[0, \infty]$	True	STEADYSTSIMULATION, p. 140
			or steady-state analysis, p. 149
Alt. steady-state	$[\infty, \infty]$	True	Is mapped to standard steady-state

algorithms known from the literature for subclasses are given in the generalized SDES setting. Due to the mapping of the different model classes to an SDES, every algorithm can be used to evaluate a model of any of the specific classes. The algorithms thus only use the definitions of SDES models as described in Chap. 2. Additional algorithms are necessary to completely derive the reward measure values; they are, however, only briefly mentioned if they do not differ from the standard application using results of the SDES-specific algorithms.

The advantage of unified algorithms for different model classes, however, has a downside. There are well-known analysis methods for specific classes that are not applicable for the general SDES case. In comparison to the SDES algorithms they are, however, often more efficient or even do not exist otherwise. Product-form solutions for classes of queuing networks, which compute performance measures without the need to analyze the whole state space, are an example. Standard model class-specific algorithms are not the focus of this text; the reader is referred to the standard literature about the model classes and their analysis algorithms for further information.

Some prerequisites for the later description of algorithms are covered in the subsequent section. Simulation algorithms for the transient and steady-state simulation (namely TRANSIENTSIMULATION and STEADYSTSIMULATION) are presented in Sect. 7.2. Some common functions are presented in Sect. 7.2.1, including a description of the syntax for algorithms used throughout this text.

Numerical analysis algorithms as opposed to simulation are covered in the subsequent Sect. 7.3. For any method therein, the first step is the generation of the reachability graph. The corresponding algorithm GENERATEFULLRG is covered in Sect. 7.3.1. Depending on whether the underlying stochastic process of the SDES model is a Markov chain or not, the appropriate numerical evaluation algorithm can be selected from the ones described in Sects. 7.3.2 and 7.3.3. Notes on background material and publications used for this section are given at the end of the chapter. Additional evaluation algorithms are presented in the subsequent chapters of this part. Application examples for the different methods can be found in Chaps. 13–16.

7.1 Prerequisites

This section describes terms and notation for the quantitative evaluation of stochastic discrete event systems used in the following. They build upon the terms introduced in Sect. 2.3 for the dynamic behavior of SDES models. It should be noted that we restrict ourself in the following to the "incomplete" states of the SDES model without action states. Recall that such a state $\sigma \in \Sigma$ associates a value to each state variable $sv \in SV^\star$, and is the same as considered by the simplified process $SProc$. The prerequisites of an action a together with one of its modes $mode$ for being enabled have been explained in the section mentioned above:

- The **initial state** of an SDES model is the starting point of any evaluation and denoted by σ_0. The values of all state variables in the initial state are specified by $Val_0{}^\star$.

$$\sigma_0 \in \Sigma \text{ such that } \forall sv \in SV^\star : \sigma_0(sv) = Val_0{}^\star(sv)$$

- An action variant v is said to be **executable** in a state σ if its action mode is enabled in it and there are no other enabled action modes with immediate delay (with a higher priority, if v is also immediate). The set of executable variants in a state σ is thus defined as

$$
\begin{aligned}
Executable(\sigma) = \{ v = (a, mode) \in &AV \mid mode \in Enabled(a, \sigma) \wedge \\
Delay^\star(v) \in \mathcal{F}^{im} \longrightarrow &\big(\forall a_i \in A^\star : Delay^\star(a_i, \cdot) \notin \mathcal{F}^{im} \\
&\vee Pri^\star(a_i) \leq Pri^\star(a) \\
&\vee Enabled(a_i, \sigma) = \emptyset \big) \\
Delay^\star(v) \notin \mathcal{F}^{im} \longrightarrow &\big(\forall a_i \in A^\star : Delay^\star(a_i, \cdot) \notin \mathcal{F}^{im} \\
&\vee Enabled(a, \sigma) = \emptyset \big) \\
\}
\end{aligned}
$$

- A state σ' is said to be **directly reachable** from another state σ $(\sigma', \sigma \in \Sigma)$, if there is an action variant $v = (a, mode) \in AV$ executable in state σ, which causes the state to be changed to σ' by its execution. We write

$$
\begin{aligned}
\sigma \xrightarrow{v} \sigma' \text{ iff } &v \in Executable(\sigma) \\
&\wedge \forall sv \in SV^\star : \sigma'(sv) = Exec^\star\big(a, mode, \sigma(sv)\big)
\end{aligned}
$$

- A state σ_n is said to be **reachable** from another state σ_1, if there exists a sequence of states $\sigma_1, \sigma_2, \ldots, \sigma_n$ that are directly reachable from the previous one each by executing arbitrary action variants. We denote this by

$$
\begin{aligned}
\sigma_1 \longrightarrow \sigma_n \text{ iff } &\exists \sigma_2, \ldots, \sigma_{n-1} \in \Sigma, \exists v_1, \ldots, v_{n-1} \in AV \\
&\text{such that } \forall i \in \{1, \ldots, n-1\} : \sigma_i \xrightarrow{v_i} \sigma_{i+1}
\end{aligned}
$$

If the executed set of action variants V that lead from σ to σ' matters, we can also write $\sigma \xrightarrow{V} \sigma'$. The simplest case is of course an empty sequence

$V = \emptyset$, and we consequently say that a state σ is reachable from itself with an empty sequence $\sigma \xrightarrow{\emptyset} \sigma$.

- The **reachability set** RS of an SDES is the set of all states that are reachable from its initial state σ_0.

$$RS = \{\sigma \in \Sigma \mid \sigma_0 \longrightarrow \sigma\}$$

States in which at least one activity with a zero delay (i.e., immediate) is enabled are called **vanishing**, because no time is spent in the state. The remaining ones are called **tangible**. Elements of the reachability set RS are thus divided into tangible states RS^{tan} and vanishing states RS^{van} as follows.

$$RS^{van} = \{\sigma \in RS \mid \exists v \in Executable(\sigma) : Delay^\star(v) \in \mathcal{F}^{im}\}$$
$$RS^{tan} = RS \setminus RS^{van}$$

- The graph that has all reachable states RS as vertices and state changes due to activity executions as edges RE is called the **reachability graph** and denoted by RG. It is a directed weighted graph. Edges of the graph correspond to state transitions, which map pairs of source and destination states to the executed action and action mode. The set of all edges is denoted by RE.

$$
\begin{aligned}
RG &= (RS, RE) \\
RE &\subseteq RS \times RS \times AV \\
&\text{with } \forall(\sigma, \sigma', v) \in RE : \sigma \xrightarrow{v} \sigma'
\end{aligned}
$$

- The reachability graph can be transformed into the **reduced reachability graph** RRG, where all states in which zero time is spent (vanishing) and all state transitions with an immediate delay are removed. The remaining graph has to be changed to capture the same dynamic behavior. The removed immediate state transitions are stored together with the previous timed state transitions such that edges of the reduced reachability graph describe one timed action variant together with any number of subsequent immediate ones.

Let $AV^{im} = \{v \mid Delay^\star v \in \mathcal{F}^{im}\}$, $AV^{exp} = \{v \mid Delay^\star v \in \mathcal{F}^{exp}\}$, and $AV^{gen} = \{v \mid Delay^\star v \in \mathcal{F}^{gen}\}$ denote the set of all action variants with immediate, exponentially, and more generally distributed delays, respectively.

$$
\begin{aligned}
RRG &= (RRS, RRE) \text{ with} \\
RRS &= RS^{tan} \text{ and} \\
RRE &\subseteq RRS \times RRS \times AV \times 2^{AV} \\
&\text{with } \forall\big(\sigma, \sigma', v_0, \{v_1, v_2, \ldots, v_n\}\big) \in RRE : \\
&\qquad (\sigma, \sigma'_0, v_0) \in RE \wedge
\end{aligned}
$$

$$(\sigma_0', \sigma_1', v_1) \in RE \wedge \ldots \wedge (\sigma_{n-1}', \sigma', v_n) \in RE \wedge$$
$$v_0 \notin AV^{im} \wedge \{v_1, v_2, \ldots, v_n\} \subseteq AV^{im}$$

All paths of immediate action variant executions from a state after a timed state change must be followed until the next tangible state has been reached. The edge set RRE of the reduced reachability graph is thus constructed as (V may be an empty sequence, in which case $\sigma' = \sigma''$)

$$\forall (\sigma, \sigma', v) \in RE, v \notin AV^{im}, V \subseteq AV^{im}, \sigma'' \in RRS \ :$$
$$(\sigma, \sigma'', v, V) \in RRE \text{ iff } \sigma' \xrightarrow{V} \sigma''$$

- The **probability of an immediate state transition** from a vanishing state $\sigma \in RS^{van}$ to a directly reachable state $\sigma' \in RS$ by executing v is denoted by $P\{\sigma \xrightarrow{v} \sigma'\}$ and can be derived from the weights of the enabled action variants.

$$\forall \sigma \in RS^{van}, \sigma' \in RS \text{ with } \sigma \xrightarrow{v} \sigma' \ :$$
$$P\{\sigma \xrightarrow{v} \sigma'\} = \frac{Weight^\star(v)}{\sum_{v_i \in Executable(\sigma)} Weight^\star(v_i)}$$

- The **path probability** $P\{\sigma \longrightarrow \sigma'\}$ denotes the overall probability of following any (possibly empty) path consisting of immediate state transitions that lead from a vanishing state σ to a state σ'. It is computed from the sum of all possible single path probabilities, for which the probability is the product over all immediate state transition probabilities.[1]

$$\forall \sigma_1 \ldots \sigma_{n-1} \in RS^{van}, \sigma_n \in RS \ : \tag{7.1}$$

$$P\{\sigma \longrightarrow \sigma'\} = \sum_{\substack{V = \{v_1, v_2, \ldots, v_{n-1}\} \subseteq AV^{im} \\ \sigma_1 \xrightarrow{v_1} \sigma_2 \xrightarrow{v_2} \ldots \xrightarrow{v_{n-1}} \sigma_n}} \prod_{i=1}^{n-1} P\{\sigma_i \xrightarrow{v_i} \sigma_{i+1}\}$$

Path probabilities and the underlying probabilities of immediate state transitions are related to the resolution of conflicts between action variants with zero delay. Action variants of this type are often used to model decisions that depend on the current state or the distribution and further paths of customers, parts or tokens. From the point of view of an analysis algorithm or a software tool, the definition of the dynamic behavior of an SDES model as given in Sect. 2.3 is sufficient, because it completely defines how such conflicts are resolved. Actions with higher priorities $Pri^\star(\cdot)$ are preferred. If there are several executable action variants with the same priority (and zero delay), a probabilistic decision is made depending on the weights $Weight^\star(\cdot)$.

---

[1] We require that the reachability graph does not contain circular state transition paths containing only immediate actions. These *vanishing loops* can be resolved numerically [66], but induce infinitely many entries in the sum of (7.1).

There are models in which the specification of weights and priorities leads to situations in which the further behavior of the model depends on the decision of which immediate action variant is executed. This decision, however, cannot be based on timing properties, because there is no first action variant if all have a zero delay. If different sequences of immediate executions lead to differing outcomes, a **confusion** has been encountered, and causal properties of the model might not be handled in a way that the modeler intended. The correct specification of priorities and weights may be a problem for a modeler, because he or she works on the structural level of the model and should not be concerned about all possible paths of states and state transitions.

Different methods have been proposed in the literature to avoid this problem. We do not go into the details of this issue; the mentioned methods could, however, be applied to SDES models as well. *Extended conflict sets* (ECS) of transitions are derived in [53] for generalized stochastic Petri nets. Such a set contains all transitions that may be in conflict during the model evolution. Weights are then regarded as relative probabilities between transitions in each ECS. Structural checks are available that ensure the independence of transition firing order between different ECS. This approach is, e.g., implemented in GreatSPN [51] and TimeNET [359]. The Möbius software tool implements a state space-based *well-defined check* that detects confusions [81, 82]. The last reference gives a good overview about the different approaches to confusions and their historical development. Confusions occur especially often in models which are interpreted in discrete time, such as discrete-time deterministic and stochastic Petri nets [337, 338]. Conditions for stochastic Petri net model to be *well defined* have been given in [70] based on an analysis of the stochastic process. One of the most recent discussions is given in [303].

7.2 Next-Event Time Advance Simulation

A simulation algorithm for a discrete event system computes sequences of states and state transitions. Section 2.3.2 defined the stochastic process of an SDES model formally. The same steps are carried out in an implementation: determine the activity to be executed, change the state according to the execution, and update the model time. The event execution is atomic and does not take time in an SDES model. States and thus accumulated rewards are constant between subsequent events and hence do not need to be simulated in more detail. The simulation time may instead jump from one event time to the next, which resulted in the name *next-event time advance*.

For a correct implementation of the formal definition, the complete state must be captured, i.e., action states together with the state variable values. While the action states are just a set of tuples from the theoretical standpoint, a reasonable implementation will use a more efficiently manageable data structure. The set of activities is stored in an **event list**, an ordered list with all currently scheduled events and their planned execution times. The event list is the algorithmic equivalence of the action state as of the complete stochastic process, and it is ordered such that entries with smaller scheduled times come first. The activity to be executed next will then always be the first entry in the list (or at least one of the first, if there are several scheduled for the same time).

Another difference between implementation and definition is that times in the event list are absolute simulation times rather than remaining activity delays as used in the formal definition of the stochastic process. If relative times would be stored, the sojourn time spent in a state would have to be subtracted from all activities after a state change. Whenever a new action variant becomes enabled in a new state, a corresponding activity is stored in the event list after the scheduled time is set to the current time plus a random value drawn from the action's delay distribution. The general approach to a simulation using an ordered event list is called **event-scheduling scheme**.

Other simulation principles include the process-oriented scheme, where individual entities are treated with individual processes. Such a process may, e.g., include creation, waiting for resources, service and departure from the model. States and events are then viewed from the side of the process, and process-individual functions implement actions and delays. The advantage of the event-scheduling technique lies in its generic applicability to a variety of systems, and its clear separation between static and dynamic model elements as well as a model-independent simulation engine. The process-oriented scheme can be used in an object-oriented program that models the real system. A class then corresponds to an object type, and instantiated objects model individual entities. Attributes and behavior of entities are naturally implemented as member variables and methods. The process view can be mapped naturally to a distributed simulation using multiple processing units (cf. Sect. 9.1).

In difference to the next-event time advance simulation scheme, another possibility is a time-driven algorithm. The time line is discretized at multiples of a constant amount of time Δt. In each step of the simulation, time is advanced to the next $t + \Delta t$, and the events that have occurred in between are executed. There is obviously an approximation error due to the rounding of all event times to multiples of Δt, and it might become complex to check which events have to be executed. This algorithm type is especially useful in the simulation of systems with continuous or hybrid state variable sets, but may become inefficient depending on the level of activity in the model. The choice of a Δt value controls the tradeoff between approximation error and algorithmic efficiency.

The simulation as well as every real system is always in one individual state. There are, however, many nondeterministic issues involved in events, e.g., when a decision between conflicting activities has to be made or when an actual delay for an activity is selected. This is where randomness comes into play, which may come from noise or system details that are below the level of model abstraction. Random number generators are thus an important element of a simulation algorithm, and their quality significantly influences the simulation. The notes at the end of this chapter point at simulation books with background information on this topic. The same applies to the following brief coverage of simulation issues.

A simulation is a stochastic experiment, and there is no guarantee that the result will be exact. Under certain weak assumptions, however, the result quality becomes better if more events are simulated or a longer simulation time is observed. There is always a tradeoff between computational effort and result quality. One of the major questions of a simulation experiment is thus when to stop the simulation run. The easiest way is to set a maximum simulation run time or computation time. While this may be useful for testing a model, it is not sufficient for a performance evaluation because the accuracy of the results remains unknown. The statistical confidence in the results can be quantified by statistical tests, e.g., Student's t-test. It is, however, preferable to continuously compute the result accuracy during the simulation and to apply a stop condition that depends on the achieved quality. Performance measures considered here are point estimations for mean values. Their accuracy can be approximately computed using a confidence interval for a given error probability.

Another problem is that standard estimation techniques are based on the assumption of independent events. This is obviously not the case for a usual simulation, because subsequent events may be causally connected and thus correlated. Covariance is a measure of such a correlation. Standard methods for steady-state simulations thus accumulate several events to batches, from which the mean is taken. The covariance in such a derived stream of events is smaller, and allows for a better statistical confidence in the estimated results. Independent replications of a simulation, each with a copy of the model and using different streams of random numbers, also lead to an improvement of estimation quality. This technique is obviously well suited for parallel simulation. Collection of different event streams from the individual and unrelated simulations substantially decrease the problematic covariance between individual events.

An additional important improvement of result accuracy detects the warmup phase (initial transient) of a steady-state simulation run, e.g., by detecting that the temporary mean value crosses the current value a predefined number of times. The initial transient phase is then ignored for the estimation of the performance measures, which leads to better estimations.

7.2.1 Common Functions

This section introduces some functions that are used later on in the algorithms.

All algorithms in this text follow the same pseudocode-like syntax. Keywords are set in a bold sans serif font like **while**. Line indentation visualizes code blocks that belong together like in a **repeat .. until**-loop to avoid explicit end statements. Some simplifications for operations as known from

---

ENABLEDMODES (Action, State)

Output: The set of enabled modes for an action in a state

ModeSet := $\emptyset$
for $\forall$Mode $\in$ $Modes^*$(Action) **do**
    if not Ena^*(Action, Mode, State) **then continue**
    Allowed := True
    for $\forall sv \in SV^*$ **do**
        if not $Cond^*\left(sv, Exec^*(\text{Action}, \text{Mode}, \text{State})\right)$ **then**
            Allowed := False
            exit
    if Allowed **then** ModeSet := ModeSet $\cup$ {Mode}
return ModeSet

Algorithm 7.1: Compute enabled action modes in a state

C-style programming languages are used (e.g., i++). Algorithm names are set in the text as ALGORITHM. Variables and parameters use a normal font (like Action), while the same symbols as introduced in Chap. 2 denote the elements of the SDES definitions, for example $Modes^*$(Action).

Algorithm 7.1 contains the function ENABLEDMODES, which computes and returns the enabled action modes for a given action in a state. Action and state are input parameters, while the results are returned as a set of action modes $\{mode_1, mode_2, \ldots\}$. The algorithm simply scans all modes of the action and checks their enabling function. If it is valid for the state, all state variables are checked if the execution of the action mode would lead to a forbidden state. If that is not the case, the mode is added to the list that is returned later on. The function thus implements the definition of $Enabled(a, \sigma)$ at page 28.

It should be noted that not every enabled action mode that the function returns may actually be executed in a state, because the enabling degree of the action in the state may forbid that. Moreover, if there are actions with a higher priority enabled, they might be executed first, thus never allowing the actual execution of another enabled action mode. These issues are considered in the algorithms that use ENABLEDMODES. In the context of stochastic Petri nets, a distinction between *being enabled* and *having concession* is made in the literature. In that sense, ENABLEDMODES returns the action modes that have concession, i.e., which are structurally enabled by the model state. The definition of executable action variants *Executable* in a state (see Sect. 7.1) does indeed only return the variants that may actually be executed.

The asymptotic complexity of the algorithm depends on the number of action modes and state variables O($|Modes^*(a)|$ $|SV^*|$).

Procedure UPDATEACTIVITYLIST shown in Algorithm 7.2 changes the event list accordingly when a new state has been reached after an activity

---

UPDATEACTIVITYLIST (State, ActivityList, Time)

---

for $\forall$Action $\in A^*$ **do**
    Enabled = ENABLEDMODES(Action, State)
    ActivityCount[$\forall mode \in$ Enabled] := 0

    ($*$ remove all nonenabled activities from list $*$)
    for $\forall$(Action, $mode, t) \in$ ActivityList **do**
        if $mode \notin$ Enabled **then**
            remove (Action, $mode, t$) from ActivityList
        else ActivityCount[$mode$]++

    for $\forall mode \in$ Enabled **do**
        ($*$ remove activities if there are too many $*$)
        while ActivityCount[$mode$] $> VDeg^*(a, mode, $ State) **do**
            Select = $\lfloor Random[0..1] *$ ActivityCount[$mode$] $- 1 \rfloor$
            $(a, mode', t) = $ first entry in ActivityList
            while Select > 0 **do**
                if $a = $ Action $\wedge mode = mode'$ **then** Select$--$
                $(a, mode', t) = $ next entry in ActivityList
            remove $(a, mode', t)$ from ActivityList
            ActivityCount[$mode$]$--$

        ($*$ add activities if the enabling degree is not reached $*$)
        while ActivityCount[$mode$] $< VDeg^*(a, mode, $ State) **do**
            Select := $\lfloor Random[0..1] * |$Enabled$| \rfloor$
            $mode' := $ Enabled[Select]
            $t' := $ Time + random value drawn from $Delay^*$(Action)
            $(a, \cdot, t) := $ first entry in ActivityList
            while $t' > t \wedge \big(t' = t \vee Pri^*($Action$) < Pri^*(a)\big)$ **do**
                $(a, \cdot, t) := $ next entry in ActivityList
            Insert (Action, $mode', t'$) into ActivityList before entry
            ActivityCount[$mode$]++

Algorithm 7.2: Update the event list

execution. It is used by the transient and steady-state simulation below. The procedure takes as input the current model state, the previous event list (ActivityList), and the current simulation time. As a result of the procedure execution, the entries in the event list are updated according to the provided state and time. This implementation is similar to the iterative definition of the action states for the stochastic process in Sect. 2.3.2.

In a big surrounding loop, all actions of the SDES model are considered. The set of enabled modes for it is derived with a call to ENABLEDMODES first. In a second step, all activities are removed from the event list for which the action mode is no longer enabled. The number of activities for every mode *mode* of the current action is counted in ActivityCount[*mode*] on the way. The subsequent for-loop adjusts the number of ongoing activities of the action to the enabling degrees in the current state. If there are too many of them present, the exceeding amount is removed after a probabilistic selection. If the enabling degree of an action variant is not yet fully utilized, the missing number of activities is created.

The selection of an action mode for which a new activity is added to the event list is done based on a probabilistic choice with equal probabilities. The time of scheduled execution is computed as the current simulation time plus a random value drawn from the delay distribution of the action.

The asymptotic computational complexity of UPDATEACTIVITYLIST is $O(|A^*| \, |\text{ActionList}| \max(Modes^*(a), Deg^*(a), |\text{ActionList}|))$. It can be substantially reduced in practice if causal relationships between actions of the SDES model can be derived from the model. The set of actions that can be disabled and enabled (or simply affected) by one action can be computed prior to an evaluation for some model classes. If then for instance an action a_i is executed in a state, the update of the event list can be restricted to check those actions that might be affected by a_i's execution. Because of the locality of usual SDES models, the number of affected actions does not necessarily increase with the model size. In that case, a much smaller number of actions need to be checked.

With the event list updated by UPDATEACTIVITYLIST, the next task is to select which activity will be executed. This may involve several steps although the activities in the event list are already ordered by completion time and priorities. Function SELECTACTIVITY as shown in Algorithm 7.3 implements this for a given event list ActivityList. It returns the selected activity consisting of action, mode, and completion time or a tuple with the time set to infinity if the list is empty. The selected activity is removed from the event list as a side effect.

The activity with the highest priority among the ones scheduled for execution at the smallest time is the first entry in the event list. The weights of all activities with the same completion time and equal priority are added after some initializations. The second step selects one of the found activities based on a probabilistic choice. The relative probability for being selected is derived from the weight divided by the sum of all weights for the mentioned activity set. The selected activity is removed from the event list and finally returned.

The algorithm has an asymptotic complexity of $O(|\text{ActivityList}|)$.

7.2.2 Estimation of Steady-State Measures

For the estimation of SDES performance measures in steady-state, procedure STEADYSTSIMULATION (Algorithm 7.4) can be applied. It takes as input the SDES model including its performance measures $rvar_i^* \in RV^*$. For each of them, an estimated value $\text{Result}_{rvar_i^*}$ is derived. We are interested in the long-term behavior of the system, and to reach the steady-state the

---

SELECTACTIVITY (ActivityList)

Input: ActivityList - the sorted event list
Output: An activity $(a, mode, t)$ that should be executed first

$(*$ if the list is empty, there is no executable activity $*)$
if $|\text{ActivityList}| = 0$ **then return** $(\cdot, \cdot, \infty)$

$(a, mode, t) :=$ first entry in ActivityList
Time $:= t$
MaxPriority $:= Pri^\star(a)$
WeightSum $:= 0$
$(*$ get the sum of weights for all possible activities $*)$
while $t = \text{Time} \wedge \text{MaxPriority} = Pri^\star(a)$ **do**
    WeightSum $+= Weight^\star(a)$
    $(a, mode, t) :=$ next entry in ActivityList

$(*$ probabilistic selection of an activity $*)$
Select $:= Random[0..1]$
$(a, mode, t) :=$ first entry in ActivityList
while Select $-= \frac{Weight^\star(a)}{\text{WeightSum}} > 0$ **do**
    $(a, mode, t) :=$ next entry in ActivityList

$(*$ delete and return selected activity $*)$
remove $(a, mode, t)$ from ActivityList
return $(a, mode, t)$

Algorithm 7.3: Select the activity to be executed next

simulation would be preferably observed until an infinite simulation time. Such simulations are called **nonterminating**, but are obviously restricted by the computational effort that is acceptable for the task.

The main simulation loop follows the formal definition of the stochastic process *CProc* closely (cf. Sect. 2.3.2). The state of the state variables σ is kept in variable State, while the activity state is held in the event list ActivityList. The simulation time SimTime corresponds to the continuous-time t that is used for the simplified process. The complete process was defined as a discrete-indexed process for simplicity.

Starting from the initial state, for every new state the corresponding event list is updated with a call to UPDATEACTIVITYLIST. Function SELECTACTIV-ITY is subsequently called to decide which one of the possible activities will be executed. In the case that a dead state has been reached, i.e., the event list is empty because there are no enabled actions, the returned execution time t is infinity. In such a case, it is required to set a maximum value for the simulation time MaxSimTime, which may also serve as a stop condition for the overall simulation. The stochastic process spends the rest of the time in the

---

STEADYSTSIMULATION (SDES)

Input: SDES model with performance measure definitions
Output: estimated values of performance measures $rvar_i^\star \in RV^\star$

(* initializations *)
for $\forall sv_i \in SV^\star$ **do** $\text{State}(sv_i) := Val_0^\star(sv_i)$
$\text{ActivityList} := \emptyset$
$\text{SimTime} := 0$
for $\forall rvar_i^\star \in RV^\star$ **do** $\text{Reward}_{rvar_i^\star} := 0$

(* main simulation loop *)
repeat
    (* get new activities *)
    UPDATEACTIVITYLIST(State, ActivityList, SimTime)

    (* select executed activity and prepare variables *)
    $(a, mode, t) := $ SELECTACTIVITY(ActivityList)
    $\text{EventTime} := \min(\text{MaxSimTime}, t)$
    $\text{Event} := (a, mode)$
    $\text{SojournTime} := \text{EventTime} - \text{SimTime}$

    (* update performance measures *)
    for $\forall rvar_i^\star = (rrate^\star, rimp^\star, \cdot, \cdot) \in RV^\star$ **do**
        $\text{Reward}_{rvar_i^\star} += rrate^\star(\text{State}) * \text{SojournTime}$
        if $t \leq \text{MaxSimTime}$ **then** $\text{Reward}_{rvar_i^\star} += rimp^\star(\text{Event})$

    (* execute state change *)
    $\text{SimTime} := \text{EventTime}$
    if $t \neq \infty$ **then** $\text{State} := Exec^\star(\text{Event}, \text{State})$
until stop condition reached (e.g., $\text{SimTime} \geq \text{MaxSimTime}$)

(* compute performance measures *)
for $\forall rvar_i^\star \in RV^\star$ **do** $\text{Result}_{rvar_i^\star} := \dfrac{\text{Reward}_{rvar_i^\star}}{\text{SimTime}}$

Algorithm 7.4: Next-event time advance simulation of steady-state behavior

---

dead state, which is handled by the algorithm by taking the minimum of the maximum simulation time and the time of the next scheduled event t. This takes effect also for the case when the next-event time is greater than the specified maximum simulation time to correctly cut the time interval. The event to be executed and the remaining sojourn time in the current state (either until event execution or maximum simulation time) are set in addition.

Now all prerequisites are available to update the performance variables. The rewards accumulated so far are stored in $\text{Result}_{rvar_i^\star}$. Rate reward earned in the current state is added first, multiplied by the actual sojourn time. If the event shall be executed before MaxSimTime, the corresponding impulse reward is added as well. Afterward the result accuracy achieved so far can be estimated with statistical techniques, but this is omitted in Algorithm 7.4.

With all state-related instructions done, the algorithm executes the event by setting all state variables to the new value as specified by the execution function $Exec^*(\cdot, \cdot)$ of the SDES model for the event.

The main simulation loop repeats until some predefined stop condition is reached. Some notes on the stop condition and its statistical significance have been given above.

A final estimation of the performance measures is computed at the end of the algorithm. Only time-averaged measures are allowed in a steady-state evaluation, thus all intermediate results for accumulated reward are divided by the overall simulation time.

The asymptotic complexity of STEADYSTSIMULATION is given by the number of times that the main simulation loop is executed multiplied by the complexity of the UPDATEACTIVITYLIST procedure (see above). The loop is carried out once for every event (or activity) being executed. The stop condition implies the number of loop executions. In the case of a predefined maximum simulation time, the expected number of events per simulation time might be estimated from the model actions, leading to an approximation of how many events will be executed.

7.2.3 Estimation of Transient Measures

Algorithm 7.5 shows the transient simulation algorithm TRANSIENTSIMULA-TION. A SDES model and its performance measures are the input parameters just like for the steady-state simulation. The output is again an estimated value for each measure in $\text{Result}_{rvar_i^*}$. In a simulation that is intended to estimate performance measures at some fixed point of time or for a known interval, there is obviously no need to continue the simulation when the simulation time exceeds the maximum of the observed time points. Such a simulation is called **terminating**, and the final estimation is computed based on the results of several terminating runs with the equal initial conditions and different random number streams (**independent replications** scheme).

The inner loop (**repeat .. until**) implements the event list update, activity selection, and state changes identical to the main simulation loop of the steady-state simulation (Algorithm 7.4). The simulation always starts with the initial SDES state at simulation time zero. The computation of rewards is, however, different for transient measures. For instant-of-time measures ($lo = hi$), the rate reward for the current state is added to the reward if the observed time point is inside the current state sojourn time. In the case of an interval-of-time measure ($lo < hi$), the intersecting time span between the observation interval and the current state's sojourn time is derived first

---

TRANSIENTSIMULATION (SDES)

Input: SDES model with performance measure definitions
Output: estimated values of performance measures $rvar_i^\star \in RV^\star$

$(\ast$ initializations $\ast)$
Runs := MaxSimTime := 0
for $\forall\, rvar_i^\star = (\cdot, \cdot, [lo, hi], \cdot) \in RV^\star$ **do**
    $\text{Reward}_{rvar_i^\star} := 0$
    MaxSimTime := max(MaxSimTime, hi)
while not stop-condition reached **do**
    for $\forall sv_i \in SV^\star$ **do** State(sv_i) := $Val_0{}^\star(sv_i)$
    ActivityList := $\emptyset$; SimTime := 0
    Runs++
    repeat $(\ast$ main simulation loop $\ast)$
        $(\ast$ get and select executed activity $\ast)$
        UPDATEACTIVITYLIST(State, ActivityList, SimTime)
        $(a, mode, t)$:= SELECTACTIVITY(ActivityList)
        EventTime := min(MaxSimTime, t)
        Event := $(a, mode)$
        SojournTime := EventTime $-$ SimTime
        $(\ast$ update performance measures $\ast)$
        for $\forall\, rvar_i^\star = (rrate^\star, rimp^\star, [lo, hi], ravg^\star) \in RV^\star$ **do**
            if $lo = hi \wedge$ SimTime $< lo \le$ EventTime **then**
                $(\ast$ instant-of-time measure $\ast)$
                $\text{Reward}_{rvar_i^\star}\ += rrate^\star(\text{State})$
            else $(\ast$ interval-of-time measure $\ast)$
                CoveredTime = min(hi, EventTime) $-$ max(lo, SimTime)
                $\text{Reward}_{rvar_i^\star}\ += rrate^\star(\text{State}) \ast \text{CoveredTime}$
                if $lo <$ EventTime $\le hi$ **then**
                    $\text{Reward}_{rvar_i^\star}\ += rimp^\star(\text{Event})$
        $(\ast$ execute state change $\ast)$
        SimTime := EventTime
        if $t \ne \infty$ **then** State := $Exec^\star$(Event, State)
    until SimTime $\ge$ MaxSimTime
$(\ast$ compute performance measures $\ast)$
for $\forall\, (\cdot, \cdot, [lo, hi], ravg^\star) \in RV^\star$ **do**
    $\text{Result}_{rvar_i^\star} := \dfrac{\text{Reward}_{rvar_i^\star}}{\text{Runs}}$
    if $ravg^\star$ **then** $\text{Result}_{rvar_i^\star} := \dfrac{\text{Reward}_{rvar_i^\star}}{hi - lo}$

Algorithm 7.5: Next-event time advance simulation of transient behavior

(CoveredTime). The rate reward associated with the current state is added to the accumulated reward, multiplied by the time span. If the next event is scheduled for execution within the observation interval, its corresponding impulse reward is added as well.

The inner main simulation loop stops when the current simulation time exceeds the maximum simulation time, which in the transient case is set to the maximal upper bound hi of all observation intervals. Dead states thus do not lead to problems, and the choice of this stop condition is independent from the statistical accuracy of the results. Every single transient simulation runs, i.e., every loop execution gives one result for each performance measure. Many runs thus have to be carried out to estimate the accuracy of the results. The number of runs is counted in variable Runs. The stop condition of the outer **while**-loop depends on the result of a statistical check of the result quality achieved so far. Some notes on this issue have been given at the beginning of this section. In the case of a transient simulation, subsequent results are, however, not correlated, because each one is the result of an independent simulation run. The random number streams obviously must be different for that reason (e.g., a different seed value needs to be set). Independent replications can be started to return more individual results in the same time. The initial transient phase naturally needs to be obeyed and not discarded.

The final estimations for the performance measures are computed as the mean over all runs after the outer loop is finished. The result is then an accumulated reward, which has to be divided by the observation interval length for time-averaged measures.

For an estimation of the asymptotic computational complexity, the number of executions of the inner loop is important. The number of processed events until MaxSimTime can be estimated from the average delays and the average number of executable variants. The complexity of one event processing is the same as for the steady-state simulation. It is hard to estimate the number of times that the outer loop is executed, because it depends on the speed of accuracy increase for the performance measures and the chosen stop condition.

7.3 Numerical Analysis

As opposed to simulation, which follows one randomly selected path through the possible behaviors of a system, numerical analysis techniques capture the whole stochastic process. The reachability graph RG is the usual approach for this task; an algorithm for its generation for SDES models is given in Sect. 7.3.1. The dynamic behavior of the model is given by the initial state and the subsequent activity executions.

Mathematical formulas are numerically solved to derive exact measure values from the process representation in a second step. A stochastic process is obviously unlimited in time at least if a steady-state analysis is required, and it is thus impossible to store all paths in a simulation-like fashion. Simplifications are necessary which allow the complete derivation and storage of the process information for any practical algorithm.

The type of process depends on the used delays and whether certain activities are enabled together in one state or not. The delays associated with action variants considered in the techniques used here [127, 130] can either be zero (immediate), exponentially distributed, deterministic, or belong to a class of general distributions called *expolynomial*. Such a distribution function can be piecewise defined by exponential polynomials and has finite support. It can even contain jumps, making it possible to mix discrete and continuous components. Many known distributions (uniform, triangular, truncated exponential, finite discrete) belong to this class. Details are given in Sect. 1.4.

The numerical analysis algorithm described in Sect. 7.3.3 only works for models with the following restriction: there must not be a reachable state in which there is more than one timed activity with nonexponentially distributed delay. This is the state of the art for practical algorithms due to numerical problems. The theoretical background for the numerical analysis of less restricted models with concurrent activities including fixed delays has been covered in [130, 224, 229]. Note that therefore the maximum allowed enabling degree of actions with timed nonexponentially distributed delay ($\mathcal{F}^{gen}$) is one, and there must only be one of them enabled in any reachable state. Formally, we require

$$\forall \sigma \in RS : \left| \left\{ v \in Executable(\sigma) \mid Delay^\star(v) \in \mathcal{F}^{gen} \right\} \right| \leq 1$$

The underlying stochastic process of such an SDES model is then a Markov regenerative (or semiregenerative) process, and an algorithm for its steady-state evaluation is presented in Sect. 7.3.3.

If the delays of all action variants that are executable in any state of an SDES model are either exponentially distributed or zero, the stochastic process is a CTMC.

$$\forall \sigma \in RS : Executable(\sigma) \subseteq AV^{exp} \cup AV^{im}$$

Algorithms for the transient and steady-state evaluation of reward measures are given in Sect. 7.3.2 for this case.

7.3.1 Reachability Graph Generation

All standard numerical analysis methods require the derivation of the complete graph of possible states and state transitions, the reachability graph RG.

Function GENERATEFULLRG shown in Algorithm 7.6 computes the full reachability graph of SDES models with the restrictions motivated above. Input to the algorithm is the SDES model – in difference to simulation, the reachability graph generation is independent of the performance measures defined inside. The function returns the reachability graph RG consisting of states and state transitions.

The initial state is set according to the SDES model specification, and it is the first element of the reachability set and the set of new states. In a second step the set of actions is sorted into an ordered list ActionList with immediate delays coming first, and among them sorted by priority. This is done to improve the efficiency of the algorithm, because only the enabled

---

GENERATEFULLRG (SDES)

Input: SDES model
Output: The reachability graph $RG = (RS, RE)$

for $\forall sv_i \in SV^\star$ **do** InitialState$(sv_i) := Val_0^\star(sv_i)$
NewStates $:= RS := \{$InitialState$\}$
ActionList $:= \emptyset$
for $\forall a \in A^\star$ **do**
    (* sorted ActionList: immediate with high priorities come first *)
    $e :=$first entry in ActionList
    while $\big(Delay^\star(a) \in \mathcal{F}^{im} \wedge Delay^\star(e) \in \mathcal{F}^{im} \wedge Pri^\star(a) < Pri^\star(e)\big)$
        $\vee\ (Delay^\star(a) \notin \mathcal{F}^{im} \wedge Delay^\star(e) \in \mathcal{F}^{im})$ **do**
        $e :=$next entry in ActionList
    insert a into ActionList before e

while NewStates $\neq \emptyset$ **do** (* main loop *)
    select State $\in$ NewStates; NewStates $:=$ NewStates $\setminus \{$State$\}$
    (* compute enabled actions *)
    Priority $:= 0$; Vanishing $:=$ False
    EnabledVariants $:= \emptyset$
    $a :=$ first entry in ActionList
    while not Vanishing $\vee \big(Delay^\star(a) \in \mathcal{F}^{im} \wedge$ Priority $= Pri^\star(a)\big)$ **do**
        Enabled $:=$ ENABLEDMODES$(a, $State$)$
        if Enabled $= \emptyset \vee Deg^\star(a, $State$) = 0$ **then continue**
        if $Delay^\star(a) \in \mathcal{F}^{im}$ **then**
            Vanishing $:=$ True; Priority $:= Pri^\star(a)$
        for $\forall mode \in$ Enabled **do**
            EnabledVariants $:=$ EnabledVariants $\cup \{(a, mode)\}$
        $a :=$ next entry in ActionList
    (* insert states and state transitions into graph *)
    for $\forall$ Event $= (a, mode) \in$ EnabledVariants **do**
        NewState $:= Exec^\star($Event, State$)$
        if NewState $\notin RS$ **then**
            NewStates $:=$ NewStates $\cup \{$NewState$\}$
            $RS := RS \cup \{$NewState$\}$
        $RE := RE \cup ($State, NewState, $a, mode)$
return $RG = (RS, RE)$

Algorithm 7.6: Generation of the full reachability graph

actions with the highest priority have to be considered in the later enabling check for each state.

The main loop is executed until the set of states which have been newly discovered and that are not yet analyzed (NewState) is empty. Otherwise one arbitrary state is taken from the set and analyzed. At the point when this set is empty, no further state can be reachable from the initial state.

As a first step, the set of enabled actions with highest priority level is derived in the **while not ..** loop. The resulting set of action variants is stored in variable EnabledVariants. The subsequent **for**-loop computes for each of the possible events the destination state NewState. It is added to the set of new states NewStates as well as the reachability set RS if it is not yet known. The corresponding state transition is added to the reachability graph RE.

The asymptotic complexity of the algorithm is $O(|RS|\,|AV|)$, because the enabling of every action variant $\in AV$ of the model has to be checked for every reachable state $\in RS$. An improvement of speed can be reached in a similar way to as it has been explained in the simulation context already. If sufficient main memory is available, we can store the set of previously enabled action variants together with every newly visited state in NewStates. When such a state is analyzed, the set of enabled events in it can be computed more efficiently if it is known from the model structure which events may have become disabled and enabled by the recent state transition. The complexity then reduces to $O(|RS|\,|AV'|)$, if we denote by AV' an upper bound on the number of action variants that are influenced (enabled or disabled) by the execution of an enabled variant. In most model classes, this number is significantly smaller than $|AV|$ because of local action influence.

The reachability graph describes the stochastic process with its reachable states properly only if the combinations of delays in a model do not inhibit state transitions that are otherwise possible. Imagine for instance two conflicting action variants with deterministic delay, which become enabled together. The one with the smaller delay will obviously be executed first and thus disable the other one in every case. The reachability graph generation algorithm, however, relies on the absence of such situations, because both executions would lead to valid state transitions. Due to the restriction of timed delays to exponential ones and a maximum of one nonexponential enabled per state, such a situation can never happen. This is due to the support $0 \ldots \infty$ of the exponential distribution, which leads to a nonzero probability for the execution of all conflicting timed action variants in any state.

The **reduced reachability graph** RRG can be constructed from the full one by removing state transitions that correspond to action variants with an immediate delay $(Delay^{\star}(\cdot) \in \mathcal{F}^{im})$. The idea is to avoid additional states in which the stochastic process does not spend any time, and which are thus of less interest during the subsequent analysis steps. Efficient algorithms for the direct generation of the reduced reachability graph (coined "on-the-fly elimination") have been proposed in the context of stochastic Petri nets [14,229,230]. Paths of immediate state transitions are followed in a recursive manner to compute all possible paths from the state after a timed state transition to the subsequent timed state. The rare case of circular state transition paths containing only immediate actions (vanishing loops) can be resolved [66].

It should be noted that the reduction is purely done for efficiency reasons; the standard numerical solution algorithms work with the full graph as well. The stochastic process as defined in Sect. 2.3.2 for SDES models does visit vanishing states, which is for instance important to cover impulse rewards related to actions with an immediate delay. Advantages and disadvantages of both approaches are, e.g., discussed in [28, 65].

Impulse rewards associated with immediate state changes cannot be captured easily if the standard approach of elimination of the vanishing states is taken. Throughputs of immediate actions, therefore, cannot be directly computed with the presented algorithms. We restrict reward variables for the numerical analysis accordingly.

$$\forall v \in AV^{im} : rimp^\star(v) = 0$$

This is, however, not a significant restriction for most applications, because it is usually possible to express the throughput of immediate actions indirectly using surrounding timed actions in the model classes. Additional remarks on this issue are given on page 150.

7.3.2 Continuous-Time Markov Chain Analysis

The reduced reachability graph of an SDES model with only immediate and exponentially distributed delays of action variants is isomorphic to a CTMC, because state sojourn times and state transitions are memoryless and the reachability set is discrete. The memoryless property then allows to compute the further evolution of the model dynamics without the knowledge of the action state – the state variable values are sufficient. This greatly simplifies the analytical treatment, and has lead to the attraction of models with only exponentially distributed (and zero) delays.

The time-dependent rate of state changes per time unit is constant for state transitions with exponentially distributed delays because it is memoryless. In fact, the parameter λ of the exponential distribution equals this rate as well as the inverse of the mean delay. For the numerical analysis of an SDES model with an underlying CTMC, we are interested in the state transition rates for the tangible reachable states. The rate q_{ij} shall denote the overall state transition rate from state σ_i to state σ_j.

As multiple rates (just like flows) add up, the rates that correspond to different activities leading to the same state change are added. This applies also to state transitions due to one action variant with an enabling degree greater than one. In a simulation, a corresponding number of activities is kept in the event list. In contrast to that, it is possible to account for multiple enabling during a numerical analysis through simply multiplying the rate by the enabling degree. Probabilistic distribution of flows between different destination states is the consequence of conflicting immediate state changes after a timed one. The corresponding path probabilities until the arrival at the next tangible state thus have to be multiplied accordingly. It should be noted that all these simplifications are only possible due to the restriction of only exponentially distributed delays.

The individual state transition rates q_{ij}; $1 \leq i, j \leq |RRS|$ are thus derived from the reduced reachability graph as follows. $\sigma_i' = Exec^\star((a, mode), \sigma_i)$ is a shorthand for the state that is reached from σ_i after the execution of action variant $v = (a, mode)$.

$$
q_{ij} = \begin{cases} \displaystyle\sum_{\substack{(\sigma_i, \sigma_j, v, \cdot) \in RRE \\ Delay^\star(v) = 1 - e^{-\lambda t}}} VDeg^\star(v, \sigma_i) \lambda P\{\sigma_i' \longrightarrow \sigma_j\} & \text{if } i \neq j \\[3ex] -\displaystyle\sum_{i \neq k} q_{ik} & \text{if } i = j \end{cases}
\tag{7.2}
$$

We are interested in every probability $\pi_i(t)$ of an SDES model to be in one of its states σ_i at time t. The actual performance measures are derived from these intermediate results later on.

The evolution of the transient state probabilities can be described by the (ordinary differential) Kolmogorov equation. It can be informally interpreted as the change of the state probabilities over time equaling the sum of all state transition rates flowing in and out of a state σ_j.

$$
\forall \sigma_j \in RRS : \frac{\mathrm{d}}{\mathrm{d}t} \pi_j(t) = \sum_{\sigma_i \in RRS} \pi_i(t) \, q_{ij}
$$

To simplify notation, a **generator matrix** $\mathbf{Q} = [q_{ij}]$ of proper dimension $|RRS| \times |RRS|$ is defined that holds all entries of the formula above. As $\mathbf{Q}$ is a rate matrix, the diagonal entries are set such that all row sums equal zero. The equation then simplifies to

$$
\frac{\mathrm{d}}{\mathrm{d}t} \boldsymbol{\pi}(t) = \boldsymbol{\pi}(t) \, \mathbf{Q}
$$

An initial condition must be set in order for the system to be fully specified. In an SDES model, we assume that the behavior starts at time zero with the initial state σ_0.

$$
\forall \sigma_i \in RRS : \pi_i(t) = \begin{cases} 1 & \text{if } \sigma_i = \sigma_0 \\ 0 & \text{otherwise} \end{cases}
$$

Transient Numerical Analysis

The transient state probabilities $\boldsymbol{\pi}(t)$ can be derived following:

$$
\boldsymbol{\pi}(t) = \boldsymbol{\pi}(0) \, e^{\mathbf{Q}t}
$$

for which the matrix exponential is defined similarly to the series expansion

$$
e^{\mathbf{Q}t} = \sum_{k=0}^{\infty} \frac{\mathbf{Q}^k}{k!}
$$

Practical numerical approaches to the derivation of $\pi(t)$ are not straightforward. The direct way using a truncated computation of the sum is numerically difficult and unstable [299]. Uniformization [185] is a commonly used technique that maps the CTMC on a discrete-time Markov chain (DTMC) as follows. A probability matrix $\mathbf{P}$ for this derived process is obtained after dividing $\mathbf{Q}$ by a value q, with q being greater or equal[2] than the biggest entry in $\mathbf{Q}$. $\mathbf{I}$ denotes the identity matrix.

$$q \geq \max_i(|q_{ii}|), \quad \mathbf{P} = \frac{1}{q}\mathbf{Q} + \mathbf{I}$$

The transient state probabilities can then be derived from

$$\pi(t) = \pi(0)\,e^{\mathbf{Q}t} = \pi(0)\,e^{qt(\mathbf{Q}-\mathbf{I})} = \pi(0)\sum_{k=0}^{\infty}\mathbf{P}^k\frac{(qt)^k}{k!}e^{-qt}$$

The $\frac{(qt)^k}{k!}e^{-qt}$ values are Poisson probabilities which can be efficiently computed [107, 130]. The advantage of this problem translation is the better numerical stability and a quantifiable approximation error for the truncation of the sum.

Performance measures that are not to be taken in steady-state can be derived from the transient state probability vector $\pi(t)$. We restrict ourself to instant-of-time measures here; measures with an observation interval length greater than zero would require the computation of an integral over the instant-of-time values. If impulse rewards are zero, it is sufficient to compute the expected sojourn times in states during the observation interval. This can be done using methods similar to the one applied in the subsequent section for the derivation of the $\mathbf{C}$ matrix.

In the restricted case of a reward variable $rvar^\star = (rrate^\star, rimp^\star, rint^\star, \cdot)$ with $rint^\star = [lo, hi]$ and $lo = hi$ (instantaneous transient analysis), the result is computed as a weighted sum of the corresponding rate rewards for all tangible states:

$$\text{Result}_{rvar^\star} = \sum_{\sigma_i \in RRS} \pi_i(lo)\,rrate^\star(\sigma_i)$$

Transient analysis with $lo < hi$ (interval-of-time or cumulative transient) is possible with an adaptation of the uniformization technique, and is presented in [59, 278].

Steady-State Numerical Analysis

The term $\pi(t)$ denotes the transient state probability vector at time t, while π corresponds to the probability vector in steady-state (if it exists). It can be

---

[2] Many references choose $q = 1.02\,\max_i(|q_{ii}|)$.

interpreted as the probability that the process is found in each state if it is observed after an infinitely long run, or as the time fractions which are spent in each of the states. The steady-state value (the **limiting state probability**) is obviously reached after an infinitely long observation of the stochastic process.[3]

$$\boldsymbol{\pi} = \lim_{t \to \infty} \boldsymbol{\pi}(t)$$

The limit exists if the CTMC of the model is ergodic. The **time-averaged limit**

$$\boldsymbol{\pi} = \lim_{t \to \infty} \frac{1}{t} \int_0^t \boldsymbol{\pi}(x) \, \mathrm{d}x$$

exists for models with fewer restrictions, and is the basis of steady-state simulation. If both limits exists, they are equal. It is not necessary to compute the transient values numerically until convergence, because the transient probability equation can be transformed into

$$\forall \sigma_j \in RRS : 0 = \sum_{\sigma_i \in RRS} \pi_i \, q_{ij} \quad \text{or} \quad 0 = \boldsymbol{\pi} \mathbf{Q}$$

which can also be interpreted that in steady-state there is no change in the transient state probabilities, and thus the flow into each state equals the flow out of it. The linear system of equations is thus coined **balance equations**. A **normalization condition** is necessary in addition, which assures that the total probability spread over the vector entries is one.[4]

$$1 = \sum_{i, \sigma_i \in RRS} \pi_i \quad \text{or} \quad 1 = \boldsymbol{\pi} \, \mathbf{1}^T$$

Both equations together have a unique solution,[5] which can be derived by standard numerical algorithms for the solution of linear systems of equations like successive over-relaxation (SOR; see, e.g., [298]).

The result for a performance measure $rvar^\star = (rrate^\star, rimp^\star, rint^\star, ravg^\star)$ for steady-state behavior can finally be computed from $\boldsymbol{\pi}$ as follows. In the steady-state case, only $rint^\star = [0, \infty]$, $ravg^\star = $ True are allowed, onto which case $rint^\star = [\infty, \infty]$, $ravg^\star = $ False is mapped (see above). Let $\lambda(v)$ denote the parameter of the exponentially distributed delay of action variant v: $Delay^\star(v) = 1 - e^{-\lambda t}$.

$$\text{Result}_{rvar^\star} = \sum_{\sigma_i \in RRS} \pi_i \left(rrate^\star(\sigma_i) + \sum_{v_j \in Executable(\sigma_i)} \lambda(v_j) rimp^\star(v_j) \right)$$

A weighted sum over all tangible states is derived. The rate reward for the state is added first, and the sum of all impulse rewards for executable action

---

[3] Assuming that the state space contains only one recurrence class.
[4] **1** is a vector of appropriate size, where all entries equal one.
[5] One equation in the system $0 = \boldsymbol{\pi} \mathbf{Q}$ is always redundant.

variants in the state multiplied by its rate λ.[6] This is done because the execution frequency of an action variant v equals $\lambda(v)$ times the probability that it is executable.

7.3.3 Steady-State Analysis of Non-Markovian Models

For the numerical analysis of SDES models containing action variants with nonexponentially distributed delays in steady-state, the vector $\boldsymbol{\pi}$ of limiting state probabilities is again to be derived as in the CTMC case. General analysis algorithms realized so far for this class of models require that at most one activity with nonexponentially delay is executable in every reachable state as described above.

State transitions between tangible states occur only with exponentially distributed delays in the CTMC case. With nonexponentially distributed delays the analysis becomes more complex, and it is necessary to distinguish between states in which only action variants with exponentially distributed delays are executable (RRS^{exp}) and the ones with nonexponential action variants (RRS^{gen}). There is always only one action variant executable in states out of RRS^{gen} because of the adopted restrictions. We denote by RRS_v^{gen} the set of tangible states in which $v \in AV^{gen}$ is executable.

$$RRS^{exp} = \{\sigma \in RRS \mid Executable(\sigma) \subseteq AV^{exp}\}$$
$$\forall v \in AV^{gen} : RRS_v^{gen} = \{\sigma \in RRS \mid v \in Executable(\sigma)\}$$

Obviously then the obtained reachability subsets are disjoint, and together they contain all tangible states.

$$\forall v_i, v_j \neq v_i \in AV^{gen} : RRS_{v_i}^{gen} \cap RRS_{v_j}^{gen} = RRS_{v_i}^{gen} \cap RRS^{exp} = \emptyset$$
$$RRS = RRS^{exp} \cup \bigcup_{v \in AV^{gen}} RRS_v^{gen}$$

In case of a CTMC, only the corresponding linear system of equations had to be solved. For models containing nonexponential action variants, an additional step is required. The underlying stochastic process is a Markov regenerative process. It is only memoryless at some instants of time, called *regeneration points*. If an action variant with nonexponentially distributed delay is executable in a state, the next regeneration point is chosen after execution or disabling this action variant. The execution time of the next exponential action variant is taken otherwise.

---

[6] This method is not applicable for impulse rewards associated with immediate action variants, which were required to be zero above. Appropriate algorithms require to keep the full reachability graph [59] or to indirectly derive the execution frequencies by considering related timed action variants.

Following results of Markov renewal theory, a discrete-time **embedded Markov chain** (EMC) is defined for the regeneration points. Its state set is a subset of the reduced reachability set RRS: included are exponential states RRS^{exp} and all general states $\sigma \in RRS^{gen}$ which are directly reachable from an exponential state.

The solution of the discrete-time EMC requires the stochastic matrix $\mathbf{P}$ of one-step state transition probabilities. An additional matrix $\mathbf{C}$ of conversion factors has to be computed for a later mapping of the EMC results to the original state set RRS as well. $\mathbf{P}$ describes the probabilities of state changes of the EMC between two regeneration points, while $\mathbf{C}$ captures the conditional sojourn times in the states between such two regeneration points.

There are some states of the original process which are not states of the EMC, specifically the states in which a nonexponential action variant is executable, and which are not directly reachable from RRS^{exp}. The time spent in those states from the enabling of a nonexponential action variant until its execution or disabling (the expected sojourn time) is kept in entries of the $\mathbf{C}$ matrix. In addition to that, the diagonal entries of the $\mathbf{C}$ matrix contain the mean sojourn times in tangible states, which are needed for a conversion at the end of the algorithm. For states with solely exponential action variants enabled, only the diagonal entry of the corresponding $\mathbf{C}$ matrix row is different from zero and can be computed directly from the reduced reachability graph.

$$\forall \sigma_i, \sigma_k \in RRS^{exp} :$$

$$\mathbf{P}_{ik} = \begin{cases} 0 & \text{for } i = k \\ \frac{q_{ik}}{-q_{ii}} & \text{otherwise} \end{cases}$$

$$\mathbf{C}_{ik} = \begin{cases} \frac{1}{-q_{ii}} & \text{for } i = k \\ 0 & \text{otherwise} \end{cases}$$

$$\forall \sigma_i \in RRS^{gen}_v, \sigma_k \in RRS :$$

$$\mathbf{P}_i = \sum_{\sigma_j \in RRS} \mathbf{\Omega}^v_{ij} \, P\{\sigma_j \longrightarrow \sigma_k\}$$

$$\mathbf{C}_{ik} = \begin{cases} \mathbf{\Psi}^v_{ik} & \text{for } \sigma_k \in RRS^{gen}_v \\ 0 & \text{otherwise} \end{cases}$$

The q_{ik} values are derived as in (7.2) while taking only exponential states into account.

To compute the entries of the $\mathbf{P}$ and $\mathbf{C}$ matrix for action variants with nonexponentially distributed firing times, the evolution of the stochastic process during the enabling of such an action variant is analyzed. At most one action variant of this type can be executable per state for this type of analysis. Therefore only exponential action variants may be executed during the enabling period, resulting in a continuous-time **subordinated Markov chain** (SMC) of the nonexponential action variant.

The SMC for an action variant $v \in AV^{gen}$ with nonexponentially distributed delay $Delay^{\star}(v) \in \mathcal{F}^{gen}$ is described by the following information. RRS_v^{gen} are the states in which v is enabled. Matrix $\mathbf{Q}^v$ of the subordinated process contains the exponential rates of action variants, which are enabled in parallel to v in states from RRS_v^{gen}. The values of q_{ij} are again computed as in (7.2).

$$\mathbf{Q}^v = \left[\begin{cases} q_{ij} & \text{if } \sigma_i \in RRS_v^{gen} \\ 0 & \text{otherwise} \end{cases} \right]$$

The transient and cumulative transient analysis of this Markov chain lead to the $\mathbf{P}$ and $\mathbf{C}$ matrix entries via the computation of $\boldsymbol{\Omega}$ and $\boldsymbol{\Psi}$ for every nonexponential variant $v \in AV^{gen}$

$$\forall v \in AV^{gen} :$$

$$\boldsymbol{\Omega}_v = \int_0^{\infty} e^{\mathbf{Q}_v t} \, Delay^{\star\prime}(v)(t) \, \mathrm{d}t$$

$$\boldsymbol{\Psi}_v = \int_0^{\infty} e^{\mathbf{Q}_v t} \, (1 - Delay^{\star}(v)(t)) \, \mathrm{d}t$$

where $\boldsymbol{\Omega}_v$ denotes the matrix of state transition probabilities of the SMC at the end of the enabling period of v, and $\boldsymbol{\Psi}_v$ the matrix of expected sojourn times of the states of the SMC process during the enabling period of v. $Delay^{\star\prime}$ denotes the probability density function, i.e., the derivative of the delay distribution. The uniformization technique, which has already been mentioned for the transient analysis of Markovian models, can be applied for both computations after some adaptations [130]. The difference is that it is not possible any more to simply integrate until a known transient time t, because the execution time of the nonexponential action variant is a random value. However, it is sufficient to change the Poisson factors accordingly.

$$\boldsymbol{\Omega}_v = \sum_{k=0}^{\infty} \mathbf{P}^k \int_0^{\infty} \frac{(qt)^k}{k!} e^{-qt} \, Delay^{\star\prime}(v)(x) \, \mathrm{d}x$$

$$\boldsymbol{\Psi}_v = \sum_{k=0}^{\infty} \mathbf{P}^k \int_0^{\infty} \frac{(qt)^k}{k!} e^{-qt} \, (1 - Delay^{\star}(v)(x)) \, \mathrm{d}x$$

An efficient combined algorithm for both is available for the case of delay distributions that are a mix of finite polynomial and exponential functions, which are sufficient to fit most practical distributions arbitrarily well. In that case the integrals over the probabilities can be derived from the Poisson probabilities $\frac{(qt)^k}{k!} e^{-qt}$. Left and right truncation points for the sum are known for a given upper bound on the approximation error [130].

The vector γ of limiting state probabilities for the EMC is computed by solving the following set of linear equations. Standard algorithms like successive over-relaxation and sparse Gaussian elimination are again applicable for this task [299].

$$\gamma\,(\mathbf{P} - \mathbf{I}) = 0, \quad \sum_i \gamma_i = 1$$

The limiting state probabilities π of the actual stochastic process can finally be obtained as the mean sojourn times in each state between two regeneration points for states $\in RRS^{exp}$. Probabilities in γ of initial general states have to be distributed over their subsequent SMC states as described in the corresponding row of $\mathbf{C}$. Formally, both transformations correspond to multiplying the EMC solution vector by $\mathbf{C}$. A normalization step ensures that the sum of probabilities in π equals one.

$$\gamma' = \gamma\,\mathbf{C}, \quad \pi = \frac{1}{\sum_i \gamma'_i}\,\gamma' \quad \text{or simply} \quad \pi = \frac{\gamma\,\mathbf{C}}{\gamma\,\mathbf{C}\,\mathbf{1}^T}$$

The user-defined performance measures are finally calculated from the state probability vector π. This is done in a way similar to their derivation for steady-state CTMC cases as described on page 150. However, there is one subtle difference: we cannot specify a $\lambda(\cdot)$ value for action variants with non-exponentially distributed delays. If such an action variant may never be disabled by the execution of another variant (which can usually be checked on the model structure with the absence of conflicts), it is sufficient to interpret $\lambda(v)$ for a $v \in AV^{gen}$ as the reciprocal value of the mean delay or the mean rate of executions per time unit, i.e.,

$$\forall v \in AV : \lambda(v) = \begin{cases} \lambda & \text{if } Delay^\star(v) = 1 - e^{-\lambda t} \\ \frac{1}{\int_0^\infty t\, Delay^\star(v)(t)\,dt} & \text{if } Delay^\star(v) \in \mathcal{F}^{gen} \\ 0 & \text{if } Delay^\star(v) \in \mathcal{F}^{im} \end{cases}$$

Immediate action variants may not carry impulse rewards in this setting; compare the footnote about the same restriction in the CTMC case above.

Notes

Quantitative evaluation of stochastic discrete event models has been an active research field for a long time now, therefore only a small selection of relevant literature can be given. Simulation algorithms and their theoretical background as well as related statistical issues are, e.g., covered in [15–17, 42, 105, 117, 150, 219].

The numerical analysis methods presented in this text are based on the work for stochastic Petri nets, which in term build upon results for Markov chains and other similar stochastic processes. Analysis methods have been presented for the numerical steady-state [6, 57, 67, 132, 224, 229] and transient analysis of DSPNs [126, 161]. Numerical analysis techniques for other non-Markovian models are, e.g., treated in [61, 126, 127, 130, 132, 133].

The most important approaches for non-Markovian models are the following. They are based on an EMC like the presentation in this text, for which efficient steady-state solution methods are available [130, 226]. The method of supplementary variables [130, 132] allows both steady-state and transient numerical analysis [125]. Another method for the transient and steady-state solution of the same class of models is based on an underlying Markov regenerative stochastic process [57]. Some restrictions of the usable preemption policies have been relaxed in [302].

In restricted cases, it is possible to solve models with more than one enabled nonexponential delay analytically. Cascaded DSPNs allow two or more deterministic delays, if they are always concurrently enabled and the delays are multiples of each other [129, 130]. This technique is based on Markov renewal theory. If all nonexponentially timed activities start at the same time, generally timed transitions may be concurrently enabled [271]. Another approach to concurrent deterministic transitions observes the generalized semi-Markov process at equidistant points, and obtains a system of integral state equations that can be solved numerically [224, 229]. It has been implemented for the special case of two deterministic transitions with equal delay in the software tool DSPNexpress [227].

Models with only immediate and exponentially distributed delays can be solved using simpler techniques [3–5, 149, 207].

Further information about stochastic processes in general and of SDES subclasses can, e.g., be found in [67, 71, 143, 170, 176, 305]. The concept of impulse and rate rewards is covered in detail in [284] and applied to multiformalism SDES models in [82]. Their computation from non-Markovian stochastic Petri nets is shown in [133].

An overview of tools that implement evaluation algorithms for stochastic discrete event models is given in Chap. 12. The underlying idea of analysis methods for an abstract model representation is the same as followed in [82].

A quite different approach is to interpret an SDES model as having a *discrete* underlying timescale (as opposed to continuous time). The abstract SDES model class in fact includes them as a special case, if all action delays are described by probability distribution functions that contain only jumps. The corresponding (generalized) probability density function is a set of weighted Dirac impulses, i.e., a probability mass function containing the probabilities of the discrete values. Figure 7.1 shows a brief overview of some relationships between Markovian models with underlying continuous and discrete timescale.

Both resulting stochastic processes *CProc* and *SProc* (as they have been defined in Sect. 2.3.2) then have state changes only at discrete points in time. The definitions and discussion of Sects. 2.3 and 2.4 are, however, valid for this special case as well. Instead of the continuous exponential distribution, the discrete geometric distribution is used preserving the memoryless property.

	Continuous time	Discrete time
Delay distribution	Exponential or zero	Geometric or zero
Stochastic process	CTMC	DTMC
Process description	Rate matrix $\mathbf{Q}$	Probability matrix $\mathbf{P}$
State equations	$\frac{d}{dt}\boldsymbol{\pi}(t) = \boldsymbol{\pi}(t)\,\mathbf{Q}$	$\boldsymbol{\pi}(t + \Delta t) = \boldsymbol{\pi}(t)\,\mathbf{P}$
Steady-state solution	$0 = \boldsymbol{\pi}\,\mathbf{Q},\ \sum \pi_i = 1$	$\boldsymbol{\pi} = \boldsymbol{\pi}\,\mathbf{P},\ \sum \pi_i = 1$
Transient solution	$\boldsymbol{\pi}(t) = \boldsymbol{\pi}(0)\,e^{\mathbf{Q}t}$	$\boldsymbol{\pi}(t) = \boldsymbol{\pi}(0)\,\mathbf{P}^{\lfloor \frac{t}{\Delta t} \rfloor}$

Fig. 7.1. Discrete vs. continuous timescale

A DTMC underlies models in which only geometrically distributed delays are allowed; discrete-time stochastic Petri nets (cf. the Notes of Chap. 5) are an example. Because in every time step the execution probabilities are given by the geometrical distribution, and the remaining activity delays (having only some discrete values) can be stored together with the state, the performance analysis does not pose mathematical problems. Moreover, as immediate and deterministic action variants are special cases of the geometric distribution, there is no problem in having any number of them enabled concurrently in a state.

A drawback is the frequent occurrence of events at the same instant-of-time, which require a subtle priority setting to avoid confusions. Section 7.1 discusses this issue. The main problem is, however, that the DTMC has to be analyzed using a discrete-time step that equals the greatest common divisor of all delays, which leads to an additional state space explosion if there are delays which are much bigger than this GCD value. The remaining delays of enabled actions lead to supplementary variables in the continuous case, which are naturally discretized in the DTMC case. Only the number of remaining discrete-time steps need to be stored in the state encoding to fulfill the Markov property. Concurrent enabling of actions with a nonmemoryless delay distribution is thus possible, but paid for with a significant increase in the state space size.

8

An Iterative Approximation Method

For many systems of real-life size, the state space is very large, which is known under the term of *state space explosion* problem. Standard numerical analysis techniques require to visit and store every individual state to produce a generator matrix to solve for the state probabilities (cf. Sect. 7.3.2). Although the matrix is sparsely inhabited by nonzero values, which is obviously exploited by all relevant algorithms, they still fail if the state space and the number of state transitions become too big to be stored in the memory of available computing hardware.

A lot of techniques have been developed to overcome this limitation. Some more technical ideas store the state space especially efficient [14], or well organized on a hard disk. Others exploit symmetrical state space structures [54,55], but are therefore restricted to models that exhibit this kind of structure. Another branch of exact analysis methods for larger state spaces avoids to store the whole generator matrix. Smaller submatrices are derived from model parts and stored such that any generator matrix entry can be computed during the execution of the solution algorithm. This is known under the term *Kronecker* (or *tensor*) *algebraic* approaches. This technique was applied to stochastic automata networks [265,267] and later adapted to stochastic Petri nets [36,91,201]. Later advances in this field try to overcome some restrictions like priorities and the use of immediate synchronization transitions among others [92]. One remaining problem is the overhead induced by unreachable states [35]. The Kronecker representation was improved by variants of matrix diagrams [62]. Overviews of efficient storage and solution techniques are given in [60,63].

However, there are still restrictions in the size of the reachability set for the individual methods. Even if the reachability graph and generator matrix can be stored efficiently using structural techniques, the size of the probability vector itself remains a problem. Simulation or approximation methods can only be used in these cases. Just like for the Kronecker technique, the idea of decomposing a model into smaller ones and solving a combination of smaller models has naturally attracted a lot of interest. The idea of these *decomposition methods* is to avoid the computation of the whole state space by dividing the original system into smaller subsystems.

This section describes an approximate evaluation method for Markovian simple as well as variable-free stochastic Petri nets described in Sects. 5 and 6.5. The technique uses ideas presented in [40, 258–260] for the analysis of GSPNs. The underlying idea is an aggregation of the state space, which keeps significant properties of the model. The results have been presented in detail in [108, 109, 112].

The approximation method requires three steps. The first decomposes the original model into a set of n smaller subsystems and is described in Sect. 8.1. It is based on a so-called **MIMO graph**, which captures paths in the model structure. The idea behind this graph-based approach is to merge sequential model elements iteratively. MIMO is short for **multiple input** and **multiple output** elements. This aggregation overcomes the decomposition boundary restrictions imposed, e.g., by approaches based on implicit places. Decomposition can thus be done in an arbitrary way, and guarantees submodels that are small enough to be analyzed.

Unfortunately in most cases, the submodels are not live in isolation and thus cannot be evaluated directly. Therefore each submodel is supplemented by an aggregation of the remaining model parts, forming a so-called **low-level system**. A **basic skeleton** is derived as well, in which all submodels are aggregated. The second step results in n low-level subsystems and the basic skeleton, which is explained in Sect. 8.3.

In the third step, the performance measures of the original model are approximately computed using the reduced models (cf. Sect. 8.4). An iterative approximation technique is used similar to the response time approximation method presented in [259]. The basic skeleton is used to take care of the interaction between the low-level system evaluations.

The benefit of the overall method is that models with a much smaller reachability graph need to be evaluated on the way, thus requiring less memory and often even smaller computational effort despite the more complex algorithm. The main advantage is thus that models which are too complex for a numerical analysis due to their state space size can be evaluated.

However, only approximate performance results are computed, and the computation of bounds on the approximation error remains an open question so far. Experiences show that the error is acceptable in most cases; it is less than 3% for the majority of examples as well as the application analyzed in Sect. 16.3. Exact structural aggregation techniques for a restricted class of GSPNs have been proposed in [110, 111]. State-dependent transition delays are computed for the aggregated models in the mentioned references. This method can be applied to GSPN models without transition synchronization to obtain better aggregations and thus lead to a smaller approximation error of the iterative method presented in this chapter.

8.1 Model Partitioning

The hierarchical structure of a vfSCPN model may guide the decomposition process. This is a significant advantage with respect to other decomposition approaches, where automated general decomposition methods are an open problem. For each substitution transition (and thus each submodel), one **subsystem** SS_i is generated. A first decomposition substitutes each of these transitions on the highest level of hierarchy by one submodel. The decomposition method is, however, not restricted to one level of hierarchy. If the state space of a resulting low-level system is still too large to be handled, it is possible to apply the decomposition to the substitution transitions at the next level automatically. An algorithm for the efficient estimation of the resulting state space size has been developed for this step, which does not require a reachability graph generation [171].

Because only transitions are refined in the net class, the model is always cut through surrounding places. From now on we call these places *buffers* of the neighboring submodel(s).

8.2 MIMO Graph-Based Aggregation

After the decomposition step, it is clear where the original model is cut. The next step is an aggregation, during which an aggregated version SS_i^* of every subsystem SS_i^* is created. A pathwise aggregation method is employed which results in a structural simplification. In [260], the aggregation method is based on implicit places. This works only if the submodels have been selected by the modeler skilfully. Moreover, aggregation based on implicit places leads to the so-called *spurious states*, which hamper the result quality. The pathwise aggregation presented here is much simpler and does not lead to spurious states. Another advantage is that an arbitrary model decomposition is possible.

The idea is that each path corresponds to a possible token flow through the system, which must not be destroyed during the aggregation. Flows of tokens are thus maintained such that for the same input both the original and the aggregated submodel have the same output. Based on a set of aggregation rules, a directed graph of joins and splits in the submodel paths (the MIMO graph) is identified and reduced. The rules are applied iteratively until no further simplification is achieved. The resulting graph is then translated back into a Petri net, resulting in one aggregated net for each decomposed submodel.

The method presented here is applicable to both simple Petri nets and variable-free colored Petri nets (vfSCPN) as described in Chap. 5 and Sect. 6.5. The reader is referred to these parts for the basic Petri net notation. The following terms are used in the following, and an uncolored model is considered for simplicity first.

The places connected to a subsystem are called **buffers**. The set of buffers is denoted with $B \subseteq P$ and the set of subsystems $S \subseteq \mathsf{SPN}$ with $S = S_1 \cup$

$S_2 \cup \ldots \cup S_n$, where $S_i \cap S_j = \emptyset$. The $x \in (P \cup T)$ are called **elements** of a Petri net. The *preset* (*postset*) of element x is denoted by ${}^\bullet x$ ($x^\bullet$).

The general idea behind the aggregation methods is to substitute complex Petri net structures by simple ones preserving some important properties of the model like, e.g., liveness. It is obvious that simple sequences of places and transitions can be aggregated without changing the token flows. Simply substituting each structure by one transition–place–transition sequence changes the dynamic behavior of a submodel, if elements are present which have multiple input and/or multiple output arcs. These elements are called **MIMO elements**. The more complex behavior of them is captured in a directed graph of MIMO elements, which is called **initial MIMO graph** and explained in Sect. 8.2.1. The MIMO graph is iteratively aggregated, which is described in Sect. 8.2.2. This yields an aggregated MIMO graph in which all neighboring MIMO elements of the same type are merged together. From the final aggregated MIMO graph, a Petri net model is derived, which is an aggregation of the original submodel (see Sect. 8.2.3).

The method and the MIMO graph are described for uncolored Petri nets first to make it easier to understand. The application to variable-free colored Petri nets is shown in Sect. 8.2.4. A more detailed presentation of the method is given in [112].

To explain the method, an example model after an imaginary decomposition is shown in Fig. 8.1. Places `buffer1` through `buffer6` are buffers connecting the submodel with the environment. The aggregation method is, however, not restricted to such a simple structure of a submodel. It is also possible that the buffers are both input and output buffers or that there are loops containing them. In the figure both the buffers and the MIMO elements are marked. Except for `P5` the elements are MIMO elements due to the fact that they have more than one input or output arc. `P5` is a MIMO element because input and output arcs have different multiplicities.

8.2.1 The Initial MIMO Graph

During the transformation of a Petri net model into a MIMO graph, only selected model elements – namely the MIMO elements – are taken into consideration. A **MIMO element** of a Petri net is defined as follows. The Petri net element $x \in P \cup T$ is a MIMO element if at least one of the following properties holds:

- $|{}^\bullet x| \neq 1 \vee |x^\bullet| \neq 1$ – the number of input or output elements differs from one
- $x \in P, \mathbf{m}_0(x) > 0$ – the initial marking is not empty
- $\exists y \in {}^\bullet x, z \in x^\bullet$ such that there are different arc cardinalities on the arcs from y to x and from x to z
- x is used in a guard function, any marking-dependent term or a reward variable definition

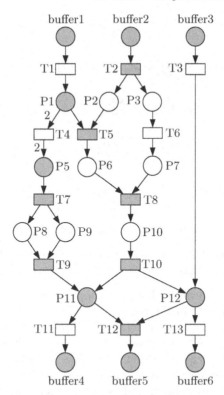

Fig. 8.1. Explanatory example

There are two types of MIMO elements: **MIMO-P** and **MIMO-T** elements, depending on whether the original model element is a place or a transition. The **MIMO graph** is a directed graph with the MIMO elements as vertexes. The arcs of the MIMO graph are related to the transitive arc relationship of the underlying Petri net. MIMO-P elements are labeled with their associated initial marking while the arcs are labeled with their corresponding multiplicity. Arc multiplicities between MIMO elements are derived as the sum of the cardinalities of all paths in the original model that connect the two MIMO elements without visiting other MIMO elements. There is exactly one cardinality in all Petri net arcs of one individual path, because a change in the cardinality would lead to an additional MIMO element following the third rule of the definition above.

It has already been stated that only MIMO elements are "interesting" for the MIMO graph-based aggregation, because all other places and transitions are part of simple sequences.[1] In the first step of the aggregation method, the Petri (sub)net is transformed into an initial MIMO graph. All MIMO elements

---

[1] They are called *SISO elements* in contrast.

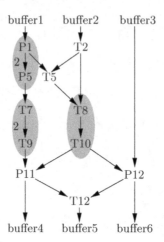

Fig. 8.2. Initial MIMO graph of the example

including the buffers are marked in the example (Fig. 8.1). The resulting initial MIMO graph is depicted in Fig. 8.2.

The algorithm to derive such an initial MIMO graph checks for every Petri net element whether the properties given in the definition hold, and marks it as being a MIMO element in that case. Take for example place P1. It is a MIMO-P element because it has more than one output element. In addition to that, there are differing input and output arc cardinalities. The associated vertex in the directed graph is labeled identically (P1). Following the left side of the model, the next MIMO element is P5 because the input and output arcs have different multiplicities. T7 is a MIMO-T element that splits the path with two output places, followed by T9 and so on. Because there are two different transitive connections (or paths) between T7 and T9 in the Petri net, the multiplicities of the two paths have to be added resulting in 2 to derive the cardinality of the connecting MIMO arc. The last MIMO element of the paths with buffer1 in the preset is P11. After derivation of the MIMO graph for all paths with buffer1 in the preset, a second root buffer2 is inserted into the MIMO graph and the associated paths are analyzed. If a MIMO element already exists in the MIMO graph, only a new arc to this vertex is inserted. The MIMO graph shown in Fig. 8.2 is derived as a result.

8.2.2 Aggregation of the MIMO Graph

The initial MIMO graph is further simplified by applying different aggregation rules. Neighboring MIMO elements of the same type can be merged if the element in the preset has no other output arcs or if the element in the postset has no other input arcs. Figures 8.3 and 8.4 show these merging rules for neighboring elements of the same type (X in the figure stands for either T or P).

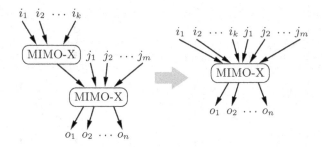

Fig. 8.3. First aggregation rule

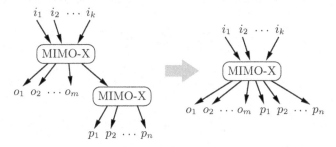

Fig. 8.4. Second aggregation rule

There are two special aggregation rules related to buffers in addition. If an input (output) buffer is connected with a MIMO-P element, both can be merged if the MIMO-P element has no other input (output) arc. Merging neighboring MIMO elements of the same type deletes only the arc connecting both while all the others are kept. An element can be deleted if it is no longer a MIMO element after an aggregation. The initial marking of MIMO-P elements is added during a merge.

See for instance the neighbors T7 and T9 in the initial MIMO graph shown in Fig. 8.2. Both can be merged into one MIMO-T element Tm1 with the input arc of the MIMO-T element T7 and the output arc of the MIMO-T element T9 (the same applies to T8 and T10 which result in Tm2). In Fig. 8.2, the elements which can be merged are marked. It is for instance not possible to merge T2 with T5 because T2 has another output arc and T5 has another input arc. P1 and P5 can be merged applying the second aggregation rule, resulting in Pm1. The aggregated MIMO graph after the first aggregation step is shown in Fig. 8.5 (left).

Possible simplifications in the resulting MIMO graph are again marked in the figure. Tm1 is not a MIMO element any more and can be deleted, leading to a direct connection between Pm1 and P11. T5 and Tm2 are merged following the aggregation rules and form Tm3. buffer1 and Pm1 can be aggregated with the special buffer rule, which leads to buffer1m. The result is shown in Fig. 8.5 (right).

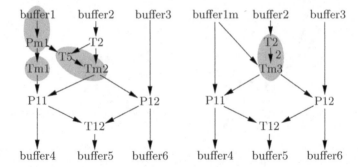

Fig. 8.5. MIMO graph after the first and second aggregation steps

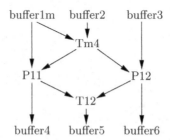

Fig. 8.6. Final aggregated MIMO graph

There is only one aggregation step possible in the shown graph: T2 and Tm3 can be merged. Figure 8.6 shows the final aggregated MIMO graph. No more MIMO elements of the same type are connected. No other MIMO-P elements can be merged with buffers, and all remaining elements are MIMO elements.

8.2.3 Translation into an Aggregated Petri Net

After the simplification of the MIMO graph, the result needs to be transformed back into a Petri net. This translation is simply done by substituting each MIMO element by the corresponding Petri net element. Figure 8.7 shows the submodel example after translating the aggregated MIMO graph. Places P11 and P12 as well as transition T12 are kept from the original model, while transition Tm4 represents transitions T2, T5, T8, and T10.

If there are neighboring MIMO elements of the same type in the aggregated MIMO graph, an additional element of the opposite type has to be inserted. Otherwise the resulting net would not be bipartite and thus not a proper Petri net. If there is for instance a buffer connected to a MIMO-P element (buffer1m, buffer3, buffer4, and buffer6 in the example), an additional transition has to be inserted for each connecting arc (see Ta1, Ta2, Ta3, and Ta4 in the example shown in Fig. 8.7). In the case of two directly connected buffers, a transition–place–transition sequence is inserted instead.

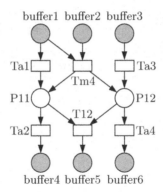

buffer1 buffer2 buffer3

Fig. 8.7. Resulting aggregated Petri net

The token flow of the submodel is preserved after aggregation. If, e.g., in a marking, only **buffer1** of both the original model and the aggregated one is marked and **buffer2** and **buffer3** are not, the only possible final marking is one in which **buffer4** is marked while **buffer5** and **buffer6** are not. The same holds for the other possible combinations of initial and final buffer markings of the submodel.

8.2.4 Aggregation of **vfSCPN** Models

The aggregation of variable-free colored Petri net models is structurally the same as for uncolored models, if every single transition mode is treated just like an individual transition. Remember that **vfSCPN** models can be seen as folded uncolored Petri nets. However, color information needs to be stored in the MIMO graph as well. The main difference in the methods is that transitions of a **vfSCPN** are MIMO elements also when they change the color of a token during firing, i.e., if the created token has a color different from the input token. Figure 8.8 shows a very simple example of a **vfSCPN** model aggregation.

The upper part of the figure shows a **vfSCPN** model example as introduced in Sect. 6.5. Arcs are inscribed with token colors: take for instance the firing of T1, which takes a token **P.empty** from place **P1** and creates one with color **A.unpr** in place **P2**.

The second graph in the figure shows the initial MIMO graph. T1 and T3 are MIMO elements because they "change the color of the tokens." They can, however, be merged during the aggregation step. The resulting aggregated MIMO graph is shown as the third part of the figure. The final result of the translation back into a **vfSCPN** model is depicted at the bottom. Only one transition **Tm** represents the aggregated behavior, with input and output token colors **P.empty** and **A.first**, respectively.

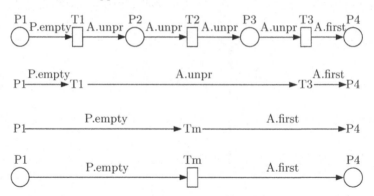

Fig. 8.8. Example aggregation of a variable-free colored Petri net

8.3 Low-Level Systems and the Basic Skeleton

The subsystems SS_i that resulted from the model partitioning as well as their aggregated versions SS_i^* are usually not analyzable in isolation. This is a consequence of the freedom of where the model can be cut, which is an important advantage with respect to other approximation methods. The reason is that the model parts are not live any more when the remaining parts are deleted.

Therefore a so-called **low-level system** LS_i is constructed for every subsystem SS_i by adding the aggregated versions of all other subsystems to it.

$$LS_i = SS_i \cup \bigcup_{j \neq i} SS_j^*$$

In addition to that, the interaction between the different submodels has to be considered in the iterative approximation algorithm later on. The **basic skeleton** BS is used for this task, which represents a model in which all subsystems are aggregated.

$$BS = \bigcup_i SS_i^*$$

The preservation of token paths during the individual aggregations ensures that all low-level systems as well as the basic skeleton are life if the original model is [112]. Example low-level systems and a basic skeleton are shown in Sect. 16.3.1 for an application.

8.4 Iterative Throughput Approximation

The low-level systems LS_i and the basic skeleton BS are used to iteratively compute an approximation of the throughput $TP(t_k)$ for all output transitions $t_k \in OT_j$ of the subsystems SS_j. Every transition with an output arc to a buffer of the subsystem is considered as such an output transition.

APPROXIMATEEVALUATION

Input: Low-level subsystems $LS_1 \dots LS_n$, Basic skeleton BS
Output: Transition throughputs
Requires: Evaluation algorithm SOLVE(model)

repeat
    for $k = 1 \dots n$ **do**
        (∗ Compute throughput of all $t \in OT_k$ ∗)
        SOLVE(LS_k)
        (∗ Adjust transition delays in the basic skeleton accordingly ∗)
        repeat
            Change transition delays $\Lambda^{BS}(t)$ for all $t \in OT_k$
            SOLVE(BS)
            (∗ until all throughput values are equal ∗)
        until $\forall t \in OT_k : TP^{BS}(t) = TP^{LS_k}(t)$
        (∗ Adjust transitions in other low-level systems accordingly ∗)
        for $i = 1 \dots n$ **do**
            for $t \in OT_k$ **do**
                $\Lambda^{LS_i}(t) = \Lambda^{BS}(t)$
until convergence: $\max_{t \in T} \Delta TP^{LS_k}(t) < \epsilon$

Algorithm 8.1: Iterative approximate evaluation algorithm

During the execution of the algorithm, the service rates of the output transitions (i.e., the inverse of their delay) are adjusted to achieve a balance between the throughput values of all low-level systems and the basic skeleton. In the nonaggregated parts, the transition's firing delays are not changed from the original model. For all application examples investigated so far, the final result was independent of the initial delay setting of the output transitions of the aggregated model parts, and convergence was reached after a few iterations only.

The method is shown in Algorithm 8.1. Informally, for each low-level system LS_i, the sojourn time for each token[2] in the preset places of all output transitions of the nonaggregated subsystem SS_i is computed. The service rates of the corresponding transitions of the basic skeleton BS are then changed such that the sojourn times of the tokens are equal to the results for the low-level system. This needs to be done in an internal iterative loop. The obtained service rates of the output transitions $t \in OT_i$ are used to change the output transition delays in all $LS_j, j \neq i$ accordingly. After that the procedure is started again with the next low-level system LS_{i+1} and so forth, beginning with the first one again when necessary. The overall algorithm is repeated until convergence is reached, i.e., the newly obtained delays do not change more than a given threshold between two consecutive steps. The algorithm

---

[2] In the case of a colored model, the different token colors are treated independently.

calls an external performance evaluation algorithm SOLVE, for which, e.g., the numerical analysis of Sect. 7.3.2 can be used.

The following notation is used in the algorithm: OT_i is the set of output transitions of low-level system LS_i (and subsystem SS_i). To differentiate between properties in the different models, superscripts are used that specify the actual model. The delay of a transition t in the low-level system LS_i is for instance denoted by $\Lambda^{LS_i}(t)$. The same applies to transition throughput results, e.g., of a transition t in the basic skeleton, which is specified by $TP^{BS}(t)$.

The presented aggregation method observes the dynamic behavior of the subsystems. There are neither structural nor dynamic restrictions for the decomposition as opposed to methods known from the literature. Very large systems can thus be analyzed. Results for an application are shown in Sect. 16.3.2.

9

Efficient Simulation of **SDES** Models

Standard simulation methods like the ones described in Sect. 7.2 work well for many practical applications. There are, however, numerous cases in which the required computational effort is very high, or may even inhibit a quantitative evaluation at all. The reason is that, independent of the model or simulation environment, there is a certain number of significant events or time intervals (samples) for which intermediate results have to be generated for a specified accuracy of the performance measures.

The basic speed of a simulation algorithm, e.g., measured in executed events per second, cannot be decreased easily: code optimization should be done in any case, and otherwise only a faster processor may help. Although it depends on the complexity model class as well, a transformation of a model into a different model class will change the number of required model elements accordingly. The use of parallel or distributed computing hardware has the potential to reduce computation time (but not effort). Section 9.1 briefly covers characteristic approaches and introduces an optimistic distributed simulation method for **SDES** models aiming at an automatic load balancing. Fine-grained model partitioning as well as a specific logical simulation time are proposed.

On the other hand, it is possible to investigate the statistical methods used for result estimation. The number of samples (significant events) to be generated may be very high with the applied estimation method. This is usually the case if the intermediate values have a high variance, or if the desired result quality (confidence interval, error probability) is very strict. The efficiency of an estimator informally states the computational effort required to achieve a certain accuracy, e.g., its mean square error. Variance reduction techniques aim at an efficiency improvement, and thus may obtain a more accurate result from the same number of samples. Well-known methods include common random numbers, antithetic variates, or control variables. Results are especially hard to estimate when only a few samples can be drawn from a large number of events.

This problem is well known under the term *rare event simulation*, and one method is described in the SDES setting in Sect. 9.2. The chapter ends with some notes on previous work and background.

9.1 Distributed Simulation of **SDES** Models

A popular approach to speedup a simulation experiment is the use of multiple processing nodes. While the overall computation effort usually increases due to organizational overhead, the waiting time reduces until statistically significant results are obtained. There are two general approaches to utilizing a set of processors. In the first one, a standard simulation algorithm with a copy of the whole model runs on each node and sends intermediate results to a central master process who collects and analyzes the results. This approach is called *independent replications*, and is for instance implemented for SPN models in the TimeNET software tool [197]. It is technically much simpler than the second approach, but the scalability is bounded by the centralized architecture.

The second approach divides the model into parts, which are simulated by communicating **logical processes** $lp_i \in LP$. Discrete event system models reflect the locality of real-world applications in that dependencies as well as the effect of executions primarily affect "neighboring" state variables. If we think of larger systems to be simulated, which are becoming ever more complex and assembled from individual modules, the second approach of *logical process simulation* is a natural choice. There are even settings in which a joint simulation of the whole model is practically impossible: if a technical system is designed by cooperating companies, and its behavior should be simulated without giving away the company's knowledge, or there are different coupled simulation engines for the model parts that run on individual hardware.[1] Modern grid or cluster environments provide the necessary environment. Another possibility is a true parallel simulation on a SIMD-operated multiprocessor, which we do not consider here further.

Just like in a standard sequential simulation, discrete event occurrences are observed in each logical process lp over a virtual simulation timescale. An event list is managed with events scheduled in the future, ordered by their occurrence time. There has to be some kind of synchronization between the logical processes that simulate parts of the model to notify others about events. The communication overhead is obviously smaller if we can find a way to separate the model such that most of the events have local effects in one logical process only. Many real-life examples are characterized by partly autonomous entities that cooperate or compete from time to time. This parallelism is inherent to the real-world and is reflected in SDES models. For its exploitation, a partitioning is required, as well as an adaptation

---

[1] Simulators in such an environment are called *federates* in the *high-level architecture* [76] framework for simulation integrating and interoperability.

of standard logical process simulation algorithms to the requirements of our abstract model class. There are different ways of synchronization possible.

A **distributed simulation** has no centralized control to synchronize the different logical processes. Synchronization is realized by exchanging messages between the nodes; a fast communication subsystem is thus important for the efficient execution. The underlying idea is that enabling of actions and event execution is often locally done in SDES models. Many events occur thus at different times, but do not affect each other. A parallel or synchronously distributed simulation would have to wait for the event executions in their time ordering due to the use of a global simulation clock. Independent execution of concurrent events is possible in an asynchronously distributed simulation, and leads to a significant speedup [103].

Distribution at the event level divides a global simulation task such that each logical process simulates a part of the global model. The model (state variables and actions) is partitioned into a set of regions that are associated to each process. *Internal events* do not affect state variables of other processes, while *external events* may do so. Nonlocal dependencies and results of events need to be propagated to the corresponding processes. Messages for state variable changes (remote events) and other notifications for management issues are exchanged via a communication system. The main problem to be solved is to guarantee causal correctness of a distributed simulation run.

A distributed simulation is obviously said to be correct with respect to the local event processing if the partial event ordering created by it is consistent with the total event order of a sequential simulation. This leads to the question which notion of time is necessary to achieve such an ordering. The *logical clock problem* [218] aims at generating clock values in a distributed system in a way that all events are ordered in a *logical time*. It was shown in [183] that this is the inverse of the problem in a distributed simulation run. Causality errors are impossible if all *LP* execute the events ordered by their time stamps. This is called the *local causality constraint* [114] and has been shown in [244].

There are several ways to ensure that this constraint is not violated. *Conservative* distributed simulation schemes guarantee causality from their algorithm and message exchange protocol. Each logical process blocks until it is safe to proceed, i.e., there will be no incoming messages with time stamps smaller than the local clock value. This method is commonly named *CMB protocol* due to the works of Chandy and Misra [43] and Bryant [33]. Inherent problems are deadlocks and memory overflows, and the performance is hampered by having to wait for other processes to catch up.

If a model has only a few events that require messages to be exchanged between logical processes, it is safe in most cases to proceed inside such a process. The idea of **optimistic logical process simulation** or **Time Warp** [183,184] is to temporarily accept the possibility of local causality violations. A violation occurs if a logical process receives a message from another one, notifying it about a past remote event execution which affected the local state of the process. Such a message is called a **straggler message**, and the causality violation is overcome by a **rollback** of the logical process to the time before the time stamp of the remote event, i.e., a state which is consistent with the received message. This obviously wastes some computation time, and the algorithm requires detailed bookkeeping of the local past (visited states and executed events), because it may be necessary to jump back in the logical time.

Logical processes communicate by exchanging positive messages to notify possibly affected processes about event executions, and negative messages that take back prior messages because the time window in which they were generated was invalidated by a rollback. All messages are time stamped. Messages with external events that are still in the future of a local simulation are simply inserted into the local event list, ordered by the time stamp and executed when the local time advanced sufficiently. Possible problems of Time Warp implementations include unnecessary or cascading rollbacks and "overoptimistic" processes that increase their local simulation times too fast, producing numerous rollbacks. Possible solutions have been proposed with *lazy cancellation* and *lazy reevaluation* as well as *optimistic time windows*. Another issue is the amount of memory required for the bookkeeping, which can be reduced from time to time by a *fossil collection*. The amount of messages exchanged between logical processes is often the most significant performance bottleneck. *Adaptive protocols* have been described in the literature which set a "level of optimism" during the simulation run for a good tradeoff between conservative and optimistic approaches. An overview can, e.g., be found in [103, 114].

A distributed simulation of SDES models requires time stamps for events that allow their unique and correct ordering. With the existing approaches it is, however, impossible to order events that are due to immediate action executions (with zero delay), or have priorities. Standard distributed simulations require a model to be decomposed into regions of logical processes in a way that there are no zero delay events to be sent, i.e., only at timed actions. It is then (practically) impossible that two events are scheduled for the same time. Our goal is, however, to improve overall performance by an automatic load balancing, which obviously can reach a better partitioning when there is no such restriction.

Global enabling functions (guards) and condition functions (capacities) are impossible without global state access, which in turn requires a complete ordering between all event times in a distributed simulation. Standard simulation times lack this feature and thus cannot be used.

We propose a new logical time scheme for SDES models that allows much better partitioning with less structural restrictions than, e.g., the approaches covered in [252]. Section 9.1.1 introduces a fine-grained model partitioning and distribution that overcomes the static association of model parts to logical processes. We show that the known logical times are not sufficient for our purposes. A **compound simulation time** for SDES models including immediate actions, priorities, and global functions is presented in Sect. 9.1.2. It allows to detect confusions in the model as a byproduct. Section 9.1.3 proves some properties of the time scheme. Algorithms for a distributed SDES simulation with the proposed techniques are shown in Sect. 9.1.4. Some notes on related issues are given in at the end of the chapter.

9.1.1 Fine-Grained Model Partitioning for Dynamic Load Balancing

A fine-grained partitioning of an SDES model has the advantage of almost arbitrary associations of model parts to computing nodes. It is a prerequisite for dynamic load balancing. The reason for this is that model parts would have to be separated and sent to another node, while it is unclear how the state and event lists would have to be updated in such a case. With a more fine-grained partitioning, there are obviously more messages to be passed among model parts; this disadvantage can, however, be simply overcome technically by avoiding communication messages between processes on one host. An additional advantage is the locality of rollbacks, which affect smaller portions if done in a more fine-grained fashion. Moreover, it is often impossible to obtain a good partitioning from the model structure. Our approach allows to achieve a good balance with simple heuristics starting from an arbitrary initial mapping to nodes. Less memory is consumed in the state lists because model parts that create few events need to store their local states only rarely.

Different to standard time-warp simulations, we propose to run one **logical process** per host, which manages several **atomic units** that are running quasiparallel in that machine. An atomic unit is responsible for the optimistic simulation of a smallest possible model part, and is created as shown below. Each atomic unit has its own local simulation time, event, and state list. This makes it possible to migrate it during runtime without touching other atomic units. An atomic unit can restore its local state accurately for a given simulation time, and send rollback messages to other atomic units that might be affected. Rollbacks are thus more precise, and unnecessary ones are avoided or canceled whenever possible. The way of scheduling the operations of atomic units inside a logical process avoids causality violations between them, reducing the number of rollbacks further. Each logical process offers a message interface to each of its atomic units, which either exchanges the information internally to another local unit or sends them over the communication network to another node. Logical processes do not share memory and operate asynchronously in parallel, e.g., on a cluster of workstations.

Formally we denote by lp one logical process that runs on its computing node; there is a one-to-one mapping between available nodes and logical processes. The set of all logical processes is called LP.

Likewise, we denote by au one atomic unit and by AU the set of all of them.

$$AU = \{au_1, au_2, \ldots, au_{|AU|}\}$$

Atomic units are distributed over the available computing nodes in an actual setting. This means a mapping of each atomic unit to a logical process lp, which we denote by the **node-distribution function** $Node$.

$$Node : AU \to LP$$

We write for simplicity AU^{lp} to denote the set of atomic units that are mapped to the logical process lp.

$$AU^{lp} = \{au \in AU \mid Node(au) = lp\}$$

The mapping of atomic units to logical processes is a key factor for the reachable speedup of a distributed simulation. Unlike other partitioning algorithms, which detect model parts that should go into one logical process and add them until there are as many model parts as logical processes, we go a different way. The smallest model parts that must be simulated together are obtained, but kept in atomic units individually. The mapping of atomic units to logical processes may be done with a simple heuristic initially, which may for instance exploit the relations defined below for connections between model parts. The goal of the proposed approach is, however, a framework in which a (near-)optimal mapping develops automatically during runtime.

Obviously should hold $AU^{lp} \neq \emptyset$ for all logical processes, at least as long as $|au| \geq |LP|$. However, the user could, in the extreme case, associate all atomic units to one logical process initially, and then let the mapping improve during the simulation by automatic migration.

An SDES model then has to be partitioned to form atomic units. The model part captured in one atomic unit au_i, the **region**, is denoted by $Region(au_i)$. It is a subset of the actions and state variables of an SDES model.

$$Region : AU \to SV^* \cup A^*$$

The main objective during the partitioning into model regions is to keep them small and to have only the minimal number of model elements in them. Communication effort might, however, become higher of the regions that are too small. Conflicts between action executions need to be solved inside an atomic unit, as it has been shown for Petri nets in [56]. Thus at least a set of input state variables of each SDES action should reside in the same atomic unit as the action itself.

For restricted model classes, it is possible to formally derive the sets of actions that may result in a conflict [53]. Such a set is called an *extended conflict set* (ECS), and has been explained in the context of simple Petri nets on page 84 as well as standard performance evaluation algorithms on page 133 already. A resulting partitioning was proposed in [50].

In the following, we assume for simplicity that there are only conflicts between actions that share state variables as their input. This disregards possible conflicts that may occur due to the change of common state variables which have a nontrivial condition function (such as a common output place with a capacity restriction), and conflicts due to the change of state variables which influence an enabling function (like a state-dependent guard).

It should, however, be noted that all these considerations only affect the execution order of activities which are scheduled at the same time. Priorities can be used by the modeler to explicitly specify a preference. This topic has been extensively studied for Petri nets in the literature [303], to which the interested reader is referred. For the "simpler" model classes such as automata and queuing networks, it is a less important issue, because conflicts can be easily found from the model structure there.

We require the SDES model to be confusion-free, as it is discussed in more detail in Sect. 9.1.2. Confusion is interpreted as a modeling error, and can be detected during runtime with the methods introduced in the following. Another restriction is the absence of *vanishing loops*, as it has been mentioned in Sect. 7.1. We slightly restrict immediate paths requiring that an ECS containing only immediate actions must not be visited more than once in one immediate path. This restriction is, however, of little practical significance.

Coming back to the partitioning, we are interested in a set of **input state variables** $SV^\star_{input}(a) \subseteq SV^\star$ of each SDES action $a \in A^\star$. Due to the general definition of the SDES model class, it is not possible to obtain this set from it directly. For a given model class,[2] it is, however, easy to do as Table 9.1 shows. Input places are for instance obvious candidates in Petri net classes.

Due to the dependence of action enabling and execution as well as conflict resolution on the input state variables, all actions that share some of these variables need to be associated to the same region. Otherwise the probabilistic choice between conflicting actions would require communication, and a distributed random choice would introduce significant overhead.

Table 9.1. Input state variables of actions from individual model classes

Model class	$\forall a \in A^\star : SV^\star_{input}(a) =$
QN	$\begin{cases} \emptyset & \text{if } a \in \mathit{Arr} \\ \{e_a \in E\} & \text{if } a \in \mathit{Trans} \\ \{q_a \in Q\} & \text{otherwise} \end{cases}$
SPN	$\{p \in P : \exists \mathbf{m} \in M, \mathbf{Pre}(p, a, \mathbf{m}) \neq 0\}$
SCPN	$\{p \in P : \mathbf{Pre}(p, a) \neq \emptyset\}$
vfSCPN	$\{p \in P : \exists \beta \in \beta^\star(a), \mathbf{Pre}(p, a)(\beta) \neq \emptyset\}$

---

[2] Automata are not considered here, because for them a distributed simulation would make no sense due to one single state variable.

Two model elements (state variables $sv \in SV^\star$ and actions $a \in A^\star$) are connected in the sense that they should be situated in the same region if the following relation $\rightleftharpoons$ holds for them.

$$\forall x, y \in (A^\star \cup SV^\star) : x \rightleftharpoons y \iff$$
$$x = y \quad \vee \quad x \in SV^\star_{input}(y) \quad \vee \quad y \in SV^\star_{input}(x)$$

Obviously $\rightleftharpoons$ is reflexive and symmetric. Now denote by $\rightleftarrows$ the transitive closure over $\rightleftharpoons$

$$\forall x, y \in (A^\star \cup SV^\star) : x \rightleftarrows y \iff \exists z \in (A^\star \cup SV^\star) : x \rightleftharpoons z \wedge z \rightleftarrows y$$

$\rightleftarrows$ is then an equivalence relation, and thus defines a partition of state variables and actions. Each of the obtained equivalence classes forms the region $Region(au)$ of an atomic unit. The set of all atomic units AU is thus given by the quotient set of $(A^\star \cup SV^\star)$ by $\rightleftarrows$.[3]

Then every part of a model is contained in exactly one region of an atomic unit, such that nothing is omitted and there are no double associations.

$$\forall au_i, au_j \in AU : au_i \neq au_j \longrightarrow Region(au_i) \cap Region(au_j) = \emptyset$$
$$\wedge \quad \bigcup_{au_i \in AU} Region(au_i) = SV^\star \cup A^\star$$

Some additional relations between actions, state variables, and atomic units are now introduced, which are used in the algorithms later on.

The set of state variables that possibly changes its value due to an action execution is for instance necessary for message generation. We denote by $SV^\star_{affected}(a)$ the set of state variables that may be affected by the execution of (a variant of) an action a.

$$SV^\star_{affected} : A^\star \to 2^{SV^\star}$$

In theory, a state variable sv is included in the set for an action a under the following condition. Assume that there are two states σ and σ' in the set of reachable states RS of the model, and σ' is directly reached by executing $(a, \cdot)$ in σ. If the value of sv is changed by this execution, it is affected by the action.

$$SV^\star_{affected}(a) = \{sv \in SV^\star \mid \exists \sigma, \sigma' \in RS, \sigma \xrightarrow{(a,\cdot)} \sigma' \wedge \sigma(sv) \neq \sigma(sv)'\}$$

This definition is, however, futile for an actual algorithm. The sets can be obtained much easier for each model class individually. Table 9.2 lists the results for the model classes used in this text.

---

[3] To avoid atomic units "running away" with their local simulation times, and thus unnecessary rollbacks, actions without input state variables are put into an atomic unit that contains another state variable which is closely connected to it in practice.

Table 9.2. Affected state variables of actions

Model class	$\forall a \in A^* : SV^*_{affected}(a) =$
QN	$\begin{cases} \{q_a \in Q\} & \text{if } a \in \textit{Arr} \\ \{e_i \in E, q_j \in Q\} & \text{if } a = (q_i, q_j) \in \textit{Trans} \\ \{e_a \in E, q_a \in Q\} & \text{if } a \in Q \end{cases}$
SPN	$\{p \in P : \exists \mathbf{m} \in M, \mathbf{Pre}(p, a, \mathbf{m}) \neq \mathbf{Post}(p, a, \mathbf{m})\}$
SCPN	$\{p \in P : \exists \beta \in \beta^*(a), \mathbf{Pre}(t, p)^\beta \neq \mathbf{Post}(t, p)^\beta\}$
vfSCPN	$\{p \in P : \exists \beta \in \beta^*(a), \mathbf{Pre}(t, p)(\beta) \neq \mathbf{Post}(t, p)(\beta)\}$

Table 9.3. Additionally required state variables of actions

Model class	$\forall a \in A^* : SV^*_{required}(a) = SV^*_{affected}(a) \cup$
SPN	$\{p \in P : \exists \mathbf{m} \in M, \mathbf{Inh}(p, a, \mathbf{m}) > 0\}$
CPN, vfSCPN	$\{p \in P : p \in G(a)\}$

Actions access not only the state variables that belong to their local region, because for instance their execution may change the value of a remote variable. The set of state variables of the SDES model that are eventually accessed by an action a is denoted by $SV^*_{required}(a)$ and defined as follows. At least the state variables that are affected by its execution are required for the action. Thus it holds

$$\forall a \in A^* : SV^*_{required}(a) \supseteq SV^*_{affected}(a)$$

Table 9.3 shows state variables which are required for read-only access in addition to the affected ones for individual model classes. For the other model classes that are considered in this text, $\forall a \in A^* : SV^*_{required}(a) = SV^*_{affected}(a)$ holds, and thus there is no need to define $SV^*_{required}(\cdot)$ specifically.

For simplicity of notation, we write $p \in G(\cdot)$ in Table 9.3, thus interpreting the guard function of a transition from a colored Petri net as an expression that depends on places.

Using the sets of required state variables for actions, we can now define the set of state variables that need to be known in an atomic unit au. We call this set **local state variables** and denote it by $SV^*_{local}(au)$. It contains the state variables of the region as well as the ones needed in any of the expressions attached to model objects in the region. Thus obviously every state variable is local to at least one atomic unit, if the model is connected.

$$SV^*_{local}(au) = \bigcup_{a \in Region(au) \cap A^*} SV^*_{required}(a)$$

We informally refer to state variables that are locally known in an atomic unit, but do not belong to its region, as **mirrored state variables**.

A **local state** σ^{au} of an atomic unit au associates a value to each of the local state variables. The set of all theoretically possible local states is denoted by Σ^{au} and obtained by taking the crossproduct over the sets of allowed values for each local state variable.

$$\Sigma^{au} = \prod_{sv \in SV^{\star}_{local}(au)} S^{\star}(sv) \quad \text{and} \quad \sigma^{au} \in \Sigma^{au}$$

A remote atomic unit au needs to be informed about the execution of an event a (i.e., au may be affected by action a), if at least one of the state variables possibly changed by the execution of a is locally known in au. We thus define the set of remote atomic units that is **affected by an action** accordingly.

$$\forall a \in A^{\star} : AU_{affected}(a) =$$
$$\left\{ au \in AU \mid a \notin Region(au) \wedge SV^{\star}_{affected}(a) \cap SV^{\star}_{local}(au) \neq \emptyset \right\}$$

9.1.2 A Logical Time Scheme for **SDES** Models with Immediate Actions and Priorities

In this section, a new logical time for SDES models is presented. It has been mentioned in the introduction already that existing time schemes for distributed simulations are not sufficient for this class of models, because the ordering of events is not ensured by a simple global clock in the presence of immediate actions and priorities. We will cover this in more detail in the following. Moreover, performance measures and conditions with global dependencies would otherwise not be possible (see below).

Causal correctness of a distributed simulation algorithm for SDES models is guaranteed if the events are processed in the same sequence that a sequential method would follow. A sequential simulation would process events in the order given by the simulation time ST, remaining delays, and by taking priorities of events into account which are scheduled for the same time (see Sect. 7.2). In the case of an optimistic approach like the one used here, possible violations of this rule are accepted temporarily, which are taken back by a rollback if the assumptions turn out to be wrong later on. Following [114,244], every atomic unit must execute events in a nondecreasing time stamp order. However, time stamps must allow a complete ordering and capture all of the problematic issues mentioned before. This section introduces a new **compound simulation time** cst, which fulfills the requirements for SDES models. The insufficiency of existing time schemes for distributed SDES simulation is shown first.

A global simulation clock alone is not sufficient, because there are actions with zero delay to be executed at the same simulation time, and there is a nonzero probability that several timed actions are scheduled at the same time as well.[4] Action priorities $Pri^\star(a)$ specify the order of execution in our definition of the dynamic behavior (cf. Sect. 2.3.2). Events to be executed at the same simulation time must be uniquely ordered at every atomic unit where their ordering matters. Priorities and causal relations between these events must be taken into account for such a decision. The question is how to manage priorities and causal dependencies in a distributed manner. There are some approaches available for causal ordering, which are briefly discussed in the following.

Lamport's algorithm [218] allows a time ordering among events [333]. A single number is associated to every event as its **logical time**, and increases with subsequent events. However, a mapping from Lamport time to real (simulation) time is not possible. In a simulation we, however, do need the actual simulation time ST, e.g., to compute performance measures. Lamport time is furthermore not sufficient to detect causal relationships between events, which is a prerequisite for models with action priorities. It is impossible to sort concurrent and independently executed events whose occurrence is based on a different priority. Lamport time would impose an artificial ordering and neglect the priorities. Moreover, it is forbidden for neighboring regions of a distributed model to exchange events that have a zero delay. In a colored Petri net, this would prevent models to be decomposed at immediate transitions, and thus restrict the formation of atomic units significantly.

A logical time that characterizes causality and thus overcomes some of the mentioned problems of Lamport time is **vector time** (or vector clocks) proposed by Mattern [235], Fidge [104], and others independently in different contexts. In our proposed setting, a vector time VT value is a vector of natural numbers, which contains one entry for every atomic unit.

$$VT : AU \to \mathbb{N} \quad \text{or} \quad VT \in \mathbb{N}^{|AU|}$$

Whenever an atomic unit executes an event or rollback, it increases the vector time entry of itself by one. The local entry of the VT vector thus always increases, even when a rollback is processed. The elementwise maximum is taken for every nonlocal entry of VT to update the local time, whenever a remote event is processed.

The informal meaning of every VT entry in an atomic unit is thus the local knowledge about the number of previously executed events, given for every atomic unit individually. Events are understood here including local action executions, execution of remote messages, and rollbacks. Vector time entries may therefore be interpreted as a version number of the atomic units' inner states. It can be used to check for causal dependency between two events. A higher-valued entry of an event denotes a causally "later" execution.

Vector time is a notion of causality and can thus be used to differ between events that depend on each other. The possible relations between vector times

---

[4] The general issue of simultaneous events in simulations and some solution methods are covered in [326].

are thus important. If event e_2 is **causally dependent** on e_1, it must naturally be scheduled after it. In that case we write $VT_1 < VT_2$.

$$\forall VT_1, VT_2 \in \mathbb{N}^{|AU|} : VT_1 < VT_2 \iff \forall au \in AU : VT_1(au) \leq VT_2(au)$$
$$VT_1 = VT_2 \iff \forall au \in AU : VT_1(au) = VT_2(au)$$

The case of all elements of two vector times being equal occurs only if two events are compared that result from actually conflicting actions in one atomic unit. Their execution sequence is then decided based on a probabilistic choice. This can, however, only happen inside one atomic unit and for events that are in the future of the local simulation time. It will never happen in the distributed simulation algorithm shown below that a remote event has the same vector time as any other locally known one, because the same event is only sent once to another atomic unit. In algorithms where this cannot be guaranteed, all equally timed events must be executed together in one step.

Unfortunately it is not the case in models with asynchronous events that every pair of events can be uniquely ordered by their vector time. Two events e_1 and e_2 are said to be **concurrent** with respect to their vector times, if there is no causal dependency found. This case is denoted by $VT_1 \parallel VT_2$.

$$\forall VT_1, VT_2 \in \mathbb{N}^{|AU|} : VT_1 \parallel VT_2 \iff (VT_1 \not< VT_2) \wedge (VT_2 \not< VT_1)$$

Vector time thus allows to detect direct and indirect dependencies by comparing VT values of events. It is possible to differ between causally dependent and truly concurrent activities.

Vector time is, however, still not sufficient for models with priorities and immediate delays. Different Petri net classes as for instance defined in Chaps. 5 and 6 allow immediate transitions with priorities. Two or more events can thus be scheduled for execution at the same (simulation) time, but the one with a higher priority must always be executed first. It may disable events with lower priorities by doing so.

A small Petri net example is shown in Fig. 9.1. Priorities of transitions are annotated in italics.[5] The correct sequence of events would be the firing of transitions T1, T3, T2, and then T4 or T5, depending on the probabilistic solution of the conflict between them.

In a distributed simulation of the model as proposed here, transitions and places are associated to atomic units $au_1 \ldots au_4$ as shown. There is no guarantee that the concurrently running atomic units process events in the mentioned order, at least if they are located on different nodes and thus in different logical processes. It might thus be that T2 fires after the firing of transition T1, and the associated event is received and processed in au_4. T4 then fires locally, which is not correct. This is detected later on, when T3 has fired and the corresponding event is received in au_4. The events of firing transitions T2 and T3 must be ordered correctly in au_4. Otherwise it cannot

---

[5] Remember that timed Petri net transitions always have a priority of zero.

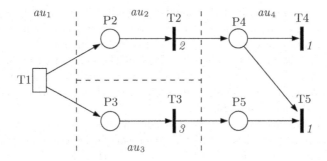

Fig. 9.1. Simple Petri net example with immediate transition priorities

be detected that T3 had to fire before T2 and that the previous local firing of T4 must be rolled back.

An ordering of events T2 and T3 in au_4 is, however, impossible based on the simulation times ST_{T2} and ST_{T3} (which are equal) or the vector times of the events. The latter are $VT_{T2} = (1,1,0,0)$ for T2 and $VT_{T3} = (1,0,1,0)$ for T3. They are concurrent following the definition $(VT_{T2} \parallel VT_{T3})$, and there is thus no hint on how to order them based on vector time. The priorities obviously need to be taken into account, which is shown in the following proposed extension of logical vector time.

The two logical time approaches described above guarantee a causally correct execution of events. However, vector time only ensures that the followed sequence of event executions is a causally possible one. Causally independent events may happen in any order [235]. This is fine as long as a truly distributed system without inherent synchronization is simulated. In a distributed simulation that is used for a performance evaluation, and where global states are important for the decision of enabling functions and performance measures, the "real" simulation time has to be used. Our model in fact has an underlying global synchronization, which is given by natural time. The stochastic delays that are chosen in an atomic unit for an event execution must be obeyed everywhere. The simulation clock, measured in model time units, therefore has to be incorporated in a time scheme for SDES. Otherwise there would be events that are simultaneous from the vector time standpoint (i.e., $VT(\cdot) \parallel VT(\cdot)$), although there is a valid precedence relation based on the execution times. Simultaneous execution in our quantitative SDES models means *at the same time*. Vector time is thus added to the simulation time to cover causal dependencies, and extended by a priority vector as described below.

Immediate Execution Paths and the Priority Vector

We have seen so far that a new time scheme is necessary for actions with priorities, which are specified by $Pri^{\star}(a)$. In a standard sequential simulation, only one global event queue is maintained, which contains the scheduled events of the whole model. It is then easily possible to detect events that are scheduled for the same time and to resolve their execution sequence in the order of the priorities (cf. Sect. 7.2.1). This is not the case in a distributed simulation, where every atomic unit manages the events of the local actions and is informed by messages about events in the neighborhood. A global detection of events that are scheduled for the same time as well as a centralized serialization is thus impossible.

However, priorities (especially of actions with a zero delay) play a significant role in many subclasses of SDES. Event serialization may only take place when different events are sorted with respect to their logical times in atomic units that process them. Thus the decision about which event has (or had) to be executed first must be taken in a distributed way, i.e., in each atomic unit that receives and sorts events by their time. A global probabilistic solution of conflicts between events is thus impossible. Therefore we required to have all actions that possibly conflict with each other to be put into one atomic unit (see Sect. 9.1.1). In the case of a Petri net, all transitions belonging to an ECS [53] are thus associated to one atomic unit.

The main reason for a new logical time is to decide the exact ordering of all incoming and local events in an atomic unit. Let us assume different paths of subsequent executions of immediate actions that lead to an event reception in an atomic unit, while a local immediate action is scheduled for execution at the same simulation time ST. Which event should be executed first, i.e., what ordering is necessary in the local event list? Obviously the priority must be taken into account, as shown earlier with the example in Fig. 9.1.

One might be tempted to think that the logical time of each event only needs to be extended by one priority value. This value then should store the minimum priority of all events on the path, because an event with a lower minimum priority in the execution path will always be executed last.

Figure 9.2 shows an example Petri net, which is partitioned into eight regions associated to atomic units $au_1 \ldots au_8$. It has been constructed to analyze the immediate event executions after the firing of the timed transition T1.

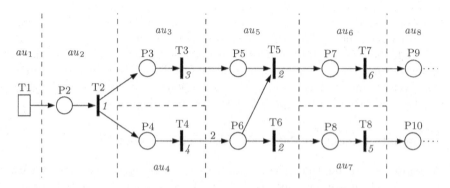

Fig. 9.2. An example for transition priorities

Transition priorities are again depicted in italics and we denote events by the names of firing transitions.[6] The correct firing sequence is T1, T2, T4, and T3, a random selection in au_5 which we assume leads to T6, T8, T5, and T7.

Assume that transitions T1, T2, T3, and T4 have fired. Associated events T3 and T4 arrive in atomic unit au_5, in which they have to be ordered. Vector times cannot be used for this, because obviously $VT_{T3} \parallel VT_{T4}$ due to the concurrent executions in au_3 and au_4. If we would store only one priority (the minimum of immediate transitions on the execution path), it would be 1 for both events, because T2 with priority one is part of both paths. Thus it is not sufficient to store just the minimum priority on the path of an event.

We introduce a **priority vector** as a supplement to the logical vector time to sort events correctly. A priority vector PV maps each atomic unit au to a natural number.

$$PV : AU \rightarrow \mathbb{N} \quad \text{or} \quad PV = \mathbb{N}^{|AU|}$$

This number stores the minimum priority of any event belonging to au, which has been executed in the current path of events that immediately followed each other. Events within the same atomic unit are ordered locally, and there is thus no need to store further information for them individually. Moreover, every atomic unit is visited at most once during one immediate path as required above, and thus there is no ambiguity in the meaning of the associated entry in the PV vector.

It should be noted that only paths of event executions between two tangible states are significant in this case, because otherwise the simple simulation time ST is sufficient for the ordering. Such a path of immediate state transitions starts with the execution of an action with nonzero delay (like transition T1 in Fig. 9.2), and ends again in a state in which some simulation time passes.[7] It should be noted that the execution of the starting timed event does, however, belong to the path itself, because the postpriorities defined in SDES models also apply to timed events that are scheduled for the same instant of simulation time. The event ordering technique described in the following is then also applicable for this case.

All event executions of one path must be considered in the priority vector, because every single one can influence the correct sequence. If two events are compared, the one with the smaller minimum priority entry must be ordered last. This is because the event with the lowest priority delays the propagation of an event until no other action with a higher priority on other paths can be executed. Entries are initialized with infinity as a neutral value of the minimum, and are again set to infinity when some simulation time ST passes (i.e., in a *tangible state*, cf. Sect. 7.1).

---

[6] Which is unambiguous here because there is only one action mode per transition in a simple Petri net.

[7] It is thus similar to the *immediate paths* for which an accumulated path probability is defined on page 132 in the context of the reduced reachability graph computation.

When an immediate action (variant) becomes executable in a local state, it is scheduled with a logical time in which vector time and priority vector are copied from the current local logical time. The priority vector entry corresponding to the local atomic unit is set to the global event priority of the scheduled action. This scheduled time is used to sort a new event into the local event list.

When an event is executed in an atomic unit, the local simulation time is advanced to a new value. The local priority vector of the atomic unit is then updated such that the entry related to the atomic unit of the event (it could be a remote or local action) becomes the minimum of the previous value and the entry in the event's priority. The remaining entries are not changed. All entries are set to the default value infinity if some time passed between the previous simulation time and the event execution. The only exception is the entry corresponding to the executed timed event, which is set to its priority.

Formal definitions of the logical time for an event that is scheduled for execution as well as the new local time after the processing of an event are given on page 188.

Table 9.4 shows vector times and priority vector values for the transition-related events in the example introduced in Fig. 9.2. The values are depicted for the events after their generation. Local vector time and priority vector are of course updated in each atomic unit upon later execution of a received remote event. This is the reason for the entries increasing by 2.

Assume that we start a distributed simulation of the model shown in Fig. 9.2 with the firing of transition T1. Vector time entries of all atomic units are thus zero at the beginning of the simulation. The entry of au_1 is set to 1 and the priority vector entry to 0 with the firing of T1. The associated event is sent to the affected atomic unit au_2. In the following we assume for simplicity that the events are executed in the right order; in reality this is achieved by rollbacks if necessary. The following happens in au_2: event T1 is received, local logical time is updated to $VT_{au_2} = (1,1,0,0,0,0,0,0), PV_{au_2} = (0,\infty,\infty,\infty,\infty,\infty,\infty)$. Firing of transition T2 is scheduled as the only entry

Table 9.4. Vector time and priority vector of events from Fig. 9.2

Event	Vector time VT	Priority vector PV
T1	$(1,0,0,0,0,0,0,0)$	$(0,\infty,\infty,\infty,\infty,\infty,\infty,\infty)$
T2	$(1,2,0,0,0,0,0,0)$	$(0,\,1\,,\infty,\infty,\infty,\infty,\infty,\infty)$
T3	$(1,2,\underline{2},\underline{0},0,0,0,0)$	$(0,\,1\,,\,\underline{3}\,,\underline{\infty},\infty,\infty,\infty,\infty)$
T4	$(1,2,\underline{0},\underline{2},0,0,0,0)$	$(0,\,1\,,\underline{\infty}\,,\,\underline{4}\,,\infty,\infty,\infty,\infty)$
T5	$(1,2,2,2,4,0,0,0)$	$(0,\,1\,,\,3\,,\,4\,,\,2\,,\infty,\infty,\infty)$
T6	$(1,2,2,2,3,0,0,0)$	$(0,\,1\,,\,3\,,\,4\,,\,2\,,\infty,\infty,\infty)$
T7	$(1,2,2,2,\underline{4},\underline{2},\underline{0},0)$	$(0,\,1\,,\,3\,,\,4\,,\,\underline{2}\,,\,\underline{6}\,,\underline{\infty},\infty)$
T8	$(1,2,2,2,\underline{3},\underline{0},\underline{2},0)$	$(0,\,1\,,\,3\,,\,4\,,\,\underline{2}\,,\underline{\infty},\,\underline{5}\,,\infty)$

in the local event list. It is executed and produces an event T2 that is sent to au_3 and au_4 with the new local time shown in Table 9.4.

The remote event T2 is received and executed in atomic unit au_4, which enables transition T4. The local time is updated, T4 is scheduled and executed locally. The associated event T4 is sent to au_5 with the logical time shown in the table. The same happens in au_3: T3 becomes enabled after the execution of the remote event T2 is scheduled and executed, and an associated event T3 is sent to au_5 with the logical time shown in Table 9.4 for T3.

The algorithm in atomic unit au_5 must be able to decide which one of the received events T3 and T4 has to be executed first, i.e., sorted into the local event list before the other one.[8] The two vector times are not sufficient, because $VT_{T3} \parallel VT_{T4}$ due to the concurrent executions in au_3 and au_4.

The priority vectors are thus used to compare the logical times of the events. However, it is obviously not adequate to compare the minimal priority vector entry. It would be zero both for T3 and T4 due to the shared event T1 on their paths. This priority must not be taken into account (as well as the priority of T2), because it belongs to an identical event on that both T3 and T4 are causally dependent. Both paths share a common initial sequence of atomic units. Cases like this can, however, be detected easily using the vector time of the events. Equal entries in the two vector times denote identical dependencies, which then have to be ignored in the comparison of the priority vectors. Priority vectors are thus compared using the minimum priority entry only for atomic units which have distinct vector time entries.

We thus define a **minimal path priority** of a priority vector PV_a to which a vector time VT_a belongs, with respect to another vector time VT_b as

$$PV_{\min}^{a,b} = \begin{cases} \min\limits_{\substack{\forall au \in AU: \\ VT_a(au) \neq VT_b(au)}} PV(au) & \text{if } VT_a \neq VT_b \\ \infty & \text{otherwise} \end{cases} \qquad (9.1)$$

Note that we define this path priority to be infinity for completeness in the case of identical vector times; this case is, however, only of theoretical interest.

Applied to our examples of T3 and T4, we obtain

$$PV_{\min}^{T3,T4} = 3 \quad \text{and} \quad PV_{\min}^{T4,T3} = 4$$

Significant entries in Table 9.4 are underlined. The minimal path priority of T3 is smaller, and thus T4 is correctly executed first. The execution of T4 already enables transition T6 but not T5 in au_5. The local time is updated to $VT_{au_5} = (1, 2, 0, 2, 1, 0, 0, 0), PV_{au_5} = (0, 1, \infty, 4, \infty, \infty, \infty, \infty)$. T6 is scheduled for execution at time $VT = (1, 2, \underline{0}, \underline{2}, \underline{1}, 0, 0, 0), (0, 1, \underline{\infty}, \underline{4}, \underline{2}, \infty, \infty, \infty)$.

---

[8] In addition to that, transition T5 might become enabled if event T3 is received and executed first. The algorithm thus has to be able to decide about the ordering of event T5 with respect to T3 and T4 as well. This case is handled below.

The algorithm managing au_5 must then decide whether remote event T3 or local event T6 is executed first. Using the formula above, we obtain

$$PV^{T3,T6}_{\min} = 3 \quad \text{and} \quad PV^{T6,T3}_{\min} = 2$$

The significant entries of T6's vectors have been underlined above. T3 is thus executed first, which is correct. If T6 already had been executed before T3 was received, it is rolled back. The execution of event T3 in au_5 enables T5 as well. T5 and T6 are in conflict now, and we assume that the probabilistic choice selects T6 to fire first. There is still one token left in places P5 and P6, and thus T5 fires afterward. Both firings lead to an event each with the logical times shown in Table 9.4 for T6 and T5.

One special case remains. It is possible that the minimal path priorities of two events are equal with the formula introduced above. This may happen if the paths share a common atomic unit where conflicting immediate actions were executed, which by chance have the smallest priority on the path. In the example from Fig. 9.2, this is the case in atomic unit au_8. Assume that the conflict in au_5 has been solved as stated above. This leads to the execution of events T6 and T5 in au_7 and au_6. Transitions T7 and T8 then fire concurrently in their respective atomic units, leading to the events and logical event times shown in Table 9.4.

The order of execution of events T7 and T8 then needs to be decided in au_8. The following minimal path priorities are then computed (significant entries are again underlined in the table):

$$PV^{T7,T8}_{\min} = 2 \quad \text{and} \quad PV^{T8,T7}_{\min} = 2$$

As it occurs, the order of the two events cannot be decided with the minimal path priorities only. The reason is that both events have au_5 as their common predecessor, in which two conflicting transitions with the same smallest priority initiated them. However, the sequence must equal the ordering in which the two events T7 and T8 were generated in the atomic unit that contained the two conflicting actions. We only need to use the vector time entries that correspond to the atomic unit in which the minimal path priorities were set to detect the causal ordering.

Two events a and b have the same minimal path priority in this case. Event a is then scheduled first if the vector time component of a that belongs to the atomic unit au_i, in which the minimal path priority was found, is smaller than the corresponding one of event b. Formally, then

$$PV^{a,b}_{\min} = PV^{b,a}_{\min} = PV(au_i) \quad \text{and} \quad VT_a(au_i) < VT_b(au_i)$$

When applied to our example of events T7 and T8, the determining atomic unit is au_5. T8 is scheduled before T7 despite its lower individual priority because $VT_{T8}(au_5) < VT_{T7}(au_5)$, which is correct. The compared values are set in a bold style in the table.

Compound Simulation Time

It has been shown above that standard logical time schemes are not sufficient to unambiguously order events in a distributed SDES simulation. Vector time must be complemented by a priority vector to manage immediate events with individual priorities.

We define a **compound simulation time** $cst \in CST$ which contains the actual simulation time ST, vector time VT, and priority vector PV of an event. The set of all possible compound simulation times is denoted by CST.

$$CST = \left\{(ST, VT, PV) \mid ST \in \mathbb{R}^{0+} \wedge VT \in \mathbb{N}^{|AU|} \wedge PV \in \mathbb{N}^{|AU|}\right\}$$

Events e mark state changes that are either executed or scheduled in the future of an atomic unit. Furthermore they are exchanged in messages. So far we have used the term *event* only informally in this chapter. With the definition of our proposed compound simulation time, a formal definition in the context of distributed simulation is given now. An **event** e comprises the executed action variant v and the corresponding compound simulation time cst. The set of all possible events is denoted by E.

$$E = \{(cst, v) \mid cst \in CST \wedge v \in AV\}$$

Not only events have a corresponding cst, but also the local current time in an atomic unit is defined as a compound simulation time. The remaining part of this section defines how compound simulation times are compared, how their values are derived for events that are newly scheduled in the event list, and how the local simulation time changes when an event is executed.

Compound simulation time is intended for an ordering between different events. For any two events e_1 and e_2, it must be clear which one has to be scheduled first. Another application is the comparison of a remote event time with the local simulation time, which is important to decide whether an event is scheduled for the future or past of an atomic unit.

The comparison is performed using compound simulation times cst_1 and cst_2 of events e_1 and e_2. While the comparison of the actual simulation times is obvious, things are more complicated when vector times and priority vectors are taken into account. This has been informally explained already above, and is now defined formally. Elements of compound simulation times are denoted by assuming that $cst_i = (ST_i, VT_i, PV_i)$.

$$
\begin{aligned}
\forall cst_1, cst_2 \in CST : cst_1 < cst_2 \iff & \\
(ST_1 < ST_2) \vee & \\
(ST_1 = ST_2) \wedge \Big((VT_1 < VT_2) \vee & \\
(VT_1 \parallel VT_2) \wedge \big[(PV_{\min}^{1,2} > PV_{\min}^{2,1}) \vee & \\
\exists au \in AU : \big(PV_{\min}^{1,2} = PV_{\min}^{2,1} = PV_1(au) \big) \wedge & \\
\big(VT_1(au) < VT_2(au) \big) \big] \Big) &
\end{aligned}
\tag{9.2}
$$

There are four cases. If simulation time or vector time allows a decision about which time is smaller, it is taken accordingly (cases 1 and 2). If $VT_1 \parallel VT_2$, the minimal path priorities are compared. The time with the greater value is then smaller (case 3). If they are equal as well, the decision is based on the vector time entry of the significant atomic unit (i.e., the one in which the minimal path priority occurred). The last line of the equation ensures that there is only one atomic unit au for which the minimal path priority is reached.

Formally it is also possible that two compound simulation times are equal, meaning that all elements are completely identical.[9]

$$\forall cst_1, cst_2 \in CST : cst_1 = cst_2 \iff \qquad (9.3)$$
$$\forall au \in AU : VT_1(au) = VT_2(au)$$

This is, however, only possible in the case of two events that belong to the same atomic unit, have zero delay and equal priority, and are being scheduled for execution in a given local state. This means that they are in conflict (otherwise they would have differing priorities and belong to other atomic units), and the order of execution will be decided locally by a probabilistic choice. Such a case will thus never be "seen" outside the responsible atomic unit, and is of less importance because it does not introduce unambiguity in event ordering. Two equal compound simulation times will otherwise never be compared in the algorithms of Sect. 9.1.4.

Another issue is the selection of the compound simulation time cst' for an event that is newly scheduled and inserted into a local event list. Assume that $cst' = (ST', VT', PV')$, current simulation time $cst^{au} = (ST, VT, PV)$, and action variant to be scheduled $v = (a, \cdot)$.

$$\text{Randomly select } RAD \sim Delay^\star(v)$$
$$ST' := ST + RAD \qquad (9.4)$$
$$\forall au_i \in AU : VT'(au_i) := VT(au_i)$$
$$PV'(au_i) := \begin{cases} \infty & \text{if } ST' \neq ST \wedge au_i \neq au \\ Pri^\star(a) & \text{if } au_i = au \\ PV(au_i) & \text{otherwise} \end{cases}$$

By doing so, the event is scheduled for the actual simulation time plus a randomly drawn delay. If it is different from zero, all priority vector entries are reset to infinity[10] except for the local value, which is set to the priority

---

[9] The simulation times ST_1, ST_2 as well as the priority vectors PV_1, PV_2 are also equal if this equation holds.

[10] It would be possible to reset all entries of the VT vector to zero every time a nonzero delay passes, just as it is done for the priority vector PV. This is, however, avoided to keep the full causal information for all messages, including rollbacks, for an improved message cancellation mechanism described in Sect. 9.1.4.

of the scheduled action. The other entries are copied. The vector time is not changed at this point; it is updated upon actual execution of the event later on.

The scheduled compound simulation time of future local events always has to correspond to the current local compound simulation time. This means that if a future event in the local event list stays executable after an event execution, its compound simulation time has to be recalculated following the equations above. The simulation time ST, however, is kept. This can be efficiently implemented without recalculation by transparently mapping the current local compound simulation time to affected future events.

When an event $e = (cst^e, v = (a, \cdot))$ (which is caused in an atomic unit $au_e, a \in Region(au_e)$) is executed in the atomic unit au, the local compound simulation time $cst^{au} = (ST, VT, PV)$ is updated to (ST', VT', PV') as follows. We assume that the elements of the event time are denoted as $cst^e = (ST^e, VT^e, PV^e)$.

$$ST' := ST^e \qquad\qquad (9.5)$$

$$\forall au_i \in AU :$$

$$VT'(au_i) := \begin{cases} VT(au_i) + 1 & \text{if } au_i = au \\ \max(VT(au_i), VT^e(au_i)) & \text{otherwise} \end{cases}$$

$$PV'(au_i) := \begin{cases} \min(PV(au_i), PV^e(au_i)) & \text{if } ST' = ST \wedge au \neq au_e \\ Pri^\star(a) & \text{if } ST' = ST \wedge au_i = au_e = au \\ PV(au_i) & \text{otherwise} \end{cases}$$

ST denotes the actual simulation time counted in model time units in this equation. Thus the system spends some time in a tangible state iff $ST' > ST$. This equation requires local causality, i.e., that $cst^e \not< cst^{au}$.

Rollbacks are necessary to overcome local violations of causality. An atomic unit then "goes back" to a state before the problem, and executes the intermediate events again. The vector time part of the local compound simulation time is, however, not set back, because the causality of event executions should cover rollbacks as well. A rollback thus leads to an increment of the vector time entry corresponding to the local atomic unit. This means that vector time and simulation time are not isomorphic in the sense that there might be two events where one has a higher simulation time, while the other has a greater vector time.

Confusions and Global Action Priorities

We require a unique and decidable ordering of different events for our algorithms to work correctly and efficiently. The *less-then* relation of compound simulation time pairs has been defined for this reason in (9.2). It worked for the examples presented so far; there are, however, exceptions possible, in which two different compound simulation times cst_1 and cst_2 cannot be ordered with the definition.

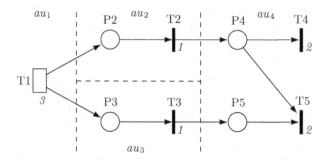

Fig. 9.3. Simple Petri net with undecidable event ordering

Figure 9.3 shows a slightly different version of the simple Petri net model depicted in Fig. 9.1; only the transition priorities have been changed. When the events of firing transitions T2 and T3 are received in atomic unit au_4 at the same simulation times $ST_{T2} = ST_{T3}$, the corresponding values are set for the vector times $VT_{T2} = (1, 2, 0, 0)$, $VT_{T3} = (1, 0, 2, 0)$ and priority vectors $PV_{T2} = (3, 1, \infty, \infty)$, $PV_{T3} = (3, \infty, 1, \infty)$. For the vector times hold $VT_{T2} \parallel VT_{T3}$, thus the minimal path priority is taken into account. However, $PV_{\min}^{T2,T3} = PV_{\min}^{T3,T2} = 1$, but there are two different atomic units au_2 and au_3 for which this value is reached in the compound simulation times.

Equation (9.2) loses its real meaning in this case, because both $cst_1 < cst_2$ and $cst_1 > cst_2$ hold at the same time. Is there anything we can retrieve additionally from the model to obtain a correct ordering of events T2 and T3? No, because a standard sequential simulation is not able to order these events as well.[11] The decision about which of the concurrent transitions T2 and T3 fires first is not determined by the model due to the equal priorities, although its further behavior obviously depends on it. This is the exact definition of a *confusion*, which we interpret as a modeling error.

Equation (9.2) defining the relation between two compound simulation times thus only holds for confusion-free models. The last line otherwise would have to be extended to

$$\left(VT_1(au) < VT_2(au) \right) \wedge \left(\nexists au' \in AU \setminus \{au\} : PV_{\min}^{2,1} = PV_1(au') \right) \quad (9.6)$$

Our definition of the compound simulation time together with the operations and comparison thus has a very nice side effect: as long as a model is confusion-free, event ordering is always uniquely determined by a comparison of the associated times. If the model behavior is simulated until a point where an actual confusion takes places, it can be detected easily on the basis of the compound simulation times. This has the advantage of only detecting errors

---

[11] The definition of the stochastic process underlying an SDES model as given in Sect. 2.3.2 is nonetheless well defined. It selects any one of the events with equal probabilities.

that actually play a role, compared to structural techniques which derive necessary but not sufficient conditions using the model structure [53].

To complement the possible relations between two compound simulation times, we formally define them to be **concurrent** as follows.[12]

$$
\forall cst_1, cst_2 \in CST : cst_1 \parallel cst_2 \iff
$$
$$
(ST_1 = ST_2) \wedge (VT_1 \parallel VT_2) \wedge (PV^{1,2}_{\min} = PV^{2,1}_{\min}) \wedge
$$
$$
\exists au_i, au_j \in AU : au_i \neq au_j \wedge VT_1(au_i) \neq VT_2(au_j) \wedge
$$
$$
\left(PV^{1,2}_{\min} = PV_1(au_i) = PV_2(au_j)\right)
$$

(9.7)

This can be interpreted as follows: If two events cannot be ordered following (9.2), then obviously the simulation times must be equal, the vector times are parallel,[13] and the minimal path priorities are the same. If there would be only one atomic unit for which the smallest path priority is found, the decision would be based on the vector time entry associated with it. The two values to be compared would be different, because only priority entries with differing vector time entries are compared for the minimal path priority. Therefore this case can be excluded, and it is clear that there must be at least two distinct atomic units in which the minimal path priority was found. The executions of the associated actions cannot be ordered causally or by using priorities; a confusion is thus encountered.

An actual implementation may either work with the confusion detection as defined above, or assume a confusion-free model for improved efficiency. If the SDES model to be simulated is confusion-free as required, conflicts can always be solved locally inside an ECS, and thus inside one atomic unit. Events of two or more actions belonging to different atomic units and having the same priority may then be executed in any order, without changing the behavior of the stochastic process. The order of execution of these events can then be fixed arbitrarily. Different priority values must, however, be valid across the borders of atomic units – the corresponding events need to be ordered by the priority of the underlying action.

We may thus introduce a **global event priority** that associates a derived priority $Pri^\star_{Global}$ to every SDES action. It leads to a unique global priority ordering as shown below, and thus makes a model confusion-free. It does not change the behavior of a model which is already confusion-free though.

$$
Pri^\star_{Global} : A^\star \to \mathbb{N}
$$

This derived priority is calculated from the actual priority as follows.

$$
\forall a \in A^\star, a \in Region(au_i) : Pri^\star_{Global}(a) = Pri^\star(a)\,|AU| + i
$$

---

[12] The definition can be obtained from (9.2) extended by (9.6) and assuming that neither $cst_1 < cst_2$ nor $cst_2 < cst_1$ holds.

[13] If the vector times are equal, an event is compared with itself.

Such a mapping has the following properties. First and most importantly, actions with a different priority keep the priority relation in the global priority.

$$Pri^{\star}(a_1) > Pri^{\star}(a_2) \iff Pri^{\star}_{Global}(a_1) > Pri^{\star}_{Global}(a_2)$$

Second, equal priorities of actions in different au regions are ordered by the arbitrary numbering of the atomic units. Third, the global priority of actions in the same atomic unit stays equal. Thus a probabilistic decision between actually conflicting events is locally possible. Finally, the associated atomic unit can be easily calculated from a given global event priority.

$$\left(i = Pri^{\star}_{Global}(a) \bmod |AU|\right) \longrightarrow a \in Region(au_i) \tag{9.8}$$

In the discussions and algorithms following below, it is possible to either use global event priorities and a simplified comparison of compound simulation times following (9.2), or the original priorities together with the confusion check based on 9.7. If there is a way to ensure absence of confusion based on the model structure for a specific model (class) a priori, original priorities and the simplified comparison are sufficient.

9.1.3 Discussion of Compound Simulation Time

In our attempt to simulate a timed synchronous system on a distributed computing hardware, a compound simulation time has been introduced earlier for a reasonable event ordering. The goal of this section is to show that (1) our algorithms associate a reasonable compound simulation time to each event, (2) the introduced time scheme complies to the nature of time, and (3) it is possible to derive global state information effectively using it.

Time can be interpreted as a set of time instances t, on which a temporal precedence order is defined [235]. Events are time stamped by a compound simulation time value as introduced in Sect. 9.1.2, and each possibly reached value equals a specific logical time.

Relation Between Events and Clock Values

Do the algorithms associate a "meaningful" compound simulation time $cst(e_i)$ to every event e_i? This issue is considered as **clock condition** in literature (see, e.g., [12, 218, 235]). If an event e_1 may somehow **affect** e_2 (which is denoted by $e_1 \rightarrow e_2$), it is mapped to an "earlier" logical time.[14] This ensures that the future cannot influence the past in the logical time scheme, as it is natural in our understanding of the passage of time.

$$\forall e_1, e_2 \in E : (e_1 \rightarrow e_2) \longrightarrow \left(cst(e_1) < cst(e_2)\right) \tag{9.9}$$

---

[14] Similar to the "happens before" relation in [218].

We consider events e_i of the distributed simulation here; they are related to the simulated SDES model thus, and not to the distributed way of computation as usually understood in the literature about distributed systems. An alternative interpretation of $e_1 \to e_2$ is thus that e_1 *is executed before* e_2 in a sequential simulation.

We examine the conditions under which e_1 affects (or is executed before) e_2 in a sequential simulation. The following cases are distinguished for a complete proof of the proposition in (9.9)[15]:

- If an event is executed later in the actual simulation (model) time, it may be affected by an earlier event.

$$\forall e_1, e_2 \in E : (ST_1 < ST_2) \longrightarrow (e_1 \to e_2)$$

Then obviously also $cst(e_1) < cst(e_2)$ because of the definition of "$<$" for compound simulation times in (9.2).

- If an event has to be executed at the same simulation time, but is causally dependent on another event, it should be executed later. Causal dependency is fully captured by vector time [235]. Thus

$$\forall e_1, e_2 \in E : ((ST_1 = ST_2) \wedge (VT_1 < VT_2)) \longrightarrow (e_1 \to e_2)$$

which obviously leads to $cst(e_1) < cst(e_2)$ due to the second line in (9.2).

- There are cases in which two events are executed at the same simulation time, but are not directly causally dependent. Both belong to individual paths of immediate executions then, which started at the same tangible state. The two paths may share some prior immediate event executions, which can be obtained from the entries of the vector time that are equal and have an associated priority vector entry smaller than infinity. The decision on which event has to be executed first must be based then on the minimal priorities of action executions that have taken place since the paths split up.

$$\forall e_1, e_2 \in E :$$
$$((ST_1 = ST_2) \wedge (VT_1 \parallel VT_2) \wedge (PV_{\min}^{1,2} > PV_{\min}^{2,1})) \longrightarrow (e_1 \to e_2)$$

This is captured exactly in the computation of the minimal path priority: the event that took the path with a smaller lowest priority will always be ordered after another event. The "$<$"-relation for compound simulation times covers this case accordingly, leading to $cst(e_1) < cst(e_2)$.

- The final case occurs, if in a setting as described above the smallest priorities of the paths occurred in their last common atomic unit. The two event executions in this common atomic unit are different ones for the

---

[15] Here and in all following proofs, we denote the compound simulation time associated to an event e_i by cst_i, and its elements simply as $cst_i = (ST_i, VT_i, PV_i)$ for notational convenience.

two paths, because their vector times would otherwise be equal, and thus their priority would not have been taken into account for the minimal path priority. Obviously there must have been a unique ordering of these two previous events that started the different paths. This ordering is simply given by the sequence of executions in the common atomic unit, which can be directly deduced from the corresponding entry in the vector time.

$$\forall e_1, e_2 \in E : \left((ST_1 = ST_2) \wedge (VT_1 \parallel VT_2) \wedge \right.$$
$$\left[\exists au \in AU : PV_{\min}^{1,2} = PV_{\min}^{2,1} = PV(au) \right] \wedge$$
$$\left. \left[VT_1(au) < VT_2(au) \right] \right) \qquad \longrightarrow (e_1 \rightarrow e_2)$$

This case is covered in the bottom part of (9.2), and ensures that $cst(e_1) < cst(e_2)$.

– There are no other cases in which two events of a confusion-free model can be in the "$\rightarrow$"-relation of a sequential simulation, as it has been discussed in Sect. 9.1.2.

With our choice of compound simulation time, even the converse condition of (9.9) is true:

$$\forall e_1, e_2 \in E : \left(cst(e_1) < cst(e_2) \right) \longrightarrow (e_1 \rightarrow e_2) \qquad (9.10)$$

Proof (Indirect). Assume we find $e_1, e_2 \in E$ such that $cst(e_1) < cst(e_2)$ and not $e_1 \rightarrow e_2$. With our assumption of global event priorities, it is always clear which event has to be executed first in a sequential simulation. Relation "$\rightarrow$" over events is thus trichotomous, thus $\neg(e_1 \rightarrow e_2) \longrightarrow ((e_2 \rightarrow e_1) \vee (e_1 = e_2))$. Obviously $cst(e_1) = cst(e_2)$ if $e_1 = e_2$, which contradicts our assumption and leaves the case $e_2 \rightarrow e_1$. From the clock condition in (9.9), it immediately follows that $cst(e_2) < cst(e_1)$, which contradicts our assumption as well (asymmetry of "$<$" is shown below).   □

Ordering based on compound simulation times of our distributed algorithm thus ensures that events are processed in exactly the same way as in a sequential simulation.

The mapping of events to compound simulation times is obviously a function: Every individual event is generated at an atomic unit, which increases its local vector time (and possibly the simulation time) during the process. The local entry thus reaches a new maximum value, which becomes a part of the new event's time stamp. There are no two events with the same vector time for the same reason. It follows that the mapping of events to compound simulation times is *bijective*, i.e.,

$$\forall e_1, e_2 \in E : \left(e_1 = e_2 \right) \Longleftrightarrow \left(cst(e_1) = cst(e_2) \right)$$

From bijectivity and the corollaries given with (9.9) and (9.10), it follows that the event set E with the "$\rightarrow$"-relation is isomorphic to the compound simulation times CST with the "$<$"-relation.

$$\forall e_1, e_2 \in E : \left(e_1 \rightarrow e_2 \right) \Longleftrightarrow \left(cst(e_1) < cst(e_2) \right) \qquad (9.11)$$

The compound simulation times can thus be used for a correct and unique decision about the ordering of events.

Properties of Compound Simulation Time

There are some conditions that any model of time should adhere to (compare, e.g., [235]), which we will check in the following for the compound simulation time. We will thus show that the "<"-relation defined in (9.2) satisfies irreflexivity, asymmetry, transitivity, linearity (more exactly trichotomy), eternity, and density.

It should be noted that it makes no sense to analyze compound simulation time entities with arbitrarily set values; we restrict ourselves to time stamps of events that could possibly be obtained during a distributed simulation of a real SDES model. Remember that such a model was required to be confusion-free, and that no atomic unit is visited more than once during a specific immediate path.

The reflexive and symmetric relation "$\|$" for compound simulation times (cf. (9.7)) denotes *simultaneity* in our time scheme. Simultaneous events may be generated, sent, and processed without problems, as long as they do not have to be ordered in an atomic unit. Section 9.1.2 concluded that the latter would be a (forbidden) case of confusion. We adopt the global event priorities as introduced in Sect. 9.1.2, because their application does not change the overall behavior in a confusion-free model. This will simplify proofs and discussions due to the fact that then never $cst_1 \parallel cst_2$.

Proof (Irreflexivity of the "<"-relation). We have to show that

$$\forall cst_1 \in CST : \neg(cst_1 < cst_1)$$

Assume that $cst_1 \in CST$ can be found such that $cst_1 < cst_1$ for an indirect proof. The elements of identical compound simulation times are of course equal; thus neither $ST_1 < ST_1$ nor $VT_1 < VT_1$ will ever be true. In addition to that, $PV_{\min}^{1,1} = PV_{\min}^{1,1} = \infty$ because $VT_1 = VT_1$. It is thus impossible to find an atomic unit satisfying the two final lines of (9.2), which is a contradiction to the assumption. □

Proof (Asymmetry of "<"). Formally,

$$\forall cst_1, cst_2 \in CST : (cst_1 < cst_2) \longrightarrow \neg(cst_2 < cst_1)$$

Indirect proof: Assume we find $cst_1, cst_2 \in CST$ satisfying $(cst_1 < cst_2) \wedge (cst_2 < cst_1)$. If $ST_1 \neq ST_2$, the decision about which time is smaller would be based on the simulation times and unique (asymmetry of "<" for real numbers), thus $ST_1 = ST_2$. With similar arguments it follows that $VT_1 \parallel VT_2$ and $VT_1 \neq VT_2$ because otherwise $cst_1 = cst_2$ and neither one would be smaller.

Due to the differing vector times, it is always possible to obtain unique minimal path priorities for cst_1 and cst_2. If we would have $PV_{\min}^{1,2} \neq PV_{\min}^{1,2}$, the "<"-relation would be true only for one comparison. It thus follows that $PV_{\min}^{1,2} = PV_{\min}^{1,2}$. Because we adopted global event priorities, there is exactly one atomic unit au for which $PV_{\min}^{1,2} = PV_{\min}^{1,2} = PV_1(au) = PV_2(au)$ holds. However, we know that $VT_1(au) \neq VT_2(au)$, because this entry would otherwise have been ignored for the derivation of the minimal path priority. Thus either $VT_1(au) < VT_2(au)$ or $VT_2(au) < VT_1(au)$. This means that only one of the "<"-relations between cst_1 and cst_2 is true, leading to a contradiction to our assumption.   □

Proof (Trichotomy of "<" for compound simulation times of events). Two values are either equal, or exactly one is smaller than the other.[16]

$$\forall cst_1, cst_2 \in CST : (cst_1 = cst_2) \oplus (cst_1 < cst_2) \oplus (cst_2 < cst_1)$$

Let us consider the case $cst_1 = cst_2$ first. The equation is fulfilled because neither $cst_1 < cst_2$ nor $cst_2 < cst_1$, which follows directly from irreflexivity.

It remains to prove that if $cst_1 \neq cst_2$, either $cst_1 < cst_2$ or $cst_2 < cst_1$ holds. There are two parts to this proof. First, we must prove that never $cst_1 < cst_2$ and $cst_2 < cst_1$, which we have already shown (asymmetry). We thus only have to show that the "<"-relation is *linear*[17]

$$\forall cst_1, cst_2 \in CST : (cst_1 \neq cst_2) \longrightarrow \big((cst_1 < cst_2) \vee (cst_2 < cst_1)\big)$$

The proof is similar to the one for asymmetry. Assume $ST_1 \neq ST_2$: then either $cst_1 < cst_2$ or $cst_2 < cst_1$ (trichotomy of "<" for real numbers). We thus only need to consider the case $ST_1 = ST_2$. We know that $VT_1 \neq VT_2$ because $cst_1 \neq cst_2$. Assume now further that $VT_1 < VT_2$ or $VT_2 < VT_1$: then obviously $cst_1 < cst_2$ or $cst_2 < cst_1$, and the proposition holds. It thus remains to show that it is also true in the case $VT_1 \parallel VT_2$.

The minimal path priorities are now inspected: If $PV_{\min}^{1,2} \neq PV_{\min}^{2,1}$, the proposition becomes true. What happens if $PV_{\min}^{1,2} = PV_{\min}^{2,1}$? There is exactly one atomic unit au for which the minimal path priority is achieved ($PV_{\min}^{1,2} = PV_{\min}^{1,2} = PV_1(au) = PV_2(au)$), because of the use of globally unique priorities. However, $PV_1(au)$ has been considered in the computation of the minimal path priority, which means that $VT_1(au) \neq VT_2(au)$. Thus one of the entries is less than the other (trichotomy of "<" for naturally numbered priorities), and the proposition is thus true in this final case as well.   □

Proof (Transitivity of "<" for compound simulation times). It has to be proven that

$$\forall cst_1, cst_2, cst_3 \in CST : \big((cst_1 < cst_2) \wedge (cst_2 < cst_3)\big) \longrightarrow (cst_1 < cst_3)$$

---

[16] $\oplus$ denotes the exclusive-or relation.

[17] Obviously only in the restricted sense of an irreflexive relation.

Application of (9.10) to the two conditions leads to $(e_1 \rightarrow e_2) \wedge (e_2 \rightarrow e_3)$. The relation "$\rightarrow$" on events $e_i \in E$ is obviously transitive: If an event e_1 has to be executed before e_2, and e_2 before e_3 in a sequential simulation, e_1 needs to be executed before e_3 as well. Thus, we conclude that $(e_1 \rightarrow e_3)$, from which $cst_1 < cst_3$ follows with (9.9).[18]   □

Proof (Eternity). This means that there is always a smaller and a greater time for any given value for our time scheme:

$$\forall cst_1 \in CST \; \exists cst_2, cst_3 \in CST : cst_1 > cst_2 \wedge cst_1 < cst_3$$

This is obviously true because already for the simulation time part ST, which is a real number, there is always a smaller and a greater value.   □

Proof (Density). There is always a compound simulation time between any pair of different times, because the real-valued simulation time entries are dense as well.

$$\forall cst_1, cst_3 \in CST \; \exists cst_2 \in CST :$$
$$(cst_1 < cst_3) \longrightarrow \big((cst_1 < cst_2) \wedge (cst_2 < cst_3)\big)$$

As a conclusion, the relation "$<$" on CST is a *strict total order*, because it has been shown to be irreflexive, trichotomous, and transitive. Moreover it is a *well-founded relation*, thus ensuring that for any nonempty subset of CST there is a unique "$<$"-minimal element. This property is necessary for each atomic unit when the future event with the smallest associated time has to be selected for execution from the event list.   □

Global States

A common problem to all distributed algorithms (and thus simulations) is to determine a global state over the numerous local states of each process. Access to global state information is necessary in an SDES model on various occasions. Guard functions of colored Petri nets are an example for a state-dependent enabling of action variants, which is captured in $Ena^\star(\cdot)$. Issues like a maximum capacity lead to a specification of a state condition $Cond^\star$ for a state variable. Finally, the derivation of performance measures requires the derivation of state variable values at certain times as well as the execution times of action variants.

An evaluation of any expression with parameters that depend on remote state variables requires a correct computation of a global state for the exact time it is requested for. This leads to an additional problem: The atomic unit in which the state variable is maintained might not have reached the

---

[18] An alternative proof based on compound simulation times and ordering possibilities of associated events has been given in [210].

simulation time for which the state is requested. In that case we follow the idea of optimistic simulation by assuming that the state will not change until the requested time. If it does so later, the affected atomic unit is notified and rolls back accordingly.

State variable access is, however, only locally possible in a distributed simulation. The issue of deriving information about remote states at a given time is known as a *global predicate evaluation problem* [12]. Due to the possibly different speed and numbers of events to be processed at each atomic unit, the local simulation times may significantly vary. Information about remote states can thus be obsolete, incomplete, or inconsistent. In the general setting of distributed algorithms, this leads to the development of methods to obtain a global state, which use only the causality relation between message sending and reception as well as the sequence of event executions of local processes [12, 44, 235].

A global state of a distributed simulation consists of a set of local states, one for each logical process (i.e., atomic unit in our approach). Every local state associates a value to each locally maintained state variable. A state is valid between two event executions at the atomic unit due to the nature of discrete event systems. It is thus possible to talk about events or states when analyzing the correctness of a global state.

Events in a distributed system are often visualized in a *space–time diagram* (see, e.g., [12, 218]), in which the event and state sequences of each atomic unit are sketched in horizontal time lines. The different lines model the spatial distribution of the processes, and messages between the processes and thus causal relationships can be drawn as arrows between the horizontal lines. If we select a local state for each atomic unit in the graph and connect all these points by a zigzag line, we have a graphical representation of a global state. Every global state that an observer may obtain can obviously be depicted in that way. The strong relationship between events and states in that aspect is clear because every state can be uniquely identified by its rightmost predecessor event.

Such a state-connecting line cuts the sets of events at each atomic unit into a past and a future set. It is thus called a **cut** C and defined as a finite subset of an event set E such that for every event e in it, all events are also included which were executed before e locally in the atomic unit producing e [235].[19]

$$C \subseteq E \text{ subject to } \forall au \in AU, \forall e \in E(au) \in C : (e' \to e) \longrightarrow (e' \in C)$$

A cut thus respects local causality: All events left of the cut in the space–time diagram are included. However, not every one of these possible observations corresponds to a consistent state.

A cut C is **consistent** if it respects global causality as well: For every event e in the cut, all events that causally precede it are included in C [235].

$$\forall e \in C : (e' \to e) \longrightarrow (e' \in C) \tag{9.12}$$

---

[19] $E(au)$ denotes the set of events belonging to atomic unit au.

In a distributed algorithm, causality is related to local event execution and message sending and reception. It is thus required that if the receive event of a message has been recorded in the state of a process, then its send event is also recorded in the state of the sender [218]. This property can be checked graphically in the time–space diagram: If there is an arrow (modeling a message transfer) which crosses the cut line backward, the cut is not consistent.

A global state in a distributed computation is consistent, if it belongs to a consistent cut. A consistent cut corresponds to a state that is *possibly* observed in a run of the algorithm, but not necessarily reached. Consistent cuts and thus states are not unique: It is possible to add or delete events that are concurrent to all other events in remote atomic units, without interfering with consistency. This is not acceptable in our environment for SDES performance evaluation, where the causal, timing, and priority relations between events must be obeyed correctly. We have already shown earlier that the introduced compound simulation time allows to order events exactly, and to clearly separate between confusions and determined behavior. It is verified in the following that our use of *CST* leads to consistent global states. Note that this would be impossible for models with immediate delays and priorities based on traditional methods.

Proof (Consistent global states). Every inquiry about remote states corresponds to a compound simulation time cst, for instance resulting from the enabling check of an action at a certain local simulation time of its atomic unit. Depending on the result, an event $e = (cst, \cdot)$ might be executed at time cst. The remote states that are valid at this time point correspond to a cut C_e. Every atomic unit can easily decide which of its own events belong to C_e based on the compound simulation times.

$$\forall au \in AU, e \in E(au) : \left(e \in C_e\right) \iff \left(e \in Past_{cst}^{EvList^{au}}\right)$$

In the equation, $Past_{cst}^{EvList^{au}}$ denotes the part of the local event list of atomic unit au that lies in the past of cst, and is defined in (9.14). Such a cut is graphically represented by a straight vertical line in a time–space diagram with simulation time as the x-axis.

The past of the event lists contains all events with a compound simulation time before cst, and the cut C_e is derived as the union of the local event sets.

$$C_e = \bigcup_{au \in AU} Past_{cst}^{EvList^{au}}$$

Such a cut is obviously unique by construction, because the membership of events to $Past_{cst}^{EvList^{au}}$ is well defined due to the trichotomy of "<" for compound simulation times. We can thus deduct that an event is in the cut iff it was executed before cst.

$$\forall e' \in E : \left(cst(e') < cst(e)\right) \iff \left(e' \in C_e\right)$$

With the isomorphism between "$<$" for compound simulation times and "$\rightarrow$" for events (cf. (9.11)), it follows directly that

$$\forall e' \in E : (e' \rightarrow e) \iff (e' \in C_e)$$

To check if a cut C_e is consistent (cf. (9.12)), we assume that it is possible to find events $e', e'' \in E$ subject to

$$e'' \in C_e \quad \wedge \quad e' \rightarrow e'' \quad \wedge \quad e' \notin C_e$$

From $e'' \in C_e$, we know that $e'' \rightarrow e$, which leads to $cst(e'') < cst(e)$. Moreover $cst(e') < cst(e'')$ because of $e' \rightarrow e''$. Thus also $cst(e') < cst(e)$, and therefore $e' \in C_e$ contradicting the assumption. □

Every global state that is constructed for a certain compound simulation time is thus consistent. It should, however, be noted that this discussion assumes a "correct" simulation that does not violate the local causality constraints; the rollback mechanism ensures that other runs are taken back.

Moreover, we remark that the theoretical discussion is based on local state variables belonging to the region of each atomic unit. In the actual implementation, a copy of the additionally required ones is maintained (mirrored state variables, cf. Sect. 9.1.4). Consistency between the copies is achieved by the event executions that are notified with messages.

9.1.4 A Distributed Simulation Algorithm for SDES

This section lists and describes the necessary algorithms for a distributed optimistic simulation of SDES models. Theoretical prerequisites have already been covered in the earlier parts of this chapter. We start with a description of the basic data structures of each atomic unit. The second subsection covers communication message types and contents. Actual algorithms are explained in the final part and have been implemented as a prototype extension of the TimeNET tool (cf. Sect. 12.1) for stochastic colored Petri nets as covered in Chap. 6.

Data Structures

Each atomic unit $au \in AU$ of a distributed simulation maintains the following basic data structures.

The **local simulation time** describes the current value of the simulated clock. It equals the time of the last processed external or internal event in the atomic unit. It is a compound simulation time (cf. Sect. 9.1.2) including simulation time, vector time, and priority vector, and is denoted by $cst^{au} \in CST$.

The **event list** $EvList^{au} \subseteq E$ is a list of events $e = (cst, v)$, i.e., pairs of action variants v to be executed, together with their scheduled compound simulation times cst. They are ordered by the cst time for efficient access.

The event list does not only hold the events that are scheduled in the future with respect to the local simulation time (as it is done in sequential simulations), but also the processed events. They are necessary to keep track of the history in case of rollbacks. The event list includes *local events* due to scheduled action executions inside the atomic unit, as well as *remote events* that inform the *au* about possible local state changes due to remote action executions. Rollback messages are processed immediately and thus not stored.

Based on a given simulation time cst, we denote the **future** and **past** of the event list with respect to cst by $Future_{cst}^{EvList}$ and $Past_{cst}^{EvList}$, respectively. The two subsets of the event list are defined as follows.

$$Future_{cst}^{EvList} = \{(cst_i, \cdot) \in EvList \mid cst_i > cst\} \tag{9.13}$$
$$Past_{cst}^{EvList} = \{(cst_i, \cdot) \in EvList \mid cst_i < cst\} \tag{9.14}$$
$$\tag{9.15}$$

In case of rollbacks, it is necessary to step back in the event list $EvList$ to a point just before the rollback time cst. To simplify notation in the algorithms, we define $LastBefore_{cst}^{EvList}$ accordingly.

$$LastBefore_{cst}^{EvList} = \begin{cases} cst_1 & \text{if} \quad \begin{aligned} &\exists(cst_1, \cdot) \in EvList : cst_1 < cst \quad \wedge \\ &\nexists(cst_2, \cdot) \in EvList : cst_1 < cst_2 < cst \end{aligned} \\ cst_{Initial} & \text{otherwise} \left(\text{i.e., } Past_{cst}^{EvList} = \emptyset\right) \end{cases}$$

The lower case $LastBefore_{cst}^{EvList} = cst_{Initial}$ covers circumstances in which there is no event before cst in the event list, i.e., it should be rolled back before the first scheduled event. The initial time of the simulation run $cst_{Initial}$ is then used.

The **local state list** $StList^{au}$ captures the past states for the state variables $sv \in SV_{local}^{\star}(au)$ of an atomic unit au. It is stored as an ordered list of states and the simulation times from when they were valid. Corresponding lists are maintained in the actual implementation both for the state variables of the region $\in Region(au)$ as well as the mirrored ones individually. This is done to improve efficiency and to differ between local and remote state variables and their changes. From the theoretical standpoint and the presentation given here, this distinction is, however, not necessary. It is thus avoided to simplify algorithms and explanations. Mirrored states are assumed to be contained in the local state list in the following. Thus a **local state** contains values for all state variables that are locally known, i.e., $\in SV_{local}^{\star}(au)$.

$$StList^{au} \subseteq CST \times \Sigma^{au}$$

The local state list is organized as an ordered list of pairs such that the local state reached after the execution of an event $e = (cst, v)$ can be obtained by $StList(cst)$. If there was no state change at the given simulation time, the term just returns the state at that time, i.e., between the two surrounding events. We thus define

$$StList(cst) = \sigma_1 \iff \exists(cst_1, \sigma_1) \in StList : cst_1 \leq cst$$
$$\wedge \quad \nexists(cst_2, \sigma_2) \in StList : cst_1 < cst_2 \leq cst$$

The current state of an atomic unit au is thus always given by $StList(cst^{au})$. This state information is used by local actions to decide about their enabling.

Mirrored and local state information is updated based on the information that is sent by the atomic units, in which actions were executed that potentially change the state. The information may not be final or up to date with the local simulation time. It is, however, assumed to be valid in all local expression evaluations until a different information might become available. A rollback is then necessary. This is exactly the same behavior as for state changes in the local state variables.

Similar to the past part of the event list with respect to a compound simulation time, we define the past of the state list as[20]

$$Past_{cst}^{StList} = \{(cst_i, \cdot) \in StList \mid cst_1 < cst\}$$

Figure 9.4 depicts some of the basic data structures of an atomic unit and their relation over time. Action variant v_i has been executed at compound simulation time cst_i, and is stored in the event list accordingly. It changed the state of the local state variables of the atomic unit to σ_{i+1}. This state was valid until the time point cst_{i+1}, at which action variant v_{i+1} was executed and led to the local state σ_{i+2}. The current local simulation time cst^{au} equals this time, which means that v_{i+1} was the last executed event. All later events in the event list, i.e., $v_{i+2}, v_{i+3}, \ldots$, are scheduled for future execution. There is thus no associated state change stored in $StList^{au}$ yet; state-dependent

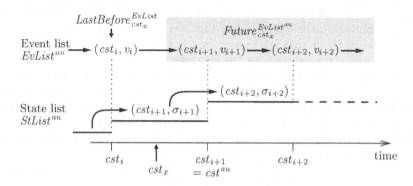

Fig. 9.4. Data structures of an atomic unit and their relation

---

[20] The future with respect to the local simulation time is always empty, because state-changing events are only processed when they are reached by the local time.

expressions in the atomic unit assume that the state σ_{i+2} is not left until further notice.

If a straggler message or rollback notification from another atomic unit is received with an associated compound simulation time cst_x, the shown au goes back to the state that was valid before cst_x, namely $LastBefore_{cst_x}^{EvList}$. Future and past parts of event and state list are used in the algorithms; the figure depicts $Future_{cst_x}^{EvList^{au}}$ as an example.

Communication Messages and Rollbacks

The distributed simulation processes need to be synchronized by exchanging messages. The implementation described here uses the *message passing interface* (MPI; [106]).

Atomic units on different nodes are managed by the corresponding logical processes, and exchange messages via them. Communication between atomic units inside one logical process is done directly without sending actual messages; this avoids the communication overhead otherwise necessary. In addition to that, copies of state variables belonging to the region of atomic units that reside on the same logical process are not kept in the local state list, because it is possible to directly access the associated state history. However, if an atomic unit is migrated to a different node, such a part of the state list has to be copied and sent together with the migrating unit.

The algorithms shown below avoid the technical details of message routing and buffering done by the logical processes. The description assumes direct message exchange between atomic units. The chronological order between messages in node–node pairs is guaranteed by the underlying MPI protocol.

The following message types are used in the approach presented here.

Event.   Messages are sent by an atomic unit to notify others of a local event execution. An event is described by simulation time and SDES action variant. The message is sent in case that the state of a remotely known state variable could change as a result. The set of atomic units to be informed about the execution of an event can be obtained from the executed action a by $AU_{affected}(a)$.

An event message is characterized by the type "EventNotify." The other parameters include source and destination atomic unit of the message as well as the executed event.

Rollback.   Messages inform other atomic units of a causality violation. They are either generated after a straggler message is received, or simply forwarded to others upon receipt.

Rollback messages are identified by the first parameter "Rollback." Source and destination atomic unit of the message are contained as well as the compound simulation time to which the time needs to be rolled back.

Logical processes in standard distributed optimistic simulations [103] keep a record of all sent messages. In case of a rollback, negative or *antimessages* are sent for all messages that were communicated in the time window that is rolled back, i.e., between the rollback time and the current local time. This leads to a lot of messages to be exchanged, and is denoted as *aggressive cancellation*. Some of the antimessages can be avoided by additional bookkeeping: It is then possible to check whether the state reached at the local time again later differs from the original state at that time. If so, no antimessage is necessary. *Lazy cancellation* employs several of these techniques. In the approach presented here, individual antimessages are not sent at all; there is only one rollback message to inform a destination atomic unit about a local rollback. This is possible because all events can be uniquely attributed to the past or future of a rollback, due to the exact ordering based on our compound simulation time.

Rollbacks can occur frequently and may be cascaded, interrelated, or even circular in conventional approaches. The reason for this is that events, which are causally dependant on the events rolled back, are not identified at the time of a causality error. The time representation of standard optimistic distributed simulation algorithms is not sufficient to causally relate events and rollbacks, because it includes the local simulation time only.

Our approach allows to rollback to an exact point in the event list because the comparison of times is transitive and trichotomous. The results of [296] can be transferred to our notion of time, which ensures that an optimal consistent state is achieved during a global rollback. It furthermore guarantees a lifelock-free simulation progress, and that there are no circular rollbacks.

The compound simulation time carries enough causal information to be exploited for an even better cancellation. The technique presented in [334] ensures a message cancellation that is shown to be optimal in the sense that no unnecessary rollback or event is processed. It is an advancement of a causality representation and cancellation mechanism for Time Warp simulations proposed in [46]. The approach of [334] has been adapted for our simulation environment. Atomic units ignore events that will be rolled back eventually, leading to significant savings in computation and communication. This is due to the fact that, assuming an aggressive cancellation strategy, the events causally dependant on the rolled-back event will be rolled back later on anyway. In addition, early recovery operations such as restoring state and ignoring events that will be rolled back can be performed for interrelated rollbacks.

A set of invalid vector time intervals is attached to each message. This helps to identify related cancellation messages and hence avoid interrelated and cascading rollbacks. It is important for this approach to causally relate rollbacks, which can be done with the vector time part of the compound simulation time. This is the reason why vector time entries are never set back to zero, which would be possible at nonzero delays if they were only used for the ordering of immediate events. An example showing the number of prevented rollbacks has been given in [211]. Details are omitted in the algorithmic description below to improve readability.

Distributed Simulation Algorithm

Algorithm 9.1 shows UPDATELOCALTIME, a function that advances the local simulation time cst^{au} to a new EventTime. This is done in case of the execution of a local or remote event from the event list, in which case the time actually increases. If a rollback is performed, the local time must be set back to the rollback time, which is also done by this function.

The three components of the local compound simulation time, namely simulation time, vector time, and priority vector, need to be set. The local vector time is processed first. For every entry it assumes the maximum value of the old local vector time and the entry of the new event. The vector time entry that corresponds to the local atomic unit, $VT_{au}(au)$, is always incremented because local causality also covers rollbacks.

The priority vector is updated next, according to (9.5). Priority values are set back to the default value infinity due to the corresponding values in scheduled events, which are copied into the local time here.

Algorithm 9.2 shows an abstract implementation of a rollback of an atomic unit, function ATOMICUNITROLLBACK. It is invoked by functions described later on either when a rollback message has been received or a rollback is caused by the reception of a straggler event. It should be noted that rollbacks

---

UPDATELOCALTIME (au, EventTime, EventPrio, Remote)

Input: Atomic unit au, time of the initiating event EventTime, priority of event (or ∞ in case of a rollback), flag Remote

(∗ Assume the following notation ∗)
$cst^{au} = (ST_{au}, VT_{au}, PV_{au})$
EventTime $= (ST_e, VT_e, PV_e)$

(∗ Local vector time is monotonic; others follow causality ∗)
$VT_{au}(au) := VT_{au}(au) + 1$
for $au_i \in AU \setminus \{au\}$ do
    $VT_{au}(au_i) := \max(VT_{au}(au_i), VT_e(au_i))$

(∗ Update priority vector ∗)
for $\forall au_i \in AU$ do
    if Remote $\wedge (ST_{au} = ST_e)$ then
        $PV_{au}(au_i) := \min(PV_{au}(au_i), \text{EventPrio})$
    else $PV_{au}(au_i) := PV_e(au_i)$
if EventPrio $\neq \infty$ then $PV_{au}(au) := \text{EventPrio}$

(∗ Set new local time ∗)
$cst^{au} := (ST_e, VT_{au}, PV_{au})$

Algorithm 9.1: Update the local simulation clock of an atomic unit

---

AtomicUnitRollback $(au, \text{RollbackTime}, au_{Source}, \text{Always})$

Input: Atomic unit au, RollbackTime, causing au_{Source}, Flag Always

(∗ Delete events after rollback time that came from au_{Source} ∗)
NoEventFound := True
for $\forall e = \big(cst, (a, \cdot)\big) \in Future^{EvList^{au}}_{\text{RollbackTime}}$ **do**
    if $a \in Region(au_{Source})$ **then**
        $EvList^{au} := EvList^{au} \setminus \{e\}$
        NoEventFound := False

(∗ Return if rollback time has not been reached or no event was deleted ∗)
if RollbackTime $> cst^{au} \vee$ (NoEventFound $\wedge$ **not** Always) **then return**

(∗ Delete local events after rollback time from the event list ∗)
NotifyAUs := $\{au_{Source}\}$
for $\forall e = \big(cst, (a, \cdot)\big) \in Future^{EvList^{au}}_{\text{RollbackTime}}$ **do**
    if $a \in Region(au)$ **then**
        $EvList^{au} := EvList^{au} \setminus \{e\}$
        NotifyAUs := NotifyAUs $\cup AU_{affected}(a)$

(∗ Delete states after rollback time ∗)
$StList^{au} := Past^{StList^{au}}_{\text{RollbackTime}}$

(∗ Update local time and event list ∗)
$cst^{au} := LastBefore^{EvList^{au}}_{\text{RollbackTime}}$
UpdateLocalTime$(au, \text{RollbackTime}, \infty, \text{True})$
UpdateActivityList$\big(StList^{au}(cst^{au}), Future^{EvList^{au}}_{cst^{au}}, cst^{au}\big)$

(∗ Notify possibly affected atomic units ∗)
for $\forall au_i \in$ NotifyAUs **do**
    Send ('Rollback', $au, au_i, \text{RollbackTime}$)

Algorithm 9.2: Rollback an atomic unit au

are never caused by causality errors between atomic units belonging to the same logical process, because events are executed in the right order there.

The function requires as parameters the affected atomic unit, the time to which it should be rolled back, the initiating atomic unit au_{Source}, and a flag denoting that the rollback should be executed in any case (see below).

The event list of the atomic unit is scanned first for events in the future of the rollback time, which came from the originating atomic unit. These events obviously have to be removed. In the case that the rollback time has not been reached by the local simulation time, nothing else needs to be done, and the function returns. The same applies to cases in which no event had to be removed before, because then au was not influenced by au_{Source} since the rollback time. If the function was called due to a straggler message, the atomic unit needs to be rolled back anyway, which is specified by setting flag Always = True.

If the atomic unit needs in fact to be rolled back, the algorithm continues by deleting all local events from the event list that are scheduled after the rollback time. The set of atomic units which need to be informed about the deleted events is derived on the fly. As every local event that has been executed already lead to an event message to all possibly affected atomic units, they must all be notified about the rollback. All states in the local state list after the rollback time are deleted as well.

The internal state of the atomic unit is finally set back to the rollback time by invoking UPDATELOCALTIME with the rollback time, and by an update of the event list for this new time via UPDATEACTIVITYLIST. The priority of the causing event needs to be set to infinity in the first call, because there is no actual event to be processed and the minimum path priority needs to be kept. Remote effects of the rollback are finally considered by sending rollback messages to all atomic units that need to be notified.

The rollback algorithm called a function UPDATEACTIVITYLIST, which can be implemented for the distributed optimistic simulation similar to the one shown on page 137 for the sequential simulation. It is thus not repeated in detail here; the differences are briefly outlined in the following.

States are only considered local to the executing atomic unit; if local or remote state variables need to be accessed, their values are taken from the state list as $StList^{au}(sv, cst^{au})$. They always exist in the atomic unit, because the local state variable set $SV^{\star}_{local}(au)$ contains all eventually used ones.

The original version of UPDATEACTIVITYLIST assumes that the event list only contains future events (as in a standard simulation), and is thus called from the distributed algorithms with $Future^{EvList^{au}}_{cst^{au}}$ to ignore past events. Another purely technical difference is that UPDATEACTIVITYLIST assumes elements of the event list as three tuples of action, action mode, and time. Elements of the distributed version are constituted of time and action variant, which in turn contains action and mode.

Additional events are scheduled for later execution in the lower part of Algorithm 7.2, if the degree of concurrency is not yet fully utilized. When elements of the event list are sorted, the compound simulation time must obviously be used instead of simple time (and priority, if applicable). This is done according to (9.2) and (9.4).

Vector time and priority vector of local events which are scheduled in the future event list have to follow the causality knowledge of the overall atomic unit (cf. explanation and equations on page 188). Both have to be updated using the local vector time and priority vector in UPDATEACTIVITYLIST for every future event which stays in the event list. Only the priority entry of the scheduled event itself must reflect its global priority. An actual implementation could improve efficiency if vector time and priority vector of future local events

---

ATOMICUNITRECEIVEMESSAGE (m)

Input: Message m

(∗ Remote event (or mirror) notification ∗)
if $m =$ ('EventNotify', au_{Source}, au, Event) **then**
$\qquad (cst, v) :=$ Event
$\qquad EvList^{au} := EvList^{au} \cup \{\text{Event}\}$
$\qquad$ **if** $cst > cst^{au}$ **then return**
$\qquad$ (∗ m is a straggler message; initiate rollback ∗)
$\qquad$ ATOMICUNITROLLBACK(au, cst, au_{Source}, True)
$\qquad$ **return**

(∗ Rollback message ∗)
if $m =$ ('Rollback', au_{Source}, au, cst) **then**
$\qquad$ ATOMICUNITROLLBACK(au, cst, au_{Source}, False)
$\qquad$ **return**

Algorithm 9.3: Receive a message and sort it into local lists

are not stored at all, but copied from the local compound simulation time whenever needed.

Whenever a message is received by a logical process, ATOMICUNIT-RECEIVEMESSAGE (shown in Algorithm 9.3) is invoked to handle it. The message is passed as the only parameter. Two cases are distinguished in the algorithm, which correspond to the types of messages (see page 203) that are exchanged.

The received event is added to the event list upon reception of an Event-Notify message. If the remote event is scheduled for a time in the future of the local simulation, nothing else needs to be done. The message might otherwise violate the local causality constraint, and is denoted as a straggler message. The atomic unit is rolled back to the time of the event by calling ATOMIC-UNITROLLBACK accordingly. The straggler event will be executed later on, because it is still in the event list.

In the case that the original action priorities are used instead of the global event priorities, confusions can be detected at the point where the new event is inserted into the event list. If the insertion point is not exactly specified, i.e.,

$$\exists (cst', \cdot) \in EvList^{au} : cst' \parallel cst$$

a confusion has been detected.

It is not necessary to remove other future events originating from the same atomic unit. In such a case the remote atomic unit has been rolled back earlier, and a corresponding rollback message would have been received. This message would then already have led to deleting all remote events from that atomic unit after the rollback time.

When a Rollback message is received, the rollback function simply needs to be invoked for the local atomic unit.

ATOMICUNITPROCESSEVENT (au)

Input: atomic unit au for which the next event should be processed

($*$ Select next event to be executed, including probabilistic selection $*$)
$(cst, v) :=$ SELECTACTIVITY($Future_{cst^{au}}^{EvList^{au}}$)
Assume $v = (a, \cdot)$
if $a \in Region(au)$ then Local $:=$ True

($*$ Execute the event $*$)
LocalState $:= StList^{au}(cst)$
NewLocalState $:= Exec^{\star}(v, \text{LocalState})$
$StList^{au} := StList^{au} \cup \{(cst, \text{NewLocalState})\}$

($*$ Update time and event list $*$)
UPDATELOCALTIME($au, cst^{au}, Pri_{Global}^{\star}(v), \neg$Local)
UPDATEACTIVITYLIST(NewLocalState, $Future_{cst^{au}}^{EvList^{au}}, cst^{au}$)

($*$ Notify affected atomic units if the event was local $*$)
if Local then
    for $\forall au_i \in AU_{affected}(a)$ do
        Send ('EventNotify', $au, au_i, (cst^{au}, v)$)

Algorithm 9.4: Optimistic distributed simulation algorithm

The actual processing of internal or remote events from the event list is done by ATOMICUNITPROCESSEVENT shown in Algorithm 9.4. Only the individual atomic unit is given as a parameter, because all other information is available in the au-specific data structures.

The event to be executed is selected from the event list by calling function SELECTACTIVITY (description given later). The algorithm checks and stores whether the event was caused by a local action. State variables are updated via the SDES execution function subsequently, which is applied to the locally known variables only. The new state is added to the state list with the compound simulation time of the state change.

Local simulation time and event list are updated using the functions UPDATELOCALTIME and UPDATEACTIVITYLIST as described earlier. Finally, all atomic units that might be affected by the execution of the event are informed about it. This is done for local events only, because atomic units that are affected by remote events are notified by the corresponding atomic units.

The next event to be executed is obtained from the future part of the event list by calling algorithm SELECTACTIVITY (see page 139). This function from the standard simulation can be reused with some changes. First of all, the tuple format of the event list elements has to be reorganized and time has to be interpreted as compound simulation time as it has been explained for UPDATEACTIVITYLIST earlier.

The priority-based search in the events that are scheduled for the same time in the original version of SELECTACTIVITY is not necessary here,[21] because action priorities are already encoded in the priority vector part of the event times, and are thus ordered correctly in the list. The only point where a decision needs to be made is when there are several immediate actions scheduled for the same time, which have the same priority. This is, e.g., the case for conflicting immediate transitions of a Petri net, and the scheduled compound simulation times are equal as defined in (9.3). The probabilistic selection is done by SELECTACTIVITY based on the individual weights $Weight^\star(\cdot)$. It should be noted that this selection will never be done between local and remote events, because they would never have equal compound simulation times, which means that there is a unique ordering in the event list. Conflicting actions have to be placed into the same atomic unit as described in Sect. 9.1.1.

Function LOGICALPROCESS is shown in Algorithm 9.5 and implements the activities of each logical process running on a node. It comprises an initialization of all corresponding atomic units and a main simulation loop.

Each atomic unit starts with a local compound simulation time equal to the overall simulation start time $cst_{Initial}$, which is provided by the calling process and usually zero. The initial states of all local state variables are set according to their SDES initial value, and the obtained state is added together with the initial time as the first entry of the state list. The event list is constructed by calling UPDATEACTIVITYLIST with a previously empty list. This call selects and schedules events due to local action enablings and adds them to the list with the adaptations described above.

The main loop of the logical process receives messages from other processes, distributes them to its atomic units by invoking ATOMICUNITRECEIVEMESSAGE, and starts event executions of its atomic units via ATOMICUNITPROCESSEVENT. The decision between these two activities might be done heuristically or even based on time lags between processes, which is not detailed here. In the algorithm it is assumed that all incoming messages are received first before any actual event is processed. The event processing in the lower part of the loop always selects the event with the smallest time of all scheduled events in the atomic units of the logical process. This avoids causality violations between them.

The computations stop and the loop is exited when some predefined stop condition is reached, e.g., a certain accuracy or a maximal simulation time. The safe time as computed using the approach [283] for the fossil collection can be used for this purpose as well, because then it is clear that all performance measures have been computed until the maximum time (see below). This does of course depend on the type of evaluation (transient or steady-state) as it was explained in more details for the standard simulation algorithm in Sect. 7.2.

---

[21] It could, however, be used with the global priority instead of $Pri^\star(\cdot)$.

LOGICALPROCESS $(AU^{lp}, cst_{Initial})$

Input: Information about atomic units AU^{lp} that are mapped to lp,
Initial simulation time $cst_{Initial}$

(∗ initializations ∗)
for $\forall au \in AU^{lp}$ **do**
$\quad cst^{au} := cst_{Initial}$
$\quad$(∗ start with initial state in state list ∗)
$\quad$**for** $\forall sv_i \in Region(au) \cap SV^\star$ **do**
$\quad\quad sv_{i,0} := Val_0{}^\star(sv_i)$
$\quad StList^{au} := \{(cst_{Initial}, (sv_{1,0}, sv_{2,0}, \ldots))\}$
$\quad$(∗ initialize event list ∗)
$\quad EvList^{au} := \emptyset$
$\quad$UPDATEACTIVITYLIST$(StList^{au}(cst^{au}), EvList^{au}, cst^{au})$

(∗ main simulation loop ∗)
repeat
$\quad$(∗ receive remote messages from other lps and sort into au_i ∗)
$\quad$**for** $\forall m \in$ ReceivedMessages **do**
$\quad\quad$ATOMICUNITRECEIVEMESSAGE(m)

$\quad$(∗ select and process next event ∗)
$\quad cst_{next} := \min_{au_i \in AU^{lp}}\left(\min_{(cst_k, \cdot) \in Future^{EvList^{au_i}}_{cst^{au_i}}}(cst_k)\right)$
$\quad$Assume $(cst_{next}, \cdot) \in EvList^{au}$
$\quad$ATOMICUNITPROCESSEVENT(au)

until stop condition reached, e.g., $\min_{au \in AU}(cst^{au}) \geq$ MaxSimTime

---

Algorithm 9.5: Logical process lp

The main distributed simulation algorithm is responsible for setting up the logical process on the available computing nodes, and for the partitioning of the SDES model according to the rules introduced in Sect. 9.1.1. The atomic units need to be distributed over the logical processes as well. The aim of this approach is, however, not a perfect partitioning obtained from the model structure, but to allow an automatic load balancing by arbitrary partitioning and migration of model regions.

The logical processes $lp \in LP$ are started one per node, and each is supplied with the necessary information about associated atomic units as well as the initial simulation time $cst_{Initial}$. The main program then waits until the logical processes have finished. Intermediate values for the performance measures are read and the results computed.

After the explanation of the central algorithms of the proposed distributed simulation algorithm, some aspects that have been left out so far are discussed briefly. This applies to message cancellation, fossil collection, performance measure computation, and load balancing.

Forwarding, execution, and *cancellation of rollback messages* work effectively when implemented as described above, and circular rollbacks are impossible. However, there are several approaches in the literature, which aim at more efficient solutions. The rollback-optimal solution presented in [334] (see page 203) has been adopted for the implementation of our approach. This required to store more elaborate information about valid and canceled time intervals in the atomic units, which is outside the scope of this text.

The amount of storage used for state and event lists of the atomic units grows as the simulation progresses, which is known under the term *limited memory dilemma* [102]. Jefferson [183] observed that there exits a **safe time** (or global virtual time) at any step of the simulation, such that all local states which are earlier than this time are confirmed and will never be invalidated by a rollback.

State and event information in the atomic units are thus not needed any more, if their associated time stamp is older than the safe compound simulation time cst^{Safe}. A standard centralized technique [283] is used to obtain cst^{Safe}. A selected node, in our case the first node of the cluster which also starts the main simulation process, requests the oldest unacknowledged message of each atomic unit. The protocol ensures that there is no pending message which would result in a rollback behind this time. The safe time is the minimum of the received times, and is broadcasted to all nodes. Every atomic unit can then discard the past of all local lists to save memory, which is called *fossil collection*

$$\forall au \in AU \ \textbf{do} \quad EvList^{au} := Future^{EvList^{au}}_{cst^{Safe}}$$
$$StList^{au} := Future^{EvList^{au}}_{cst^{Safe}}$$

The fossil collection never removes events such that *LastBefore* becomes invalid.

The computation of *performance measures* is done as follows. They can be associated to an atomic unit in which the state variables or actions reside which contribute rate or impulse rewards to them. Another possibility would be to form additional atomic units which only contain performance measures.

In any case, all state variable changes and action executions that contribute to a reward measure have to be forwarded to the managing atomic unit, to compute intermediate reward results as it is done in Algorithm 7.4 (STEADYSTSIMULATION). This is efficiently done at the time of a fossil collection: All discarded events and states are final, and can be used for a measure computation. It is not practical to compute intermediate values before, because the values could be invalidated by a rollback later on.

The final goal of the overall approach is a distributed simulation of SDES models with an automatic *load balancing*. The fine-grained model partitioning described in Sect. 9.1.1 is an important prerequisite. Atomic units manage their local data structures independently, and can thus be migrated from one computing node to another one easily. The logical processes only need to update their information about the mapping of atomic units and logical processes, and the data structures of the migrating *au* have to be transferred efficiently. The underlying message passing protocol has to update its routing information as well.

The migration of atomic units has already been implemented in the software environment TimeNET. Different heuristics are currently investigated for the decision when an atomic unit should be migrated and to which logical process. Such an algorithm could relate to the time difference of logical processes between global safe time and local *au* simulation times. A logical process with a local time close to the safe time seems to be a bottleneck, and should give away one of its atomic units to a logical process that has a very high local time. Communication overhead should also be taken into account; past message exchanges can be used as an approximation. Atomic units that exchange a lot of messages should obviously be located in the same logical process. Another issue to be resolved is the frequency (or reaction time and sensitivity) of migrations, because thrashing could occur otherwise, i.e., atomic units migrate between nodes back and forth all the time. Corresponding heuristics are the topic of current work [212].

9.2 Simulation of Models with Rare Events

There are many examples of technical systems in which the designer is interested in the performance under failures, or the probability of reaching a state of potential catastrophic behavior. If a quantitative model of such a system exceeds the restrictions of symbolic or numerical analysis techniques, simulation is the only applicable evaluation method. Very low probabilities of reaching a state of interest requires a vast amount of events to be generated. This case is called *rare event simulation* due to the low ratio of significant samples with respect to the overall event number. There are numerous applications in which probabilities in the range of 10^{-8} or less need to be quantified. Standard simulation methods are not applicable in practice because they would require prohibitively long run times before achieving statistical accuracy. Variance reduction techniques such as control variables, common random numbers, and others do not solve this problem, because it may easily happen that no significant event is ever generated in an acceptable time.

Several approaches have been investigated in the literature to overcome this problem; overviews are e.g., given in [141, 147, 159, 160, 237]. They have the common goal to make the rare event happen more frequently in order to gain more significant samples out of the same number of generated events.

Importance sampling [144] changes the model in a way that lets the rare event happen more often. Its main drawback is that it usually requires deep insight into the model, and is thus considered to be useful for simple models only [147].

The second main technique for rare-event simulation is *importance splitting*, which has been introduced in the context of particle physics [191] and later used in simulation [23]. Its underlying idea to generate more samples of the rare event is to follow paths in the simulated behavior that are more likely to lead to an occurrence of the event. It thus requires an algorithm to decide which paths to take and which to discard; Sect. 9.2.1 briefly shows how this is usually done.

There are two different approaches in the literature, one under the name of *splitting* [140, 141] and the other called *RESTART* [317, 318, 320, 321] (which is short for *repetitive simulation trials after reaching thresholds*). Splitting is applied to estimate the (small) probability of reaching a set of states out of an initial state before the initial state is hit again. It requires the system to return to this state infinitely many times. RESTART is considered to be less restrictive than splitting with respect to the types of performance measures to be computed. Moreover, measures can be estimated both in transient and steady-state. It has been extended significantly in [321] by the authors that introduced this version of splitting techniques. The application to arbitrary models (e.g., in a software tool or a model framework such as SDES) is easier because it can use the model as a black box, as long as the measure of interest is defined as described below.

We consider the RESTART method out of the mentioned reasons here; its implementation for SPN models in the software tool TimeNET (cf. Sect. 12.1) is used for the application example considered in Chap. 14. The technique is explained in Sect. 9.2.1, and its application to the simulation of SDES models covers Sect. 9.2.2 including an algorithm.

9.2.1 The RESTART Method

Assume that the goal of a simulation is to estimate the probability $P\{A\}$ of being in a set of state A in steady-state, and that significant samples are generated only rarely due to the model. Let the set of all reachable states of a model be denoted by B_0, and the initial state of the system by σ_0. This situation is sketched in the Venn diagram in Fig. 9.5.

A standard simulation would require a very long run time until A has been visited sufficiently often to estimate $P\{A\}$. The idea of importance splitting techniques in general is to let A be visited more frequently by concentrating on promising paths in the state set. The line in Fig. 9.5 from the initial state into A depicts one possible "successful" path. A standard simulation would usually deviate from this and take paths leading "away" from A. If we can find a measure of "how far away" from A a state is, it becomes possible to decide which paths are more likely to succeed and should be followed more frequently. Such a measure can often be deduced from the actual application: rare failures may be the result of a continuous wear-and-tear, and unavailability or losses

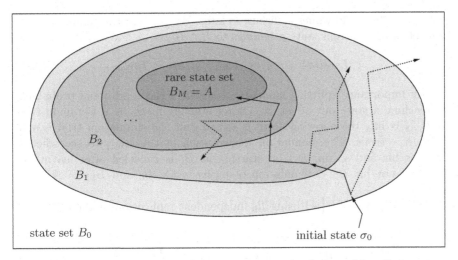

Fig. 9.5. State sets and paths in a rare-event simulation with splitting

due to blocking happen after a buildup in buffers. Corresponding regions of the state set with increasing probability of reaching A can then be defined. The samples resulting from such a guided simulation must of course be corrected by the ratio with respect to the probability that the path would have been taken in a standard simulation.

Formally, define M subsets $B_1 \ldots B_M$ of the overall state space B_0 such that

$$A = B_M \quad \text{and} \quad B_M \subset B_{M-1} \subset \ldots \subset B_1 \subset B_0$$

The conditional probabilities $P\{B_{i+1} \mid B_i\}$ of being in an enclosed set B_{i+1} under the precondition of being in B_i are much easier to estimate than $P\{A\}$, because every one of them is not rare if the B_i are chosen properly. The measure of interest can then be obtained from the product of the conditionals.[22]

$$P\{A\} = \prod_{i=0}^{M-1} P\{B_{i+1} \mid B_i\}$$

States that are visited during a simulation must be mapped to the respective sets B_i. An **importance function** f_I returns a real value for each state. Guidelines on how to choose such a function are given in [320].

$$f_I : B_0 \to \mathbb{R}$$

A set of **thresholds** (denoted by $Thr_i \in \mathbb{R}, i = 1 \ldots M$) divides the range of importance values such that the state set B_i can be obtained for a state.[23]

$$\forall i \in \{0 \ldots M\} \quad : \quad Thr_{i+1} > Thr_i$$
$$\sigma \in B_i \iff f_I(\sigma) \geq Thr_i$$

---

[22] Obviously $P\{B_0\} = 1$.

[23] We assume $Thr_0 = -\infty$ and $Thr_{M+1} = \infty$ here to simplify notation.

Setting of threshold values is discussed later. We say that the simulation is in a **level** i if the current state σ belongs to $B_i \setminus B_{i+1}$.

$$Level(\sigma) = i \iff Thr_i \leq f_I(\sigma) < Thr_{i+1} \tag{9.16}$$

An importance splitting simulation measures the conditional probability of reaching a state out of set B_{i+1} after starting in B_i by a Bernoulli trial. If B_{i+1} is hit, the entering state is stored and the simulation trial is split into R_{i+1} trials. The simulation follows each of the trials to see whether B_{i+2} is hit and so on. A trial starting at B_i is canceled after leaving B_i if it did not hit B_{i+1}. Simulation of paths inside B_0 and $B_M = A$ is not changed.

An estimator of $P\{A\}$ using R_0 independent replications is then [140, 310]

$$\widehat{P\{A\}} = \frac{1}{R_0 R_1 \ldots R_{M-1}} \sum_{i_0=1}^{R_0} \cdots \sum_{i_{M-1}=1}^{R_{M-1}} \mathbf{1}_{i_0} \mathbf{1}_{i_0 i_1} \cdots \mathbf{1}_{i_0 i_1 \ldots i_{M-1}}$$

if we denote by $\mathbf{1}_{i_0 i_1 \ldots i_j}$ the result of the Bernoulli trial at stage j, which is either 1 or 0 depending on its success.

The reduction in computation time results from estimating the conditional probabilities $P\{B_{i+1} \mid B_i\}$, which are not rare if the sets B_i are selected properly. Even more computational effort is saved by discarding paths that leave a set B_i, and which are therefore deemed unsuccessful. The optimal gain in computation time is achieved if the sets are chosen such that [321][24]

$$M = -\frac{1}{2} \ln(P\{A\})$$
$$P\{B_{i+1} \mid B_i\} = e^{-2}$$
$$R_i \approx \frac{1}{P\{B_{i+1} \mid B_i\}} = e^2$$

It should be noted that the optimal conditional probabilities as well as number of retrials do not depend on the model.

The numbers of retrials R_i can only be set approximately, because they have to be natural numbers. Apart from that, the optimal setting of the RESTART parameters for a given model and performance measure is not trivial. The first problem is that, following the equations above, the final result needs to be known before the simulation has been started. The second problem is how the importance function and especially the thresholds should be set. The problem of finding good thresholds is e.g., discussed in [310, 320]. Finally, even if the optimal settings are known, the model structure might require to set e.g., different thresholds.

---

[24] Later results of the same authors [319] recommend an alternative setting such that $P\{B_{i+1} \mid B_i\} = 1/2$, if it is possible to set the thresholds dense enough.

However, even if the optimal efficiency might not be reached easily, experiences show that the technique works robustly for a wide range of applications [318, 319]. If it is at least possible to specify the optimal number of thresholds, they should be set according to

$$\text{minimize} \sum_{i=0}^{M-1} \frac{1}{\sqrt{P\{B_{i+1} \mid B_i\}}} \text{ subject to } \prod_{i=0}^{M-1} P\{B_{i+1} \mid B_i\} = P\{A\}$$

which means that every $P\{B_{i+1} \mid B_i\}$ should be chosen as close as possible to e^{-2}. The advantage of RESTART compared to standard simulation actually becomes bigger for lower probabilities of visiting A, and speedups of several orders of magnitude have been reported. Formulas for the asymptotical speedup to be achieved with a RESTART simulation using optimal parameters are derived in [318, 319].

Several variants of RESTART have been considered in the literature [118]. We follow the approach taken in [198, 310], which can be characterized as *fixed splitting* and *global step* according to [118]. The first aspect corresponds to the number of trials into which a path is split when it reaches a higher level. The second issue governs the sequence in which the different trials are executed. Global step has the advantage to store fewer intermediate simulation states. Following [194], it is also possible in a global step RESTART to adjust thresholds during the simulation, which overcomes the disadvantage of possibly setting fixed nonoptimal thresholds.

Following the presentation in [310], the steady-state value of our example measure $P\{A\}$ is for a standard simulation given by

$$P\{A\} = \lim_{T \to \infty} \frac{1}{T} \int_0^T \mathbf{1}_A(t) \, \mathrm{d}t$$

if we denote by $\mathbf{1}_A(t)$ the indicator variable that is either one or zero, depending on whether the current state of the simulation at time t is in A.

An estimator for this steady-state measure for a RESTART implementation needs correction factors that take into account the splitting. We adopt the method of [310], where *weights* ω are maintained during the simulation run, which capture the relative importance of the current path elegantly.

The weights are computed as follows: A simulation run starting from the initial state $\sigma_0 \in (B_0 \setminus B_1)$ has an initial weight of 1, because it is similar to a "normal" simulation run without splitting. Whenever the simulation path currently in level i crosses the border to an upper level u, the path is split into R_u paths, which are simulated subsequently. The weight is obviously divided by R_u upon splitting. Paths leading to a level $< i$ are discarded, except for the last one, which is followed further using the stated rules. The weight of the last path is multiplied by R_i when it leaves level i downwards. The weights of the previously discarded paths are thus taken back into consideration, to maintain an overall path probability of one. This procedure is repeated until

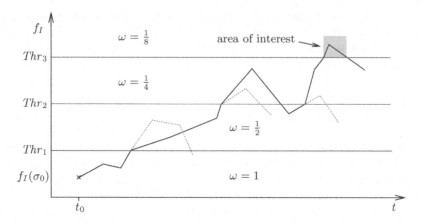

Fig. 9.6. Simulation runs in a RESTART algorithm

the required result quality is achieved. This technique has the advantage of allowing "jumps of levels" over more than one threshold with respect to the original method.

Figure 9.6 depicts a sample evolution of a RESTART simulation with three thresholds. Every path that crosses the border of a threshold upwards is split into two paths in this example; the ones that drop below a threshold are canceled, unless they represent the last trial.

Based on the weight factors, an estimator for the steady-state probability of A is

$$\widehat{P\{A\}} = \frac{1}{T} \int_0^T \omega(t)\, \mathbf{1}_A(t)\ \mathrm{d}t \tag{9.17}$$

with a large T. T counts in this context only the time spent in final paths, i.e., in the last path of each split.

9.2.2 RESTART Simulation of **SDES** Models

Approaches in the rare-event simulation literature estimate the probability of a rare state set A in transient or steady-state. This is, however, a significant restriction in the context of SDES. We apply the RESTART technique in the version described in [310] to the simulation of SDES models here, and extend it such that a complex SDES reward measure as defined in Sect. 2.4 can be estimated.

When we transfer the RESTART method to SDES models, B_0 is given by the set of all reachable states RS of such a model[25]. Instead of estimating $P\{A\}$, the goal is to obtain an estimation of a reward variable $rvar^\star$ specified as described in Sect. 2.4.1. This extension is useful for all

---

[25] Compare Sect. 7.1

performance measures that significantly depend on rewards gained in areas of the state space which are only visited rarely. For simplicity of notation, we restrict ourself to one measure which is assumed to be analyzed in steady-state.

Following the notation in Sect. 2.4.1, the reward variable $rvar^\star$ is then characterized by $lo = 0, hi = \infty, ravg^\star = $ True (steady-state, averaged). The "rareness" of event sets that are significant for a measure are obviously related to individual reward variables; it would thus only make sense to estimate several measures together if they depend on similar rare state sets. Otherwise it would be hard to find an importance function and thresholds that are meaningful for both.

Using the definition of the value of an SDES reward variable $rvar^\star$ given in (2.2) leads to an estimator $\widehat{rvar^\star}$ in the sense of (9.17).

$$\widehat{rvar^\star} = \frac{1}{T} \int_0^T \omega(t)\, R_{inst}{}^\star(t)\, \mathrm{d}t \qquad (9.18)$$

where T is the (sufficiently large) maximum simulation time spent in final paths, and $R_{inst}{}^\star(t)$ denotes the instantaneous reward gained at time t which is derived by the simulation, and $\omega(t)$ is the weight as managed by the algorithm and described in Sect. 9.2.1

An importance function $f_I : \Sigma \to \mathbb{R}$ for SDES states needs to be specified together with threshold values Thr_i that correspond to the different levels (or subsets of RS). There is no method known yet that obtains an importance function automatically for general models such as SDES (see e.g., the discussion in [310] and the references therein). In many model classes it is however easy to find one for the modeler. If for instance in a queuing model a high number of customers in a particular queue is seldom reached, lower numbers of the same value are natural thresholds. In the case of a reward measure defined as the probability that a certain number of tokens is exceeded in a place of a Petri net, an automatic method is for instance known [195, 198]. In the following, we assume that thresholds are either specified by the user or obtained with a set of presimulations, as it is e.g., done in the implementations of SPNP [310] and TimeNET [198]. They can even be adjusted during the simulation run without wasting the previous results [194].

The RESTART method for SDES models is started with RESTARTSIM-ULATION shown in Algorithm 9.6. Its input is the SDES model to be simulated, and a single reward variable $rvar^\star$ as a part of it. The initial state of the simulation is set with the initial SDES state and empty activity list. Accumulated reward and simulation time start with zero. The actual simulation function RESTARTPATH is called with level zero[26] and weight one. The call returns when the stop conditions are reached; the estimation of the performance measure is finally computed. It should be noted that SimTime is not the sum of

---

[26] We assume that $Level(\sigma_0) = 0$ for simplicity.

RESTARTSimulation (SDES)

Input: SDES model with performance measure definition $rvar^\star$
Output: estimated values of performance measures $rvar_i^\star \in RV^\star$

(* initializations *)
for $\forall sv_i \in SV^\star$ **do** $\sigma_0(sv_i) := Val_0{}^\star(sv_i)$
ActivityList $:= \emptyset$
$t_0 := 0$
Reward$_{rvar^\star} := 0$

(* call path simulation *)
$(\cdot, \cdot, \mathrm{SimTime}) := \mathrm{RESTARTPATH}(0, 1.0, \sigma_0, \mathrm{ActivityList}, t_0)$

(* compute performance measure *)
$\mathrm{Result}_{rvar^\star} := \frac{\mathrm{Reward}_{rvar^\star}}{\mathrm{SimTime}}$

Algorithm 9.6: Startup of RESTART simulation

all simulated time spans as in a standard simulation. It equals the simulation time spent in all final paths, compare (9.18).

Algorithm 9.7 implements RESTARTPATH, the actual simulation of a path in the RESTART approach. The two functions are divided mainly to allow recursive execution of the latter. The main simulation loop is an adapted copy of Algorithm 7.4 for the standard steady-state simulation, thus only the differences are explained here. Its parameters include RESTART-specific level and weight as well as the complete state of the simulation itself, namely state, activity list and simulation time. This is necessary to start simulation paths at splitting points. The reward that is accumulated in a state and upon state change is multiplied by the weight factor ω, to compensate for the splitting procedure.

Level control according to the RESTART rules and (9.16) is done at the end of the loop. If the next state is on a higher level lvl' than the current simulation state, the path is split into $R_{lvl'}$ paths that are started with recursive calls to RESTARTPATH. Each new path starts at the current new state, with a weight ω divided by the number of paths. The final state of the simulation that is reached with the last of the paths is followed thereafter by copying the returned state information.

The main simulation loop repeats until a new state belongs to a lower level; the current path is then aborted by a return to the upper level of recursion. In addition to that, some stop condition depending on the simulation time or the accuracy achieved so far is used. Corresponding formulas for the estimation of result variance in a RESTART simulation can be found in the literature, see e.g., [318, 319].

RESTARTPATH $(lvl, \omega, \sigma, \text{ActivityList}, t)$

Input: Level lvl, Weight ω, state σ, activity list, time t
Output: Final state of the simulation: new $(\sigma, \text{ActivityList}, t)$

$(*$ main simulation loop $*)$
repeat
    $(*$ get new activities $*)$
    UPDATEACTIVITYLIST$(\sigma, \text{ActivityList}, t)$

    $(*$ select executed activity $*)$
    $(a, mode, t') :=$ SELECTACTIVITY(ActivityList)
    Event $:= (a, mode)$

    $(*$ update performance measure $rvar^* = (rrate^*, rimp^*, \cdot, \cdot)$ $*)$
    Reward$_{rvar^*}$ $+= \omega * \left((t' - t) * rrate^*(\sigma) + rimp^*(\text{Event})\right)$

    $(*$ execute state change $*)$
    $t := t'; \sigma := Exec^*(\text{Event}, \sigma)$

    $(*$ RESTART level control $*)$
    $lvl' := Level(\sigma)$
    if $lvl' > lvl$ **then** $(*$ split $*)$
        for $i = 1 \dots R_{lvl'}$ **do** $(\sigma', \text{ActivityList}', t') :=$
            RESTARTPATH$(lvl', \frac{\omega}{R_{lvl'}}, \sigma, \text{ActivityList}, t)$
        $(*$ Continue the final path $*)$
        $\sigma := \sigma'; \text{ActivityList} := \text{ActivityList}'; t := t'$

until $(lvl' < lvl)$ **or** (stop condition reached, e.g. $t \geq \text{MaxSimTime}$)
return $(\sigma, \text{ActivityList}, t)$

Algorithm 9.7: Main RESTART algorithm for steady-state simulation

Notes

Section 9.1 is based on joint work published in [209–211, 348]. The distributed SPN simulation implemented in TimeNET has been described in [196, 197, 213].

There is an immense number of publications on distributed simulation. Some books covering the subject include [15, 17, 115, 151, 332]. Different algorithms for the distributed simulation of Petri net models are compared in [252]. Recent results in this area are reported in [116, 330]. An overview of distributed simulation algorithms can be found in [114, 280].

Section 9.2 draws on the application of the RESTART method to the example of Sect. 14, which has been published in [346, 347]. More recent results including the treatment of extended reward measures are reported in [357, 358].

The evaluation of examples is based on a Petri net application of RESTART in the TimeNET software tool [195, 196, 198]. Other implementations of the RESTART technique include SPNP [310] for stochastic Petri nets and ASTRO [317]. Importance sampling techniques are, e.g., included in SAVE [148] and UltraSAN [253, 254], the latter for stochastic activity networks. Different ways of implementation and their characteristics are compared in [118], and an overview of software tools for reliability is given in [189].

References to literature relevant to RESTART have been given in Sect. 9.2 already. Recent treatments of its theoretical background can be found in [318–320]. Surveys of the more general field of rare-event simulation include [141, 147, 159, 160, 237].

10

System Optimization

During the design of a technical system, one of the basic tasks is to choose between different options, such that an optimal behavior is achieved as closely as possible. Optimization is the problem of finding specific parameter values for a given system such that an optimal behavior is reached. It is a key issue in the design of complex technical systems because even comparably small performance improvements can lead to huge savings. This chapter presents an efficient optimization technique based on stochastic Petri net models.

Parameters (or decision variables) are values in the model, such as a machine speed, a buffer capacity, or the number of pallets, that have some degree of freedom in the design and need to be decided. We denote the number of those parameters for a specific optimization problem by D in the following. Each **parameter set** $\mathbf{x}$ thus consists of D elements $\mathbf{x}_1 \ldots \mathbf{x}_D$, which are assumed to be real values for simplicity $\mathbf{x} \in \mathbb{R}^D$. Actual parameters may be integers, real values, or enumerations, all of which can be mapped to a real number. The **search space** $\mathbf{X}$ of possible solutions thus has D dimensions, and is constrained by the **restrictions** of the parameters $\mathbf{x}^{min}$ and $\mathbf{x}^{max}$ such that

$$\mathbf{X} = \{\mathbf{x} \mid \forall i \in \{1, 2, \ldots, D\} : \mathbf{x}_i^{min} \leq \mathbf{x}_i \leq \mathbf{x}_i^{max}\}$$

The goal of the optimization is to find a parameter set $\mathbf{x}$ for which the designed system behaves in the best way. To define how good a certain behavior is, an **optimization function** (or cost/profit function) needs to be specified. It returns a real value $\text{cost}(\mathbf{x})$ for a parameter set $\mathbf{x}$. Any of the methods described in Part II for the quantitative evaluation of a model can be used to compute the result. Throughout this chapter it is assumed that the value of this function should be minimized. If in later chapters a profit function is used, it is clear that the maximum should be found.

Equation (10.1) depicts the optimization task in mathematical terms. The parameter for which the minimum cost is achieved corresponds to the optimal set of decision variable settings.

$$\text{Optimal cost} = \text{minimum } \text{cost}(\mathbf{x})$$
$$\text{subject to } \mathbf{x}_i \geq \mathbf{x}_i^{min} \ \forall i \in \{1, 2, \ldots, D\} \quad (10.1)$$
$$\mathbf{x}_j \leq \mathbf{x}_j^{max} \ \forall j \in \{1, 2, \ldots, D\}$$

The subsequent section describes why only **indirect optimization** techniques are applicable for the kind of system we are interested in. There are several methods available for this task, of which **simulated annealing** has been chosen here. The underlying idea and algorithm is explained in Sect. 10.1.1. The problem is, however, that because of the number of required evaluations runs, the overall computational effort is very high.

The aim of the work presented in Sect. 10.2 is to reduce this computational cost, which is done by dividing the optimization into two phases and the fast estimation of performance measures. The first phase, **fast preoptimization**, tries to get a reasonably good initial solution very fast. The cost function is approximately computed using *performance bounds* as explained in Sects. 10.2.2 and 10.2.3. The second phase, **fine grain optimization**, is started once the fast preoptimization is finished. The aim of this phase is to improve the approximate solution found in the first phase using an accurate performance measure computation. It can be considerably accelerated due to the initial starting point. The overall speedup typically reaches about two orders of magnitude.

More technical information about the method can be found in Sect. 12.1. Application to an example and the achieved efficiency is demonstrated in Sect. 13.

10.1 Indirect Optimization

The cost function in the sense that has been used above may look like a simple mathematical formula. For the examples that we are interested in here, it is in fact not. It has to incorporate all possible effects of parameters and requires the parameterized system model as well as a quantitative evaluation technique. Such an evaluation may be quite complex in itself, as has been shown in Part II. From now on we will denote by a cost (or profit) function only a quantitative measure of an SDES model. The dependence of the model will not be explicitly mentioned, because only the changeable parameters are of real interest. Section 13.3 for instance covers profit function elements for the example application area of manufacturing systems.

The optimization problems arising for discrete event system design can be solved using *evaluative* techniques, leading to an algorithm with iterative computations. *Generative* techniques obtain a solution directly for a given cost function and set of constraints. Among the latter are problems that fit into well-known mathematical programming templates like linear programming problems. The algorithms that are available for these optimization problems are in practice very efficient and mature. They unfortunately require to describe the dependency of the cost function from the parameters and the model itself in a linear algebraic way.

Models of complex technical discrete event systems are nonlinear in principle. This is especially the case for selection problems where parameters determine different system layouts, strategies, or machine types. The relation between parameters and cost function value can not be described as a linear equation. It requires a quantitative evaluation by analysis or simulation. Thus only an evaluative method can be used. The optimization algorithm iteratively generates new parameter sets and controls the search for the optimum. Formally speaking, it is not an optimization in mathematical terms, but a heuristic search for a near-optimal parameter set. It is, however, possible to show that the real optimum will be found with some statistical probability.

Typical optimization problems involve integer, boolean, and real variables (compare, for example, the types of design issues in manufacturing systems, Sect. 13.1). Modern optimization techniques are able to cope with these issues. They approach the problem using some kind of search heuristic, among which are *tabu search* [142, 277], *genetic algorithms* [119, 145], and *simulated annealing* [2, 313, 314].

Figure 10.1 depicts the principal iterative optimization process of an evaluative method. The heuristic search generates new parameter sets for which the cost function value is derived until a stop condition is reached. Simulated annealing [2] is adopted here as the optimization technique. It is explained in

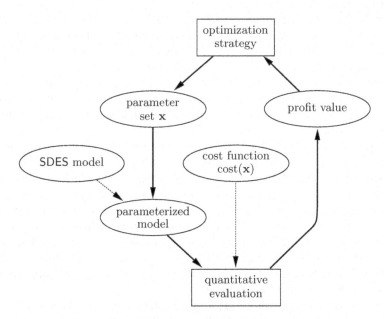

Fig. 10.1. Indirect optimization

more detail in the subsequent section. Other heuristic search techniques such as the ones mentioned above could be used instead. The general idea of the two-phase optimization method (cf. Sect. 10.2) as well as the approximation technique for performance measures (Sect. 10.2.2) are not restricted to it.

10.1.1 Simulated Annealing

For the indirect optimization a heuristic method is necessary that guides the search for a near-optimal solution. **Simulated annealing** has been chosen because of its robustness and applicability to mixed optimization problems involving integer and continuous parameters. It is moreover known to be useful in cases where local optima lead to problems with simpler algorithms.

A *Monte Carlo simulation* is a technique to explore a search space using random moves that start from the previous point in the space. This method was improved by the introduction of a "temperature" of the system, such that the Boltzmann average of the system energy (based on the Boltzmann distribution law of energy) can be obtained. This modified method is called *Metropolis Monte Carlo simulation* [241] after its first author. It was originally used as an importance sampling technique for integral derivations in statistical physics. The application of simulated annealing as a technique for optimization was introduced in the early 1980s [206].

The algorithm is named "simulated annealing" because it imitates the physical process of a hot material that cools down slowly, allowing the atoms to assume a regular arrangement. This process is called annealing and leads from a system at a high temperature (and thus in a highly disordered state of atoms) to an ordered state. Crystal growing or production of hard glass is a real-life example for this process, where irregularities can be avoided by a proper high initial temperature and a sufficiently slow cooling.

In terms of an optimization, a state of the physical system corresponds to a possible solution (i.e., a decision variable setting). The energy at the current state is mimicked by the cost function value at that point. A state change is a step to a neighboring solution in the search space, where the distance of the step is randomly sampled based on a distribution that depends on the current temperature. The distance is longer with a higher probability if the temperature is high. Paths to intermediate states with a higher energy may be followed by using a probabilistic Boltzmann acceptance criterion. Local optima can therefore be escaped by temporarily accepting worse solutions.

The temperature decrease follows a cooling scheme, which acts as a control parameter of the optimization and directly influences the tradeoff between speed and probability of reaching the global optimum. The final result of the process is a frozen state, which corresponds to the statistically optimal solution of the optimization problem.

The speed of the algorithm for a given computational complexity of each cost function derivation is mainly determined by the temperature reduction scheme. The original method (called *Boltzmann annealing*) uses a temperature for the kth annealing step according to the formula

$$T_k = \frac{1}{\ln k} T_0 \tag{10.2}$$

It has been shown that for this temperature scheme the global minimum of the cost function is statistically found. The weak point of this method is that the scheme is quite slow. Improvements have been found that show how under weak restrictions much faster cooling schemes are possible.

Fast annealing [301] uses a Cauchy distribution instead of the original Boltzmann form, resulting in a temperature scheme

$$T_k = \frac{1}{k} T_0 \tag{10.3}$$

For the special case of finite known intervals of the parameters as in the multidimensional optimization problems with D parameters that are considered here, an even faster scheme can be applied. The so-called *very fast annealing* [173, 174] uses a temperature scheme of

$$T_k = T_0 e^{-c\,k^{1/D}} \tag{10.4}$$

A freely available implementation of the latter [175] has been used for the computations in this text.

Algorithm 10.1 SIMANNEAL shows how simulated annealing works. The syntax follows the format introduced in Sect. 7.2.1. It is shown simplified for one parameter x only to improve readability. The parameters are normally vectors $\mathbf{x}$ with dimension D, and all variable accesses have to be performed correspondingly. Additional exit tests are applied in actual implementations, considering a maximum number of generated and accepted states or stop if the cost improvement or acceptance rate become too small from parameter set to parameter set. The recalculation of the temperatures are, e.g., done every 100 times that a new value is accepted and every 10 000 times a parameter set has been generated. In the algorithm it is shown as if done every time for simplicity.

The cost temperature T^{cost} influences the probability with which worse solutions are accepted. Its default initial value T_0^{cost} is 1. The parameter temperature T^{par} controls how far away from the last accepted parameter the new parameter value is selected; the default starting value T_0^{par} is also 1. The speed of "cooling down" the temperatures is controlled by the constant c. Its default value depends on the number of parameters D and is, e.g., 0.115 (3.64) for one (four) parameter(s). It is computed with TRatioScale = 0.00001 and TAnnealScale = 100 in the applied algorithm [175].

$$c = -\ln \text{TRatioScale} * e^{-\ln \text{TAnnealScale}/D} \tag{10.5}$$

SimAnneal (x_0, cost)

Input: Initial parameter x_0, Cost function cost($\cdot$)
Output: parameter set x_{best} and cost value cost$_{best}$ of optimum

$T_0^{cost} := T_0^{par} := 1$
accepted :=generated :=0
$x_{best} := x_{current}$
cost$_{best}$:= cost($x_{current}$)
loop
    generated := generated + 1
    if $T^{par} = T_0^{par} e^{-c\,\text{generated}^{1/D}} < \epsilon$ **then exit**
    if $T^{cost} = T_0^{cost} e^{-c\,\text{accepted}^{1/D}} < \epsilon$ **then exit**
    $r := \text{random}[-1 \ldots 1]$
    $d := x_{max} - x_{min}$
    repeat
        $x_{new} := x_{current} + \text{sign}(r)\, T^{par} \left[(1 + \frac{1}{T^{par}})^{|2r-1|} - 1 \right] d$
    until $x_{min} \leq x_{new} \leq x_{max}$
    cost$_{new}$:= cost(x_{new})
    if cost$_{new}$ < cost$_{best}$ **then**
        $x_{best} := x_{new}$
        cost$_{best}$:= cost$_{new}$
    $\delta := \text{cost}_{new} - \text{cost}_{current}$
    if $\delta < 0$ **or** random$[0 \ldots 1] < e^{-\delta/T^{cost}}$ **then**
        $x_{current} := x_{new}$
        cost$_{current}$:= cost$_{new}$
        accepted := accepted + 1
end loop
return x_{best}

Algorithm 10.1: Simulated annealing for one parameter

10.1.2 Avoidance of Recomputations with a Cache

The computational effort for computing the value of the cost function cost($\mathbf{x}$) from a parameter set $\mathbf{x}$ is high for our cases because every cost function is computed by a simulation or numerical analysis of an SDES. For the stochastic Petri net applications considered in Sect. 13.4, for example, the simulation can take some minutes to complete, depending on the model complexity and the confidence interval. During the first preoptimization phase this is not such a big problem. However, a certain number of linear programming problems have to be solved for every step, which is the computational bottleneck also for this phase.

When the simulated annealing algorithm generates a new parameter set to evaluate, no book-keeping of already searched areas is adopted (as, e.g., in a tabu search). Moreover, the "distance" between subsequent parameter sets statistically becomes very small when the algorithm comes to its end due to a low temperature. The probability of evaluating the same or very similar parameter set again is thus very high. Hence, the overall efficiency of the optimization can be increased a lot by avoiding recomputations.

Every result that the cost function returns is thus stored in a cache-like data structure together with its corresponding parameter set. The desired accuracy of the overall optimization is used to check each parameter individually to decide whether it is considered equal to a cached result. This is especially necessary for real-valued parameters. If $\Delta_1 \ldots \Delta_D$ are the accuracies for the D parameters $\mathbf{x}_1 \ldots \mathbf{x}_D$, a parameter set $\mathbf{x}$ is considered to be equal to a set $\mathbf{y}$ that is stored in the cache iff

$$\forall i \in \{1, 2, \ldots, D\} : |\mathbf{x}_i - \mathbf{y}_i| \leq \Delta_i$$

In this case the already stored result for $\mathbf{y}$ is adopted for $\mathbf{x}$ as well, assuming that with sufficiently small accuracy ϵ

$$|\mathrm{cost}(\mathbf{x}) - \mathrm{cost}(\mathbf{y})| \leq \epsilon$$

Section 12.1 covers the cache integration together with the implementation details of the optimization algorithm. In the examples considered so far the simple supplement of a cache to the cost function call lead to a reduction of the computational complexity by almost one order of magnitude. Chapter 13 demonstrates results for an application model.

10.2 A Two-Phase Optimization Strategy

The main problem of model-based indirect optimization lies in the computational effort required for a series of long simulation runs. We propose a two-phase optimization method that aims at finding an approximation of the best parameter set during a fast preoptimization. Speed is more important than accuracy during the first phase preoptimization phase. This phase is inspired by *ordinal optimization* ideas [165]. Cost function values need to be obtained using some fast approximation method. The idea of the two-phase optimization can be combined with any efficient quantitative estimation method for SDES.

An approximation technique for stochastic Petri nets using bounds on the performance measures of the model has been developed and is taken for the preoptimization phase. Petri nets enable the application of linear programming techniques for the approximate analysis as shown in the subsequent Sect. 10.2.2.

For the preoptimization phase the default initial parameters of simulated annealing are used without changes. This results in a standard simulated annealing procedure, for which with some statistical probability the algorithm will find (at least the area of) the global optimum. This "optimum" is of course obtained for the approximated profit function, not for the original one. During this step it is important that the region of the solution will be found, while the accuracy of the result is not.

The second phase then takes as input the best parameter set of the first one as the starting point of the optimum search. The cost function derivation from the model and the current parameter set is done with a standard performance evaluation technique, e.g., simulation as described in Chap. 9. It can be accelerated significantly, because the promising region of the search space is already known. The speed of the algorithm (leading to the number of parameter sets for which the model has to be analyzed) depends essentially on the temperature management. For an acceleration of the second phase the following heuristic changes have been introduced (cf. Sect. 10.1.1):

- The initial cost temperature T_0^{cost} is decreased from 1 to 0.1, thus reducing the acceptance probability of worse solutions already at the beginning.
- The cooling speed control constant c is made smaller by reducing the value of TAnnealScale from 100 to other values (20, 10, 5, 1), resulting in a faster temperature reduction process. Depending on the parameter space dimension D, c is thus adjusted such that only a certain percentage of the originally generated parameter sets are being analyzed until the temperature reaches the stop condition value ϵ.

An important question is when to change from the preoptimization phase to the second one. As each calculation in the preoptimization phase takes less than a second in our prototype implementation, the number of annealing iterations during this phase does not play a big role for the overall effort. Therefore, the default parameters have been used, resulting in a standard simulated annealing procedure in the preoptimization. When the annealing algorithm has reached its end, the second step is started with adjusted parameters and temperatures.

It should be noted that for the annealing algorithm the duration can be more or less arbitrarily changed by choosing different temperature schemes. However, the question is then how good the found solution will be. There is obviously a tradeoff between speed and statistical accuracy. Simulation and annealing both contain stochastic behavior that makes it hard to exactly assess the general quality of the heuristic temperature settings. However, the examples considered in previous papers [281, 351–353] as well as in Sect. 13 show that the method leads to significant speedups for temperature settings that are still robust.

The following sections show how quantitative measures can be approximated to obtain an estimation of the cost function. Section 12.1 contains more implementation-oriented information about the two-phase optimization method.

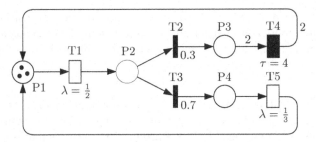

Fig. 10.2. Small Petri net example

10.2.1 Preliminary Notes on Petri Nets

We first recall some basic structural properties of Petri nets that are used in the sequel to make the text self-contained. The example model shown in Fig. 10.2 is used in the following to explain formulas and intermediate results. Firing weights of immediate transitions are annotated as numbers, while for timed transitions the rate λ of the exponential distribution or the deterministic firing time τ is shown.

Remember that the pre- and post-incidence functions of a Petri net, **Pre** and **Post**, map pairs of transitions and places to the respective arc cardinality, possibly depending on the current marking (cf. Sect. 5.3). We assume in this chapter that the arc cardinalities are not marking-dependent. In that case **Pre** and **Post** can be interpreted as matrices with dimension $|P| \times |T|$ containing natural numbers. Matrices **Pre** and **Post** have the following values for the example in Fig. 10.2:

$$\mathbf{Pre} = \begin{pmatrix} 1 & 0 & 0 & 0 & 0 \\ 0 & 1 & 1 & 0 & 0 \\ 0 & 0 & 0 & 2 & 0 \\ 0 & 0 & 0 & 0 & 1 \end{pmatrix} \quad \mathbf{Post} = \begin{pmatrix} 0 & 0 & 0 & 2 & 1 \\ 1 & 0 & 0 & 0 & 0 \\ 0 & 1 & 0 & 0 & 0 \\ 0 & 0 & 1 & 0 & 0 \end{pmatrix} \tag{10.6}$$

The *token flow matrix* **C** captures the flows of tokens between transitions and places,[1] and can be obtained from the incidence matrices.

$$\mathbf{C} = \mathbf{Post} - \mathbf{Pre} = \begin{pmatrix} -1 & 0 & 0 & 2 & 1 \\ 1 & -1 & -1 & 0 & 0 \\ 0 & 1 & 0 & -2 & 0 \\ 0 & 0 & 1 & 0 & -1 \end{pmatrix}$$

A marking vector $\mathbf{m}'$ that is reached from a source marking $\mathbf{m}$ by firing transition t_k can thus be expressed by

$$\forall i \in \{1 \ldots |P|\} : \mathbf{m}'[i] = \mathbf{m}[i] + \mathbf{C}[i, k]$$

---

[1] Often called incidence matrix in the literature, although it only contains the full structural information if the net is *pure*, i.e., does not contain self-loops.

If we denote by $\boldsymbol{\sigma} \in \mathbb{N}^{|T|}$ a vector that counts for each transition t of the Petri net the number of firings that occured in some firing sequence starting at the initial marking $\mathbf{m}_0$, the finally reached marking $\mathbf{m}$ is given by the *state equation*

$$\mathbf{m} = \mathbf{m}_0 + \mathbf{C} \cdot \boldsymbol{\sigma}$$

It follows that if it is impossible to find a $\boldsymbol{\sigma}$ for a given $\mathbf{m}$, this marking is unreachable. This is, however, only a sufficient condition: for many Petri net variants there are markings with a solution of the state equation, which are not reachable. They are called *spurious* in the literature [293].

There are two important types of linear invariants that can be obtained from the state equation. A vector $\mathbf{y}$ is called a **P-flow**[2] iff

$$\mathbf{y} \cdot \mathbf{C} = 0 \quad \longrightarrow \quad \mathbf{y} \cdot \mathbf{m} = \mathbf{y} \cdot \mathbf{m}_0 + \mathbf{y} \cdot \mathbf{C} \cdot \boldsymbol{\sigma} = \mathbf{y} \cdot \mathbf{m}_0 = \text{constant}$$

because the weighted sum of tokens specified by $\mathbf{y}$ for every reachable marking $\mathbf{m}$ is constant (see e.g. [293]).

Nontrivial and nonnegative P-flows (i.e., if $\mathbf{y} \geq \mathbf{0}$) are called *P-semiflows*. They correspond to *conservative* parts of a model, where tokens are neither lost nor created.

It is possible to derive transition invariants similar to place invariants: A vector $\mathbf{x} \in \mathbb{N}^{|T|}$ is called a **T-flow** if the following equation holds.

$$\mathbf{C} \cdot \mathbf{x} = 0 \quad \longrightarrow \quad \mathbf{m}' = \mathbf{m} + \mathbf{C} \cdot \mathbf{x} = \mathbf{m}$$

A nonnegative, nontrivial T-flow is called **T-semiflow**, and describes multi sets of transitions that, when fired in a sequence, always lead back to the first marking. This does, however, not guarantee that such a sequence is in fact always executable.

P- and T-semiflows are said to be **minimal** if there is no smaller P-semiflow. An alternative condition is that there is no other semiflow of the same type that has a strictly smaller support. The set of minimal semiflows is unique for a model; it represents the basis of a vector space containing all P- or T-semiflows. Efficient algorithm for their computations exist, see e.g. [73, 293, 304].

The example shown in Fig. 10.2 has the following semiflows:

$$\mathbf{y}_1 = (1, 1, 1, 1) \quad \text{with } \mathbf{y}_1 \cdot \mathbf{m} = 3$$
$$\mathbf{x}_1 = (2, 2, 0, 1, 0)$$
$$\mathbf{x}_2 = (1, 0, 1, 0, 1)$$

The number of tokens in the model is thus constant, and there are two minimal firing sequences that always lead back to any originating marking: firing T1 two times, T2 two times, and T4 once as well as firing T1, T3, and T5.

---

[2] Sometimes informally called *Place invariant* in the literature.

10.2.2 Computation of Performance Bounds

Stochastic Petri nets (cf. Chap. 5) are used as the modeling formalism for the systems to be optimized. The computation of approximated quantitative values can be done through several techniques, e.g., *response time approximation* [190, 260], where relatively accurate results are achieved with a computational effort that is still high. The goal of most approximation techniques is not a faster computation, but to cope with models that are too complex to be analyzed exactly.

Opposed to this, a rough computation of performance measures is sufficient for the approach presented here. For certain classes of Petri nets, efficient algorithms based on linear programming problems (LPP) exist for the computation of upper and lower bounds of performance measures. In the following, we denote with $\chi_+[t_i]$ and $\chi_-[t_i]$ the upper and lower bound of the real throughput $\chi[t_i]$ of transition t_i, and with $\mathbf{m}_+[p_i]$ and $\mathbf{m}_+[p_i]$ the upper and lower bound of the real mean number of tokens $\overline{\mathbf{m}}[p_i]$ in place p_i in steady-state.

Upper and lower bounds for the throughput of transitions as well as mean numbers of tokens in steady-state can be determined based on a linear algebraic description of a Petri net that was introduced in the previous section very briefly. The formulas are applicable to any kind of Petri net, but they are more exact if the models are restricted to the class of **FRT-nets** (see below; FRT stands for freely related T-semiflows). For the exact definition of FRT nets the reader is referred to the references at the end of the chapter.

For an application of the following equations we thus restrict ourselves to models with the following properties.

– The net has to belong to the class of FRT nets. This requires mainly that if there are transitions in conflict, the probabilities of firing each of them has to be computable from the net structure (as shown below). Hence conflicts are only allowed between transitions that are in *equal conflict relation*, that is, their pre-incidence function is the same: $\mathbf{Pre}[\cdot, t_i] = \mathbf{Pre}[\cdot, t_k]$. Additionally, there must not be different T-semiflows for which the relative throughput cannot be computed from the net structure. This is, e.g., the case in nets that are not connected. However, there are connected nets that do not comply with this restriction as well. If for instance the routing strategy for parts in a flexible manufacturing system model depends on the current state of the subsequent machines, the algorithm is not applicable, and less exact bounds can only be obtained.

– Conflicts are restricted to immediate transitions for the implemented algorithm. This requires to separate conflict resolution and timing conditions, which makes sense in many applications, but prohibits, e.g., modeling of deadlines. Most net structures can be changed to comply with this condition by adding immediate transitions before conflicting timed transitions. The firing probabilities corresponding to these additional transitions can be computed from the mean firing times of the timed transitions.

However, it is not easily possible to replace conflicts of timed transitions if their firing delay is not exponentially distributed and they do not have *age memory* firing policy.

– The net has to be *structurally bounded* and *structurally life*, which should be the case for most correctly modeled systems.

– Priorities and inhibitor arcs are not allowed for the bounds computation. By ignoring them a net could still be analyzed with the algorithm. There would, however, be no guarantee that the actual values are inside the computed range, making the results approximate and their exactness unknown. While for some systems the use of priorities or inhibitor arcs is just a way of expressing behavior conveniently, it is known that the description power of Petri nets is improved by adding either one of them. However, for the considered subclass of structurally bounded nets, it is always possible to substitute them by adding some net elements.

– Timed transitions are considered as having *infinite server* firing semantic. If a transition should have single (or fixed multiple) server semantic, its firing parallelism can be restricted by adding net elements such that an additional place invariant ensures this restriction.

The example of Fig. 10.2 does not violate any of the restrictions given above. The model is a FRT net because it is completely covered by its two T-semiflows $(2, 2, 0, 1, 0)$ and $(1, 0, 1, 0, 1)$. Moreover, the relative firing probabilities of the structurally conflicting transitions T2 and T3 can be obtained from the net structure as 0.3 and 0.7. The conflict between these two transitions, which are in equal conflict relation, "connects" the two semiflows. Their relative throughputs can thus be obtained directly.

The first step for the bounds computation is to calculate routing rates at conflict of the Petri net system. For this we first consider all pairs of transitions t_i, t_j, which are in **equal conflict relation** (denoted by $EC(t_i, t_j)$). This is the case if their input arcs come from the same places and have identical corresponding multiplicities, i.e., their pre-incidence function $\mathbf{Pre}$ is equal.

$$\forall t_i, t_j \in T : EC(t_i, t_j) \quad \text{if} \quad \mathbf{Pre}(t_i) = \mathbf{Pre}(t_j)$$

It is obvious that a transition is in equal conflict relation with itself, and that the relation is both symmetric and transitive. The relation is thus an equivalence relation, which divides the set of transitions T into a set of nonempty sets $T_1 \ldots T_k$ of transitions, which have no common elements and together form the set T. Every transition t belongs to exactly one of the sets T_i. For the example shown in Fig. 10.2 the two immediate transitions T2 and T3 are in equal conflict. The corresponding sets are $\{T1\}, \{T2, T3\}, \{T4\}, \{T5\}$.

Transitions that are in equal conflict may only be immediate ones following the restrictions. Thus every timed transition is in one subset T_i on its own, while immediate transitions may be part of a set with several elements. Immediate transitions t have a firing weight $W(t)$ associated to them, which uniquely defines the probability of their firing in our case due to the equal conflict restriction.

The relative **visit ratios** of transitions are computed in the following. The term $\mathbf{v}^{(1)}[t_i]$ denotes the number of times that transition t_i fires in steady-state in relation to transition t_1. Note that the restriction of the method to FRT nets is only due to the computability of the visit ratios for this kind of nets [38]. The relative visit ratios between two transitions can be determined easily if they belong to the same T-semiflow. It can also be computed if the transitions are in equal conflict, i.e., belong to the same transition subset T_i. In that case it is possible to find routing rates r that satisfy

$$\forall T_i, \forall t_1 \ldots t_k \in T_i : \quad \begin{aligned} r_2\mathbf{v}[t_1] - r_1\mathbf{v}[t_2] &= 0 \\ r_3\mathbf{v}[t_2] - r_2\mathbf{v}[t_3] &= 0 \\ &\cdots \\ r_k\mathbf{v}[t_{k-1}] - r_{k-1}\mathbf{v}[t_k] &= 0 \end{aligned} \quad (10.7)$$

where r_k denotes the relative routing rate at conflict of transition t_k. These rates are given in the model definition for the immediate transitions by their *weights* (relative firing probabilities). They are computed iteratively following

$$r_1 = 1; \; r_{i+1} = r_i \frac{W(t_{i+1})}{W(t_i)} \quad (10.8)$$

because of the conflict solution probabilities that are given by the firing weights.

Because of the definition of the FRT net class, all transitions are a member of a T-semiflow, and all T-semiflows are related by relative visiting rates set by the conflict probabilities. The **vector of visit ratios** $\mathbf{v}^{(1)}$ is therefore uniquely defined.

The homogeneous system of linear equations shown in the previous equation can be expressed in matrix form as $\mathbf{R}[T_i] \cdot \mathbf{v}^{(1)} = 0$, where $\mathbf{R}[T_i]$ is a $|T_i| \times |T|$ matrix and combined to the **routing matrix R**:

$$\mathbf{R} = \begin{pmatrix} \mathbf{R}[T_1] \\ \vdots \\ \mathbf{R}[T_n] \end{pmatrix} \quad (10.9)$$

The information that transitions belonging to the same T-semiflow need to have corresponding visit ratios is added by extending the routing matrix $\mathbf{R}$ by the token flow matrix $\mathbf{C}$. This results in the following system of linear equations normalized for transition t_1:

$$\begin{pmatrix} \mathbf{C} \\ \mathbf{R} \end{pmatrix} \cdot \mathbf{v}^{(1)} = 0, \quad \mathbf{v}^{(1)}[t_1] = 1 \quad (10.10)$$

Solving this system of equations results in the visit ratios of all transitions of the net with respect to transition t_1. An important side effect of this computation is the following: it can be shown that if the model belongs to the class of FRT nets, the above system of linear equations has exactly one solution [15, 38]. Hence if this is not the case for a specific net, it is clear that the model is not FRT and the algorithm stops. An additional previous check of the net, which would require a computation of the T-semiflows and their conflict relations, is therefore not necessary.

The vector of visit ratios for the transitions in the example is obtained as $\mathbf{v}^{(1)} = (1.0, 0.3, 0.7, 0.15, 0.7)$. The correctness of these values can be easily checked because for every firing of transition T1, transitions T2 and T3 fire 0.3 and 0.7 times due to their firing weights. Based on that it is obvious that transition T5 fires as often as T3 does, while transition T4 fires once for every two firings of T3.

Little's Law (see p. 70) holds for stochastic Petri nets and can be applied to each subnet of a model that consists of a timed transition t_1 and (one of its) input place(s) p. Remember that timed transitions are not allowed to be in conflict. We denote in the following by $\overline{m}[p]$ the mean number of tokens in place p, the average token waiting (or residence) time in p by $\overline{r}[p]$, and by $\chi[t_1]$ the throughput of transition t_1. Applying Little's Law (The mean number of customers in a subnet is equal to the mean interarrival rate into it multiplied by the mean delay to traverse it) to such a subnet leads to

$$\overline{m}[p] = (\mathbf{Pre}[p, \cdot] \cdot \chi)\, \overline{r}[p] \tag{10.11}$$
$$= \mathbf{Pre}[p, t_1]\, \chi[t_1]\, \overline{r}[p] \tag{10.12}$$

The equation also holds for immediate transitions, because then both the firing time (and thus the token waiting time) as well as the marking is zero.

$$\overline{m}[p] = \overline{r}[p] = 0 \tag{10.13}$$

It is obvious that the mean token waiting time $\overline{r}[p]$ in place p is at least as long as the mean service time $\overline{s}$ of the output transition t_1 (which is given by its firing delay specification Λ).

$$\overline{r}[p] \geq \overline{s}[t_1] \tag{10.14}$$

which can be inserted into (10.12) leading to

$$\overline{m}[p] \geq \mathbf{Pre}[p, t_1]\, \chi[t_1]\, \overline{s}[t_1] \tag{10.15}$$

$\mathbf{Pre}[p, t_j] = 0$ for $j \neq 1$ holds because there are no conflicting timed transitions. Thus

$$\overline{m}[p] \geq \sum_{j=1}^{m} \mathbf{Pre}[p, t_j]\, \chi[t_j]\, \overline{s}[t_j] \tag{10.16}$$

Multiplication with the average interfiring time (the inverse of the throughput) $\Gamma[t_1] = 1/\chi[t_1]$ leads to

$$\Gamma[t_1]\, \overline{m}[p] \geq \sum_{j=1}^{m} \mathbf{Pre}[p, t_j]\, \Gamma[t_1]\, \chi[t_j]\, \overline{s}[t_j] \tag{10.17}$$

and since $\Gamma[t_1]\,\chi[t_j] = \frac{\chi[t_j]}{\chi[t_1]} = \mathbf{v}^{(1)}[t_j]$,

$$\Gamma[t_1]\,\overline{\mathbf{m}}[p] \geq \sum_{j=1}^{m} \mathbf{Pre}[p, t_j]\,\mathbf{v}^{(1)}[t_j]\,\overline{s}[t_j] \tag{10.18}$$

We define **average service demands** $\overline{\mathbf{D}}^{(1)}$ relative to transition t_1 as

$$\overline{\mathbf{D}}^{(1)}[t_i] = \mathbf{v}^{(1)}[t_i]\,\overline{s}[t_i] \tag{10.19}$$

By multiplying the visit ratio with the service demand of a transition, a measure of the relative workload due to the transition corresponds to the service demand. It should be noted that because of the dependence of the visit ratios of the arbitrary selection of transition t_1, the service demand vector also depends on the selection of t_1.

Service times and service demands for the transitions in the example of Fig. 10.2 are $\overline{s} = (2, 0, 0, 4, 3)$ and $\overline{\mathbf{D}}^{(1)} = (2, 0, 0, 0.6, 2.1)$.

The service demands simplify (10.18) in matrix notation together for the whole net to

$$\Gamma[t_1]\,\overline{\mathbf{m}} \geq \mathbf{Pre} \cdot \overline{\mathbf{D}}^{(1)} \tag{10.20}$$

It can easily be checked that this equation holds for the case of immediate output transitions as well because of

$$\overline{\mathbf{m}}[p] = \mathbf{Pre}[p, \cdot] \cdot \overline{\mathbf{D}}^{(1)} = 0 \tag{10.21}$$

Following the definition and properties of P-semiflows [73] we know that $\mathbf{y} \cdot \mathbf{m} = \mathbf{y} \cdot \mathbf{m}_0$ for all markings $\mathbf{m}$ of the Petri net reachable from the initial marking $\mathbf{m}_0$. It can be concluded that for the average marking $\mathbf{y} \cdot \overline{\mathbf{m}} = \mathbf{y} \cdot \mathbf{m}_0$ holds. Using this together with (10.20) results in a lower bound for the average interfiring time of transition t_1:

$$\Gamma[t_1] \geq \max_{\mathbf{y} \in \{\text{P-semiflows}\}} \frac{\mathbf{y} \cdot \mathbf{Pre} \cdot \overline{\mathbf{D}}^{(1)}}{\mathbf{y} \cdot \mathbf{m}_0} \tag{10.22}$$

The search for a lower bound in (10.22) can be formulated as a fractional programming problem:

$$\Gamma[t_1] = \text{maximum } \frac{\mathbf{y} \cdot \mathbf{Pre} \cdot \overline{\mathbf{D}}^{(1)}}{\mathbf{y} \cdot \mathbf{m}_0}$$
$$\text{subject to } \mathbf{y} \cdot \mathbf{C} = 0 \tag{10.23}$$
$$\mathbf{1} \cdot \mathbf{y} > 0$$
$$\mathbf{y} \geq 0$$

which can be rewritten because for life systems $\mathbf{y} \cdot \mathbf{m}_0 > 0$ holds:

$$\Gamma[t_1] = \text{maximum } \mathbf{y} \cdot \mathbf{Pre} \cdot \overline{\mathbf{D}}^{(1)}$$
$$\text{subject to } \mathbf{y} \cdot \mathbf{C} = 0 \tag{10.24}$$
$$\mathbf{y} \cdot \mathbf{m}_0 = 1$$
$$\mathbf{y} \geq 0$$

where $\mathbf{y}$ is a P-semiflow and $\mathbf{C}$ denotes the token flow matrix. Equation (10.24) constitutes a standard linear programming problem (LPP), which can be solved efficiently in practice. An interpretation of the LPP is to search for the "slowest subsystem" among the ones defined by P-semiflows in isolation, similar to a bottleneck analysis.

Throughput upper bounds χ_+ can now be computed from the mean inter-firing time $\Gamma[t_1]$ of t_1

$$\chi_+[t_1] = \frac{1}{\Gamma[t_1]} \tag{10.25}$$

The bounds for all other transitions are directly calculated from the result and the relative corresponding visit ratios.

$$\chi_+[t_i] = \chi_+[t_1]\,\mathbf{v}^{(1)}[t_i] \tag{10.26}$$

The values computed for the example are shown and compared in Table 10.1 in the following subsection.

A pessimistic upper bound for the average interfiring rate of transition t_1 is computed by assuming that the worst case for firing this transition again is after having fired all other transitions the number of times that their visit ratio specifies:

$$\Gamma[t_1] \leq \sum_{t \in T} \mathbf{v}^{(1)}[t]\,\bar{s}[t] = \sum_{t \in T} \overline{\mathbf{D}}^{(1)}[t] \tag{10.27}$$

which leads to obvious lower bounds χ_- for the transition throughputs:

$$\chi_-[t_1] = \frac{1}{\sum_{t \in T} \overline{\mathbf{D}}^{(1)}[t]} \tag{10.28}$$

$$\chi_-[t_j] = \chi_-[t_1]\,\mathbf{v}^{(1)}[t_j] \tag{10.29}$$

Upper and lower bounds of transition throughputs are used to estimate the throughput values $\chi_\sim$ of the transitions as a part of the profit function estimation as shown in the subsequent section. Mean number of tokens $\overline{\mathbf{m}}$ in places need to be estimated as well, based on upper $\mathbf{m}_+$ and lower bounds $\mathbf{m}_-$ of the mean markings as described in the following.

Recall that for the mean interfiring times holds

$$\Gamma[t_1]\,\overline{\mathbf{m}} \geq \mathbf{Pre} \cdot \overline{\mathbf{D}}^{(1)} \quad \text{and} \quad \chi[t_1] = \frac{1}{\Gamma[t_1]} \tag{10.30}$$

which is also true if we exchange any transition t_i for t_1. The mean marking $\overline{\mathbf{m}}$ is thus constrained by

$$\forall t_i \in T : \overline{\mathbf{m}} \geq \mathbf{Pre} \cdot \overline{\mathbf{D}}^{(i)} \cdot \chi_-[t_i] \tag{10.31}$$

and a lower bound $\mathbf{m}_-$ can be obtained by using the maximum

$$\mathbf{m}_- = \max_{t_i \in T}\left(\mathbf{Pre} \cdot \overline{\mathbf{D}}^{(i)} \cdot \chi_-[t_i]\right) \tag{10.32}$$

These lower bounds may be heuristically improved by substituting the estimated throughput $\chi_{\simeq}$ for the lower throughput bound χ_- in the formula as explained in the subsequent section.

The next step is the determination of an upper bound of the mean marking. Consider a P-semiflow $\mathbf{y}$ whose support includes place p_i. Then

$$\mathbf{y}^T \cdot \mathbf{m}_0 = \mathbf{y}^T \cdot \overline{\mathbf{m}} \tag{10.33}$$

and thus

$$\mathbf{y}^T \cdot \mathbf{m}_0 \geq \mathbf{y}_i \cdot \overline{\mathbf{m}}[p_i] + \sum_{i \neq j} \mathbf{y}_j \cdot \mathbf{m}_-[p_j] \tag{10.34}$$

$$\mathbf{y}_i \cdot \overline{\mathbf{m}}[p_i] \leq \mathbf{y}^T \cdot \mathbf{m}_0 - \mathbf{y}^T \cdot \mathbf{m}_- + \mathbf{y}_i \cdot \mathbf{m}_-[p_i] \tag{10.35}$$

$$\overline{\mathbf{m}}[p_i] \leq \mathbf{m}_-[p_i] + \frac{1}{\mathbf{y}_i} \mathbf{y}^T \cdot (\mathbf{m}_0 - \mathbf{m}_-) \tag{10.36}$$

An upper bound $\mathbf{m}_+$ for the mean number of markings $\overline{\mathbf{m}}$ can thus be computed for a place p_i as

$$
\begin{aligned}
\mathbf{m}_+[p_i] = \text{minimum} \quad & \mathbf{m}_-[p_i] + \mathbf{y} \cdot (\mathbf{m}_0 - \mathbf{m}_-) \\
\text{subject to} \quad & \mathbf{y} \cdot \mathbf{C} = 0 \\
& \mathbf{y} \cdot \mathbf{e}_i = 1 \\
& \mathbf{y} \geq 0 \\
\text{with} \qquad & \mathbf{e}_i[k] = \begin{cases} 1 & \text{for } i = k \\ 0 & \text{otherwise} \end{cases}
\end{aligned}
\tag{10.37}
$$

Upper and lower marking bounds for the example are shown in Table 10.2 below.

A possible improvement of the marking upper bound can be found in the special case of transitions t_j with only one input place p_i. Let K denote the multiplicity of the arc going from place p_i to t_j. Little's Law ensures that the mean marking $\overline{\mathbf{m}}[p_i]$ of place p_i is equal to the flow of tokens into the place times the token residence time, i.e., the delay of transition t_j in our case (infinite server firing semantic). The flow of tokens is balanced in steady-state, which means that the number of tokens flowing into the place must be equal to the number of tokens flowing out of it. The token flow out of the place obviously equals the arc multiplicity K times the throughput $\chi[t_j]$ of transition t_j. Thus

$$\overline{\mathbf{m}}[p_i] = K \chi[t_j] \overline{s}[t_j] \tag{10.38}$$

$$\mathbf{m}_+[p_i] \leq K \chi_+[t_j] \overline{s}[t_j] + (K - 1) \tag{10.39}$$

$$\text{with } K = \mathbf{Pre}(p_i, t_j)$$

because $(K - 1)$ tokens do not enable transition t_j, while K might.

10.2.3 Approximate Derivation of Profit Values

It is clear that $\chi_- \leq \chi \leq \chi_+$ and $\mathbf{m}_- \leq \overline{\mathbf{m}} \leq \mathbf{m}_+$,[3] and the approximated values $\chi_\simeq$ and $\mathbf{m}_\simeq$ should of course be within the range of their bounds as well. The actual values for throughput and mean number of tokens are estimated from the bounds after the latter have been computed using the methods described in the previous section. The following weighted sums of upper and lower bounds are therefore used.

$$\chi_\simeq = \alpha\,\chi_+ + (1-\alpha)\,\chi_- \qquad (10.40)$$
$$\mathbf{m}_\simeq = \beta\,\mathbf{m}_+ + (1-\beta)\,\mathbf{m}_- \qquad (10.41)$$

Experiments with numerous examples showed that the throughput upper bound is much better (nearer to the actual value) in most cases. This is, however, not surprising from the "trivial" formula of the throughput lower bounds. Hence the value of α has been set to 0.9, which often results in a reasonable approximation of the throughput.

For the examples considered so far, the marking bounds were in general not very close to the actual values. No observation could be made whether the upper or lower bound is systematically better. Therefore, β has been chosen as 0.5, resulting in an equal importance of marking lower and upper bound.

It has already been stated that the lower bounds for the transition throughputs are usually not very exact. The marking lower bounds are computed using them and the marking upper bounds depend on the marking lower bounds. Hence, the marking bounds are also not very tight in most cases. However, we are mainly interested in a good approximation of the throughput and mean marking. An approximation of the marking bounds can then be computed by assuming that $\chi_\simeq$ (as computed in (10.40)) equals the correct throughput values. The bounds on the throughput that are being used in (10.32) and (10.39) can then be substituted by $\chi_\simeq$.

The mean marking approximations are not guaranteed to be within their theoretical bounds after this change. If for instance values of $\chi_\simeq$ are bigger than the actual throughput values, this technique leads to an overestimation of the lower marking bound and an underestimation of the marking upper bound. In extreme cases no result might be found for the LPP (10.32) because the system of equations is contradictory. The implemented algorithm detects this case, adjusts the α value, and retries from the point where the first approximation was adopted. By making α smaller, the approximated throughputs are not bigger than the actual ones at some point, and the problem will be solved. This kind of problem did, however, not occur in any of the examples that have been analyzed so far.

Applying the formulas to the example in Fig. 10.2 leads to a vector of upper and lower bounds of transition throughputs. The obtained values are

---

[3] If the heuristic improvement of the throughput approximation is not used, see further.

Table 10.1. Approximation of throughput values for the example

	Transition				
	T1	T2	T3	T4	T5
Lower Bound	0.213	0.064	0.149	0.149	0.032
Upper Bound	0.566	0.170	0.396	0.396	0.085
Approximation	0.531	0.159	0.371	0.080	0.371
Real Value	0.465	0.140	0.326	0.070	0.326

Table 10.2. Approximation of mean marking values for the example

	Place			
	P1	P2	P3	P4
Lower Bound	0.425	0.000	0.255	0.447
Upper Bound	2.298	1.872	2.128	2.319
Improved Upper Bound	1.132	0.000	1.679	1.188
Approximation	0.779	0.000	0.967	0.818
Real Value	0.931	0.000	1.090	0.978

shown in Table 10.1 together with the approximated value, compared to the real value that was computed by numerical analysis. The throughput approximation works very well for the example, the relative error is in all cases only about 14%.

The same result analysis was done for the marking approximation of the example. Table 10.2 shows the results. The quality of the approximation is good, which among others is due to the improvement after (10.39) (see row Improved Upper Bound). Relative errors range between 11 and 16% except for the perfect value for place P2. The good approximation quality is based on the very simple net structure, which keeps the approximation error small.

The approximation error is typically bigger for models with a high influence of synchronization, because waiting times are introduced, which are not captured in the bounds equations.

Algorithm 10.2 COMPUTEAPPROXIMATION gives an overview of the steps that are carried out for an approximate computation of a cost function value for a stochastic Petri net. Equation numbers that describe the basis of the individual steps are shown in brackets.

The profit value as defined by a profit function can be computed based on the estimated mean marking and throughput values. Section 13.2 shows for the exemplary application area of manufacturing systems how typical profit function elements can be mapped on a Petri net performance measure. The mathematical terms of these profit functions can be approximated as shown above. The approximation quality of the overall profit function value might suffer in some cases from the different errors in the individually estimated elements. The quality of the approximations is analyzed in Sect. 13.6 for an application

---

COMPUTEAPPROXIMATION (SPN)

---

derive $\mathbf{C} = \mathbf{Post} - \mathbf{Pre}$ and P-semiflows $\mathbf{y}$
derive routing equations r_i for conflicting transitions (10.8)
derive visit ratios $\mathbf{v}^{(1)}[\cdot]$ (10.10)
compute service demands $\overline{\mathbf{D}}^{(1)}[t_i]$ (10.19)
solve LPP (10.24)
derive transition throughput upper bounds for t_i (10.25), (10.26)
compute transition througput lower bounds (10.28), (10.29)
obtain place marking lower bounds (10.32)
for $\forall p_i \in P$ **do**
    solve LPP (10.37) for marking upper bound
    improve place upper bound if possible (10.39)
compute throughput approximation (10.40)
compute marking approximation (10.41)
return result approximation using throughput and marking results

---

Algorithm 10.2: Computation of approximate cost function value

example, and shows that a sufficiently good estimate for a promising region in the optimization parameter space can be efficiently detected with the presented technique.

It should be noted that the approximation error of the profit value is not important for the two-phase method as long as the multidimensional shape of the real profit function is sufficiently well estimated. Let $\mathbf{X}$ denote the n-dimensional solution space of all possible decision variable values $\mathbf{x} \in \mathbf{X}$ of an optimization problem with the given constraints. Any approximation function $Profit_\sim(\mathbf{x})$ for the real profit $Profit(\mathbf{x})$ would be perfect for our purposes if it satisfied

$$\forall \mathbf{x}_1, \mathbf{x}_2 \in \mathbf{X} : \mathrm{Profit}_\sim(\mathbf{x}_1) > \mathrm{Profit}_\sim(\mathbf{x}_2) \iff \mathrm{Profit}(\mathbf{x}_1) > \mathrm{Profit}(\mathbf{x}_2)$$

In that case the optimization running on the approximated values will come to the same optimal parameter set, although probably not with the real profit value.

Notes

The chapter is an extended presentation of joint work published in [351–353]. Additional heuristics for the further improvement of the two-phase optimization methodology have been reported in Rodriguez et al. [281]. Among them are temperature reduction schemes for the second phase that exploit the intermediate results of the first phase. Another proposal uses the two-phase optimization technique together with a simulation-based preoptimization phase, and could thus be directly extended to any SDES.

The method used to obtain the different performance measures involved in the objective function is based on the results obtained by Campos et al. [15, 37–40, 49, 261]. The computational effort is only linear in the size of the net structure, and efficient solvers for linear programming problems are freely available.

11

Model-Based Direct Control

After a technical system has been modeled, its performance evaluated and possibly improved, the final step is to bring the specified behavior into reality. The model itself needs to be interpreted in order to create an interface to the environment. Only one model should be used throughout the whole design process. It is easily possible to integrate control rules in an SDES model. Afterwards, the influence on the system behavior (liveness, performance) can be analyzed. The behavior specified in the model is directly executed finally.

The control interpretation proposed in the following associates control (output) and sensor (input) signals to transitions, which are exchanged between model and "outside world." In the applications considered so far, this approach turned out to be very natural in the Petri net understanding, and did not require the exchange of state information. It is, however, easy to define state inputs and outputs as well, as it is done in Sect. 11.2 for general SDES models. For a state output, the most general way is similar to the definition of a marking-dependent boolean expression, the result for which is available to the environment. State inputs can be used in guard functions of transitions.

Many control-related publications explicitly distinguish between the controlled system (often called *plant* or *process*) and the *controller* itself. The plant model describes the uncontrolled behavior, i.e., all possible states and state changes. The controller model is connected to the plant model such that the intended behavior is achieved. This is often only defined as a set of forbidden states. There is a significant amount of literature about automatically finding such a controller, which should have some properties like deadlock-free operation and smallest possible restriction of behavior [100, 152]. With the control interpretation described below it is also possible to describe plant and controller independently, and to connect both models to check for the results. In a strict sense, input and output signals may then only be connected to the plant part of the model. The generation of controller models is beyond the scope of this work.

The chapter is structured as follows. A control interpretation for Petri nets is explained using an example in Sect. 11.1. The transformation of the general idea for abstract SDES models is covered in Sect. 11.2, and some notes are given about the influence of the control interpretation on the dynamic behavior of the model. Related work is pointed at in the notes at the end of the chapter. The control method is applied to a more complex example in Sect. 16.4, and some implementation details are given in Sect. 12.1.

11.1 A Control Interpretation for Petri Nets

The discussion below applies to simple and colored Petri nets in principle. It has been implemented for vfSCPN models in the TimeNET tool (compare Sect. 12.1). The way of interpretation could, however, be easily transferred to any kind of Petri net or even a general SDES model, see Sect. 11.2.

To allow the control of a real-world process using a Petri net model, possibilities for its interaction with the outside world have to be added to the otherwise autonomous model. From the model's point of view, input and output signals are necessary. They should be added in a simple way, naturally following the meaning of the model elements. This is possible without problems if the modeler follows the real structure of the system in the modeling process.

Only "active resources" like machines, transport facilities, etc. are controllable. Their activities are modeled by transitions, who can either move tokens (transport) or change token attributes in a colored model (processing). A transition becomes enabled when its guard function evaluates to true, the necessary input tokens are available, and enough space for added tokens is free in the output places (if they have a restricted capacity). This is the point of time at which the firing time of the transition starts to run if the dynamic behavior is evaluated. The actual firing with the corresponding marking change takes place when the firing delay has elapsed. It then appears to be natural to assign controllable activities to transitions. When the transition becomes enabled, a **control signal** (or output signal) is sent from the model to the technical process (e.g., a motor is switched on). After termination of the activity (e.g., a sensor detects the stop position), a **sensor signal** (or input signal) from the process is sent to the model, which then initiates the instantaneous transition firing. This model interpretation only changes the model behavior by adopting the unknown delays of external activities. Therefore, results of the qualitative analysis still hold, and quantitative results should be comparable if the actual delays are similar to the modeled ones.

Transitions with associated input signals are called **external**, all others **internal**. Associating control signal(s) to a transition does not change its firing semantic in the model – a signal is sent to the outside world if the transition becomes enabled, which does not influence the model itself. Opposed to that, external transitions fire if and only if they are enabled *and* receive their sensor signal. The firing delay specified in the model for an external transition is thus ignored during the online control. The modeler should be careful with cases in which external transitions can be disabled by the firing of other transitions, because their input signal could then be lost. Conflicts of this type can be automatically detected from the model structure. However, they are not generally forbidden, because there are cases in which this behavior is useful.

It is possible to assign any number of input and output signals to a transition. All control signals are being sent when the transition becomes enabled, while the arrival of any one of the sensor signals triggers the firing of the transition. Transitions with output signals but without input signals (or vice versa) are allowed as special cases. An example of an activity that can be finished at any time without having been started before is the failure of a machine or the arrival of a customer. A sensor that detects this event can trigger the firing of an associated transition in the model. The firing time of internal transitions keeps its semantics from the autonomous model. Timing issues like delays or deadline violation detection can thus be easily achieved.

11.1.1 An Example

The control interpretation is explained using a ficticious example. Imagine a street crossing over a waterway on a drawbridge, as it is sketched in Fig. 11.1. For simplicity we assume a one-way street; vehicles approach from the left, may be stopped at a traffic signal, and cross the bridge if the signal is off. The drawbridge should work as follows: after 1 h of road traffic, the bridge is opened for boats for 10 min. There is no dependency on the amount of traffic or waiting boats to simplify the model. We furthermore assume that boats just wait in front of the drawbridge and pass under it, always finishing within the 10-min interval. Vehicle sensors are installed at both driveways to ensure that the bridge is opened only if it is empty. The red signal may be switched

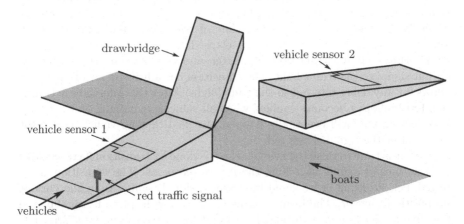

Fig. 11.1. Drawbridge example

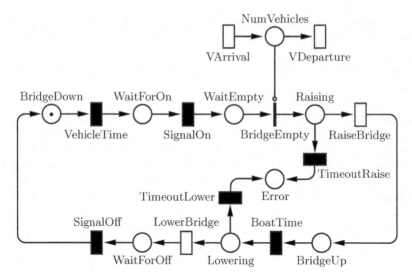

Fig. 11.2. Petri net model for the control interpretation example

on and off, and the drawbridge can be controlled by setting the motor to raise or lower it. There is a stop-position sensor at the up and down positions.

Figure 11.2 depicts a possible simple Petri net model solution to the control task of the drawbridge example. The top part of the model describes the arrival and departure of vehicles with timed transitions VArrival and VDeparture, respectively. Both are external transitions and fire only when the corresponding sensor signal is received from the real application. Provided that the sensors work properly, the number of tokens in place NumVehicles always equals the number of vehicles between the sensors.

The lower part of the model describes the bridge states and control actions. In the figure the initial marking has one token in place BridgeDown, which is the case for 1 h (the firing time of transition VehicleTime) to allow road traffic. After that time has elapsed, the token is moved to place WaitForOn, enabling transition SignalOn, and thus sending the associated output signal to the traffic light. The transition has a firing delay of 1 s, which allows for any delay internal to the traffic light. After that, immediate transition BridgeEmpty becomes enabled when all vehicles remaining on the bridge have left it. This will eventually happen provided that no vehicle enters the bridge after the signal shows red.

Now it is time to raise the bridge, which is done by transition RaiseBridge with its output signal, switching the bridge motor on. The transition is external, i.e., its firing is triggered by a sensor signal coming from the "bridge up position" sensor. The bridge status token is now in place BridgeUp, and the time for boats passes during the firing delay of transition BoatTime. The bridge is lowered and the red signal switched off afterwards in a similar way,

Table 11.1. Transition details of the drawbridge control model

Transition	Delay	Control output	Sensor input
VArrival	External	–	Vehicle sensor 1
VDeparture	External	–	Vehicle sensor 2
VehicleTime	1 h	–	–
SignalOn	1 s	Red light on	–
BridgeEmpty	0	–	–
RaiseBridge	External	Raise bridge	Bridge up position
BoatTime	10 min	–	–
LowerBridge	External	Lower bridge	Bridge down position
SignalOff	1 s	Red light off	–
TimeoutRaise	5 min	–	–
TimeoutLower	5 min	–	–

arriving back at the initial marking from where the behavior starts again as described.

For the case of a motor failure, the actions of raising and lowering the bridge are monitored if they finish in time. Transitions `TimeoutRaise` and `TimeoutLower` have a fixed delay of 5 min, which is assumed as a safe upper boundary on any correct bridge movement. If the movement did not come to an end during that time, i.e., the corresponding sensor signal did not arrive, the timeout transition fires and puts the token into place `Error`. This stops the bridge operation and could be used to signal the error state to a supervisor.

Table 11.1 summarizes transition details of the example. Every possible combination of input and output signal attachments to transitions was used in the example, as well as timeout supervision transitions.

11.2 Model-Based Control of SDES

In the general context of an abstract SDES model, the control interpretation explained above for Petri nets can be applied as well. In the following the interface between model and the "real world" environment is explained. Input and output for SDES models is defined and its impact on the dynamic behavior sketched afterwards.

There are four ways of possible information exchange between model and environment: instantaneous signals and piecewise constant states, both for the input and output direction. Because of the similarity between Petri net transitions and SDES actions (and place markings and states), input and output control signals are naturally associated with actions and their modes. When an action (mode) becomes enabled, the associated output signal is sent to the controlled system (if there is any). The sets of all **output signals** of a model is denoted by $Signals_{Out}$ and **input signals** by $Signals_{In}$. To every possible action variant $v = (a, mode) \in AV$, there is an associated subset of output signals $Trigger(v) \subseteq Signals_{Out}$, which are the ones that are activated when that action variant becomes enabled. This subset can of course be empty. The same mapping exists vice versa for input signals. There is no type or any other elaborate information associated to a signal, it just happens at certain time instants.

$$Trigger : \begin{cases} AV \to 2^{Signals_{Out}} & \text{associated output signals} \\ Signals_{In} \to 2^{AV} & \text{associated input signals} \end{cases}$$

State information is exchanged between model and environment via **state inputs** $States_{In}$ and **state outputs** $States_{Out}$. State inputs and outputs are not attached to any individual model element as signals are. They always have some value $Value(\cdot)$ depending on the current state $\sigma \in \Sigma$ of the model. Although in many cases a boolean type will be sufficient, we do not restrict the state information. The value of a state output $o \in States_{Out}$ is given by the evaluation of the associated function $Value(o, \sigma)$ in a state σ as specified in the SDES model, for which the sort must be fixed. We extend the SDES sort function accordingly.

$$\forall o \in States_{Out}, \sigma \in \Sigma : Value(o, \sigma) \in S^{\star}(o)$$

The value of the state inputs depends on the environment state in a way that is unknown inside the model. We only require the set of possible values to restrict to some predefined sort, which must be the same for all observation times t.

$$\forall i \in States_{In} : Value(i, t) \in S^{\star}(i)$$

An action variant $v \in AV$ is called **external** if there is at least one triggering input signal, i.e., if $v \in \bigcup_{i \in Signals_{In}} Trigger(i)$. An enabled action variant (i.e., activity) is executed when it is triggered by such a signal. This extension of the Petri net concept described in the previous section to models with different action modes is not problematic, because in the simple standard case a controllable entity may only be active under one mode and should therefore be modeled with single server semantics. This ensures that there will be only one activity at a time, resulting in a clear correspondence between controlled resource and model activity. There are, however, cases in which a controlled resource needs to be modeled with infinite server semantics, think, e.g., of a machine that executes different processing steps depending on the type of workpiece.

The definition of an SDES model that should be used to control an environment thus needs to be extended as follows.

- The interface between model and environment needs to be specified first. The sets of input and output signals and states ($Signals_{Out}$, $Signals_{In}$, $States_{Out}$, $States_{In}$) have to be defined.

-   Mappings between SDES model elements and interface parts have to be set in a second step. State output functions $Value(\cdot, \cdot)$ map model states to state outputs. Interactions between model actions and input/output signals are specified using $Signals_{Out}$ and $Signals_{In}$.

For the Petri net example described in Sect. 11.1, no state information was exchanged, leaving the corresponding sets empty $States_{Out} = States_{In} = \emptyset$. The association of transitions to output signals and of input signals to triggered transitions is listed in Table 11.1.

11.3 Behavior of a Control-Interpreted SDES

The dynamic behavior of an SDES model as defined in Sect. 2.3.2 is influenced by a control interpretation. The behavior of the environment adds to the previously independent model. Only the interface can be taken into account for the behavior, because the environment is not directly visible from the model. If we restrict the environment to a stochastic discrete event system,[1] it can be understood as a model part with its own internal state and functions that govern its state transitions as well as the emitted state information and signals.

A full definition of the dynamic behavior of a control-interpreted SDES is avoided here. In short, one possibility is to set the remaining activity delay of external action variants to infinity when they become enabled. Execution of such an activity then takes place only at the point in time when one of the associated input signals arrives from the environment.

The main problem for the definition of the underlying stochastic process is that events may happen in the environment that are thus unknown to the model. In a real control application, the actual time of a signal from the environment can not be known in advance. It thus makes no sense to use remaining activity delays and the minimum of them to select the first activity to be executed. The sojourn time in a state needs to be changed such that it accounts for the first signal that arrives from the environment before the current minimum RAD might have elapsed. Input state changes need to be taken care of as well. This requires to insert an additional state of the discrete-parameter stochastic process $CProc$ every time that an input signal arrives or one of the input states changes its value. All elements of the interface would therefore need to be included in the process definition.

The way of handling output signals and state output should be obvious from the definition of the dynamic behavior, especially because they do not change the behavior directly.

---

[1] As far as it can be seen through the interface, "invisible" continuous states are allowed.

Notes

A theory of discrete event system control based on automata and formal languages was proposed in Ramadge and Wonham [273], see [274] for a survey.

Petri nets as an example of SDES have often been considered for the control of manufacturing systems [166]. The various approaches to interpret a Petri net model for control use different methods to exchange information between model and environment. The use of places to output state information to the controlled environment and transition enabling based on environment information has been proposed in Silva [290]. A similar approach is taken in GRAFCET [79,80] and related methods. The enabling of transitions that depend on external information can also be done with control places [167]; the models are then called controlled Petri nets. Another possibility is the association of control procedures with transitions [234] or places [192]. However, the internal behavior of the associated program parts can not be analyzed using the model. An in-depth analysis of the different approaches and references is contained in Zhou and Venkatesh [336], and a good introduction can be found in Cassandras and Lafortune [42].

Colored Petri nets are used for the control of a manufacturing system, e.g., in [192, 233]. In Feldmann [101], transitions modeling processing steps are hierarchically refined and input/output signals are associated with the subtransitions. Generally, control design based on a Petri net model is advantageous with respect to a state machine description, because they are able to capture parallelism in a much clearer way.

Places are proposed for state output and transitions for state input in other work on control-interpreted Petri nets [214]. With the extension of state input and output as described in Sect. 11.2, the exchange of information between model and environment described here is similar to the interface understanding of *net condition/event systems* [297].

This chapter is partially based on work previously published in [342, 344, 356]. The application of the proposed control interpretation to an example is described in Sect. 16.4. Its implementation in the software tool TimeNET is presented in Zimmermann and Freiheit [342] and briefly covered by Sect. 12.1.

12

Software Tool Support

Modeling and evaluation of nontrivial technical systems are only possible in practice with the support of appropriate software tools. A graphical user interface should be provided to efficiently enter the graphical representations of discrete event system models. The task of improving a model until it behaves as expected can be supported by qualitative checks on the model as well as by an interactive simulation or visualization of the behavior. Finally, evaluation algorithms and other aspects such as code generation or direct control have to be implemented. Modular algorithms and code reuse might become more important in the future, especially with the *Petri net markup language* (PNML, [27]) exchange format. Abstract model descriptions like SDES of this text or the one used in the Möbius [81,82] software tool have the potential for an integration of models and tools. Until then, however, most research groups and obviously commercial companies design and implement a complete tool individually, usually only for one specific model class.

This chapter briefly covers the software tool aspect in the context of the SDES modeling environment. An overview of selected existing software tools is given in Sect. 12.2. The subsequent Sect. 12.1 describes TimeNET, a software tool that is being designed and implemented in the group of the author. The techniques described in Part II of this text have been implemented in TimeNET, and it has been used for the application examples presented in Part III.

12.1 TimeNET

This section presents TimeNET, a software tool for the modeling and performability evaluation using stochastic Petri nets. The tool has been designed especially for models with nonexponentially distributed firing delays. TimeNET has been successfully applied during several modeling and performance evaluation projects and was distributed to more than 300 universities and other organizations worldwide at the time of this writing. Its functions are being continuously enhanced with the inclusion of results from Ph.D. theses at the modeling and performance evaluation group of TU Berlin. Most of the implementation work is done by master students.

The development of TimeNET started around 1994. It was based on an earlier implementation of DSPNexpress [227–229], which was also performed at Technische Universität Berlin. It contained all analysis components of the latter at that time, but supports the specification and evaluation of extended deterministic and stochastic Petri nets (eDSPNs). The graphical user interface was based on the X Athena Widget toolkit, and not easily adaptable to extensions. It was completely rewritten in 1997 using the Motif toolkit to capture several modeling environments within the same tool. Corresponding extensions included variable-free colored Petri nets, fluid stochastic Petri nets, discrete-time stochastic Petri nets, and modular blocks of SPNs, which were added later on. Other extensions deal with specialized analysis algorithms for existing model classes. A list of model classes and available evaluation algorithms is given in Sect. 12.1.1.

Previous publications describing the tool TimeNET include [124, 130, 131, 197, 199, 213, 345, 349, 359, 360], and several applications have been considered in [83, 85, 134, 341, 342, 344]. References to publications covering analysis algorithms and background of the tool are given in the section where appropriate.

TimeNET is in a transition phase at the time of this writing. Version 3 had been stable for some time now, while development of the next major revision took place. TimeNET 4 [349] features a new graphical user interface written in JAVA to allow platform-independent use of the tool. It is described in Sect. 12.1.3. Another significant improvement is the availability of colored stochastic Petri nets as described in Chap. 6. Not all model classes and evaluation algorithms of TimeNET 3 have been integrated into the new architecture by now. We describe the new version in the following, and point out differences where necessary.

TimeNET runs under Solaris 5.9 and Debian Linux 3, and TimeNET 4 under Windows as well. The tool is available free of charge for noncommercial use from its home page at `http://pdv.cs.tu-berlin.de/~timenet`, where a user manual [362] can be found as well.

12.1.1 Supported Net Classes and Analysis Methods

Model classes and corresponding evaluation algorithms included in TimeNET are briefly explained in the following.

The classic main model class of TimeNET are *extended deterministic and stochastic Petri nets* (eDSPNs, cf. Notes on p. 96). Firing delays of transitions can either be zero (immediate), exponentially distributed, deterministic, or belong to a class of general distributions called *expolynomial* in an eDSPN.[1] Such a distribution function can be piecewise defined by exponential polynomials and has finite support. It can even contain jumps, making

---

[1] Compare Sect. 1.4.

it possible to mix discrete and continuous components. Many known distributions (uniform, triangular, truncated exponential, finite discrete) belong to this class.

Under the restriction that all transitions with nonexponentially distributed firing times are mutually exclusive, stationary numerical analysis is possible [67, 132, 226] as described in Sect. 7.3.3. If the nonexponentially timed transitions are restricted to have deterministic firing times, transient numerical analysis is also provided [126, 161]. For the case of concurrently enabled deterministically timed transitions, an approximation component based on a generalized phase type distribution has been implemented [127]. If there are only immediate and exponentially timed transitions, the model is a GSPN and standard algorithms for steady-state and transient numerical evaluation based on an isomorphic Markov chain are applicable (cf. Sect. 7.3.2).

Structural properties of eDSPNs like extended conflict sets and invariants can be obtained with TimeNET. They are displayed and checked by the modeler to examine the correct model specification [171, 344]. For a steady-state or transient analysis of a simple stochastic Petri net model of any kind, the reachability graph is computed (cf. Sect. 7.3.1). Structural properties are exploited for an efficient generation of the reachability graph [52, 53]. Subnets of immediate transitions are evaluated in isolation, following the ideas presented in Balbo et al. [14].

The transient analysis of DSPNs is based on supplementary variables [123, 126, 161], which capture the elapsed enabling time of transitions with nonexponentially distributed firing delays. TimeNET shows the evolution of the performance measures from the initial marking up to the transient time graphically during a transient analysis.

The tool also comprises a simulation component for eDSPN models [196], which is not subject to the restriction of only one enabled non-Markovian transition per marking. Steady-state and transient simulation algorithms are available with the techniques described in Sect. 7.2. Results can be obtained faster by parallel replications [197], using control variates [193], or with the RESTART method in the presence of rare events [195, 198] as described in Sect. 9.2. During the simulation run, intermediate results of the performance measures are displayed graphically together with the confidence intervals.

The automatic optimization method described in Chap. 10 has been implemented as a prototypical extension of TimeNET, which is not part of the distributed version at the moment. More details can be found in the mentioned section as well as [281, 351–353].

Simple stochastic Petri nets in TimeNET can either be interpreted in continuous time as an eDSPN or as a *discrete deterministic and stochastic Petri net* (DDSPN [337,343]). DDSPNs allow geometric distributions, deterministic times, and discrete phase type distributions as delays. Steady-state and transient numerical analysis as well as efficient parallel simulation are available for this type of model. The notes on p. 154 give an overview of the solution methods.

A TimeNET model class capturing UML Statecharts is currently under development [307–309]. Such a model will then be translated into a stochastic Petri net for a later evaluation as described in Sect. 3.5.

Variable-free colored Petri nets vfSCPN as briefly covered in Sect. 6.5 represent another model class of TimeNET 3.0. They were initially developed for manufacturing systems [339–341, 344, 354, 355, 361], but later used for workflow systems as well [83–85]. Firing delays of transitions have the same range as in eDSPNs.

Structure and work plans of a manufacturing system[2] are modeled with slightly adapted versions of vfSCPNs. Templates from a library of common submodels can be parameterized and instantiated to ease the description of large systems. Function blocks can be used to model a system instead [350], which are later automatically translated into a Petri net. The separate models of structure and work plans are later on automatically merged to a complete model.

Qualitative analysis can be used to derive structural properties [171, 344]. Different performance evaluation techniques are available for this net class in TimeNET: numerical steady-state analysis as described in Sect. 7.3, the approximation method of Sect. 8, and standard simulation as covered in Sect. 7.2. The modeled system can be directly controlled using TimeNET [342] with the method shown in Chap. 11. A wavelet-based approach for the approximate numerical analysis of vfSCPN models with large state spaces is currently investigated.

Online control of vfSCPN models is implemented in TimeNET 3.0 as follows: during a token game with activated control, an additional process monitors the model state. If a state change enables a transition, the associated output signals are sent to the controlled process. The sensor states are checked additionally by the software process. The corresponding transition(s) are fired when a sensor state change is detected, which is associated with an input signal. This is done only if they are enabled in the model, and leads to an update of the displayed model state. Internal transitions (the ones without associated input signals) may fire independently, only depending on their associated firing time. The correspondence between transitions and input/output signals is described in a file. Data exchange between the software tool and the example application (cf. Sect. 16.1) is realized via a RS 232 serial link.

Stochastic colored Petri nets as introduced in Chap. 6 have been added recently to TimeNET version 4 [172, 331, 363]. Because of the inherent complexity of the models, a requirement of only one nonexponential transition per marking was decided to be too restrictive. Thus only simulation has been implemented for the performance evaluation of SCPN models so far, following the algorithms in Sect. 7.2. Currently the distributed simulation method introduced in Sect. 9.1 is being implemented [209–211], which allows the efficient

---

[2] Or structural and object-process related information for a model from another application area.

simulation of complex models on a cluster of workstations. Section 12.1.2 covers some details of the software architecture that became necessary for an efficient evaluation of SCPN models.

Other features of TimeNET 3.0 include fluid stochastic Petri nets (FSPNs) with steady-state and transient analysis algorithms [328,329], and SPNL models that combine SPN with concepts of modular programming languages [128].

The *token game*, i.e., an interactive simulation of the behavior of a model, is available for all main model classes in TimeNET. Enabled transitions are highlighted in the graphical user interface and are fired by clicking them with the mouse. The state is updated and shown in the window.

12.1.2 Software Architecture of TimeNET 4.0

Software development for TimeNET is usually done by students as part of their diploma or Ph.D. theses. It is thus of high importance to keep all analysis components modular with well-defined interfaces. The overall tool architecture as well as the graphical user interface have to be extendable and adaptable to new net classes and analysis algorithms. A major goal in the development towards TimeNET 4 12.1.2 is platform-independency, in order to allow Windows user's access to the tool. It has thus been completely rewritten in JAVA; more details are given in Sect. 12.1.3. Efficient computations are the main goal for analysis algorithms, which are therefore implemented in C++.

The remainder of this section briefly describes the software modules involved in a sequential simulation of SCPN models (cf. Chap. 6 and Sect. 7.2) as well as the optimization of SPN models (Chaps. 10 and 5) as examples for the software architecture. Other parts of TimeNET's software architecture are explained in some of the publications that have been mentioned at the beginning of Sect. 12.1.

The main constituents of TimeNET are the graphical user interface (GUI) and analysis algorithms. The latter are usually started as background processes from the GUI, but can be run from the command line prompt as well. Data exchange between GUI and analysis algorithms is mainly done with data files, while sockets are used between processes for efficiency.

Figure 12.1 shows the interaction between programs for the simulation of stochastic colored Petri nets. The model is edited with the GUI according to the SCPN model class. Models as well as model class descriptions are stored in XML format. The tool architecture allows to run the graphical user interface on a client desktop PC, while the computationally expensive simulations run on a remote server. Both parts may reside on the same host as well. This aspect is implemented with some kind of a simple middle-ware, which is denoted as *remote system* in the figure. It allows to start and stop programs, transfer input and output data, and other functions independently of whether the interacting programs are located on machines running Unix or Windows.

The program code implementing the SCPN transition's functionality is generated at the beginning of a simulation run by an automatic code generation

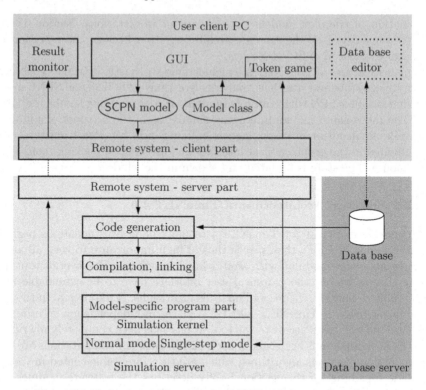

Fig. 12.1. Software architecture for SCPN modeling and evaluation

module. This takes the model as input and writes functions for each transition that has been changed since the last simulation. By doing so, efficiency of the later simulation program increases substantially compared to a standard simulation that "interpretes" the model parts during the run. The resulting code is then compiled forming a model-specific program, which is linked to a simulation kernel to generate the actual simulation program. All this is done on the remote simulation server. Because of the complexity and required adaptability of SCPN models in a research project [331, 363], a database can optionally be combined with the tool to store model information. A SCPN model can thus be parameterized using data from the database. The interface between model generation and database follows the ODBC standard, thus allowing the majority of modern data base systems to be connected to it.

The simulation program has two working modes: normal simulation, which is intended for an efficient computation of performance measures, and a single step mode used in conjunction with the GUI for an interactive visualization of the behavior (token game). Results of the simulation run are graphically displayed on the client PC during the simulation in a result monitor program (see Sect. 12.1.3). They are also stored in files that can be analyzed after a completed simulation.

The implementation of the optimization technique described in Chap. 10 as a module of TimeNET is briefly explained now. It combines the user interface of TimeNET for the specification of a model, an evaluation algorithm like simulation, and a software package implementing the simulated annealing method. We decided to use the ASA (adaptive simulated annealing [173,174]) tool for the latter, which can be used for optimizing multivariate nonlinear complex systems. It is especially useful for adaptive global optimization of complex stochastic systems. Another advantage is its adaptability to new applications by programming a cost function and specifying some additional parameters.

Figure 12.2 sketches the interaction of the different program parts. Boxes depict actions, ellipses data, and arcs show control and data flow. Dotted lines correspond to steps of the second optimization step in contrast to solid lines for the preoptimization. The box containing the interface procedure implements the main algorithm for the interaction between ASA and TimeNET. Other than that, only the initialization routines of ASA had to be reprogrammed for the combined optimizer. Every result is stored in a cache-like table together with its corresponding parameter set to avoid recomputations.

Before an optimization can be started, the model to be optimized and some parameters like the decision variables and their search ranges have to be specified. The model contains the definition of a cost or profit function as a performance measure. During the simulated annealing algorithm a quantitative evaluation algorithm for the underlying model is called with a parameter set, for which the resulting profit function value is computed. In the next step it is tested whether convergence is reached and if so the algorithm exits with the final optimization result. A new parameter set is generated otherwise and a subsequent iteration begins.

In the case that the parameter set has not been evaluated, the interface procedure prepares a parameterized model from the original Petri net model by substituting the actual parameter values in the model description. Depending on the optimization phase, either the approximation component or the TimeNET simulator is called afterwards. For the computations of the bounds, the software tool lp-solve is used for the solution of linear programming problems. The resulting file with the computed value of the profit or cost function is read by the interface procedure. The new value is stored in the queue together with the parameter set and afterwards returned to the ASA optimizer. In the next step, ASA tests whether convergence is reached and exits with the final optimization result. A new parameter set is generated otherwise and a new iteration begins. The computed optimal parameter set is written into the configuration file for the fine-grain optimization at the end of the preoptimization phase. Afterwards, ASA is started again for the second phase with a new starting point and temperature adjustments as explained in Chap. 10.

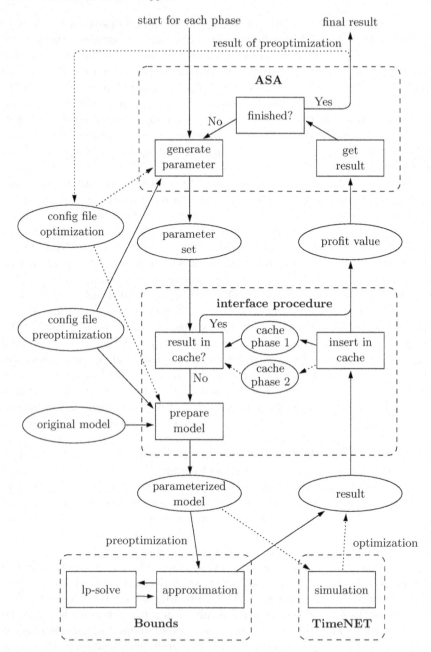

Fig. 12.2. Interaction of ASA, bounds computation and TimeNET

12.1.3 A Model-Class Generic Graphical User Interface

The graphical user interface for TimeNET 4 has been completely rewritten in JAVA and can therefore be run on both Unix- and Windows-based environments. The new GUI retains the advantages of the former one, especially in being generic in the sense that any graph-like modeling formalism can be easily integrated without much programming effort. Nodes can be hierarchically refined by corresponding submodels. The GUI is thus not restricted to Petri nets, and is already being used for other tools than TimeNET. As a stand-alone program it is named PENG, which is short for *platform-independent editor for net graphs* [178].

Model classes are described in an XML schema file, which defines the elements of the model. Node objects, connectors, and miscellaneous others are possible elements. For each node and arc type of the model the corresponding attributes and the graphical appearance is specified. The shape of each node and arc is defined using a set of primitives (e.g., polyline, ellipse, and text). Shapes can depend on the attribute value of an object, making it possible to show tokens as dots inside places. Actual models are stored in an XML file that must be consistent with the model class definition, which can be checked automatically with library toolkits for XML. Editing and storing a model can already be done after the corresponding schema is available.

Program modules can be added to the tool, which implement model-specific algorithms. A module has a predefined interface to the main program. It can select its applicable net classes and extend the menu structure by adding new algorithms. All currently available and future extensions of net classes and their corresponding analysis algorithms are thus integrated with the same "look-and-feel" for the user.

Figure 12.3 shows a sample screen shot of the GUI during an editing session of the model considered in Chap. 15. There are standard menus with the necessary editing commands in the top row, e.g., **File** to open, close, save, or create a new model. Commands under the entries **Edit** (cut, copy, paste, undo, ...), **View** (grid, zoom, go up or down in the hierarchy, ...) and **Window** (iconify, arrange, ...) should be self-explanatory and follow usual GUI style. There is a set of icons below the menu bar where the modeler can access menu commands which are most commonly used.

The main window contains the editing area. Models can be edited with the left mouse button like using a standard drawing tool with operations for selecting, moving, and others. There might be different windows opened at the same time. The lower icon bar shows all model elements that can be added in the model class. The contents of this bar are automatically derived from the model class description of the currently opened model. Clicking one of them changes the mouse pointer into a tool that creates the corresponding object. Arcs are added by clicking and holding at the source element, and then drawing the mouse to the destination.

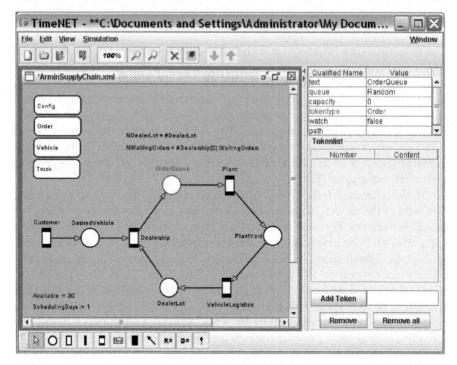

Fig. 12.3. Sample screen shot of the graphical user interface

Individual attributes of an element can be edited by selecting it in the drawing area and then changing the values in the right tab. There is one entry in this window for every object attribute as defined in the model class for that object. Place OrderQueue is currently selected in Fig. 12.3, and the attributes of a place are shown. The initial marking of a place can be specified in the lower part of the right tab.

The dynamic behavior of a Petri net model can first be checked with an interactive simulation, the token game. A background simulation without a visualization can be started as well for SCPN models. Data exchange between GUI and simulation takes place via the remote subsystem as explained earlier. Once the actual simulation program is linked, it starts and initializes its internal information. The result monitor starts up as well. This JAVA program runs on the user PC, receives all result measures from the simulation during run time and displays them graphically in windows. A sample screen shot is shown in Fig. 12.4.

12.2 Software Packages for Stochastic Discrete Event Systems

Numerous commercial, research, and prototype software tools have been developed for stochastic discrete event systems in the past; overviews are

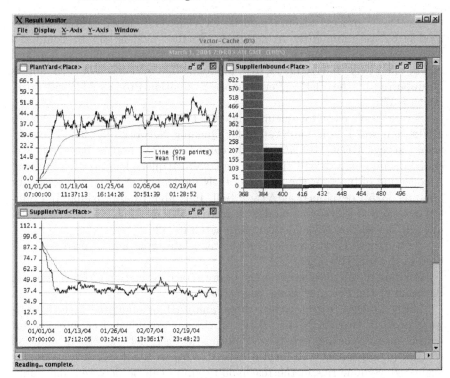

Fig. 12.4. Graphical result output

given in [156, 158, 306]. Most tools provide a graphical user interface for the convenient drawing and editing of the model. They usually include analysis components for one or several classes of models. The following list of brief descriptions for tools related to our subject is by no means exhaustive.

Commercial general-purpose simulation tools do not necessarily use a formal underlying model, but have their own kind of graphical or textual description language for systems instead. Graphical user interfaces allow to combine predefined templates from libraries as building blocks. An example is the tool Arena [200], successor of SIMAN. The simulation tool Visual SLAM [269] is able to model discrete, continuous, and combined models. Another example from this category is Extend [217]. Many simulation tools have been specially designed for an application area like production systems or logistics.

Stochastic automata networks (SANs) are parallel and synchronized Markovian stochastic automata. The PEPS software tool [266] analyzes large SAN models analytically.

The StateMate tool [153, 155] implements state charts modeling and evaluation, and is distributed by I-Logix. Other applications include model validation, test case generation, and code generation.

Stochastic process algebra models (SPA) extend classic process algebra models by assigning rates to activities. Smaller models can be composed with operators. The underlying stochastic process is then Markovian, and can be efficiently analyzed [138,162]. Modeling and evaluation of such models is supported by the PEPA (Performance Evaluation Process Algebra) workbench [137].

Several tools exist for queuing models. PEPSY-QNS is the Performance Evaluation and Prediction SYstem for Queuing NetworkS [30,31]. The software packege RESQME [45] was developed at IBM and is based on extended queuing networks. Utilization and average queue lengths are obtained. A collection of tools for the analysis of QN models is QNA (queuing network analyzer [325]) developed at AT&T Bell. It is able to approximately evaluate general non-Markovian models. Selected software tools for the analysis of queuing models are covered with more detail in [42].

A structured notation based on abstract data types and layered design is used in the tool HIT [324].

Pointers to software tools for the performance evaluation of Petri nets can be found at the Petri net home page [264].

GSPN models are graphically edited, analyzed, or simulated with the software package GreatSPN [48,51]. It provides a graphical user interface for editing and evaluating generalized stochastic Petri net models. In addition to GSPNs, stochastic well-formed nets (SWN) with colored tokens can be evaluated. Symmetries in the reachability graph can be detected and lumping techniques facilitate analysis algorithms with less computational effort in many cases. Components for direct numerical analysis and simulation are contained in GreatSPN. Algorithms for the computation of performance bounds based on linear programming techniques have been added later.

The software package DSPNexpress [227,228] provides a graphical interface running under X11 and is especially tailored to the steady-state analysis of deterministic and stochastic Petri nets (DSPNs). For the class of generalized and stochastic Petri nets, steady-state and transient analysis components are available. A refined numerical solution algorithm is used for steady-state evaluation of DSPNs [229], facilitating parallel computation of intermediate results. Isolated components and isomorphisms of subordinated Markov chains of deterministic transitions are detected and exploited. A recent update under the name DSPNexpress-NG allows transient and steady-state analysis of DSPN with two equally timed concurrent deterministic transitions, and the evaluation of UML models.

The tool WebSPN [29] analyzes non-Markovian stochastic Petri nets and has a web-enabled user interface. A discrete time approximation of the stochastic behavior overcomes some of the restrictions for numerical analysis techniques.

The software package SPNP [64,164] has been developed at Duke University. It contains components for the transient and steady-state analysis of stochastic reward nets (SRNs), which are comparable to GSPNs. Models are specified alphanumerically as C functions. Very general performance measure specifications can be given and a sensitivity analysis module is available. Transient and cumulative reward measures are obtained using randomization. The tool has been extended with a Tcl/Tk based graphical user interface, which is referred to as iSPN. SPNP has been combined with other analysis tools in an integrated framework called IDEA [113].

At the University of Arizona the software package METASAN was developed. The provided class of models is called stochastic activity networks (SANs), a variant of stochastic Petri nets. The successor of the tool is known under the name UltraSAN [75, 285] (later developed at the University of Illinois at Urbana-Champaign). It contains a graphical user interface and the possibility of hierarchical model construction. Rate and impulse reward measures can be specified. Components for the transient and steady-state analysis are provided. The algorithms can take advantage of symmetries by lumping of states. SANs extended with deterministic delays can be analyzed in steady-state. Transient and stationary simulation modules with importance sampling for rare event handling are also available.

Colored Petri nets are supported by the CPN-Tools [24] (formerly DesignCPN), offering a limited time model that is not compatible with the usual understanding of time in SPN. Performance evaluation and stochastic delays are not supported.

ALPHA/Sim [249] is a general-purpose, discrete event simulation tool with Petri net support. ALPHA/Sim allows a user to graphically build a simulation model, enter input data via integrated forms, execute the simulation model, and view the simulation results within the graphical environment.

Object-oriented Petri nets are used in PNtalk [179]. Tokens are similar to objects of a programming language in this class. Programming of transition properties is, however, required and only simulation with simple statistics are available.

Other approaches to tool implementations have been towards the integration of other tools in one common user interface and by combining the results of the individual tools.

One example is the SHARPE tool (Symbolic Hierarchical Automated Reliability and Performance Evaluator [163, 282]). It can be used for specifying and analyzing performance, reliability, and performability models. The toolkit provides a specification language and solution methods, e.g., for fault-trees, queuing networks, (semi-)Markov reward models, and stochastic Petri nets. Steady-state, transient, and interval measures can be computed and used in other models.

Models can be expressed as a combination of parts of different model classes in other tools. Additional remarks on this topic are given in the notes for the unified SDES model class at p. 42.

The software package HiQPN [21, 22] has been developed at the University of Dortmund. It uses the model class queuing Petri nets (QPNs) as well as their hierarchical combination (HiQPNs), including colored Petri net tokens and queuing places. Several analysis techniques for the steady-state performance evaluation are provided, namely decomposition approaches based on tensor algebra. The hierarchical structure of the model is used for a structured description of the stochastic generator matrix. SimQPN [216] is a recent Java-based simulator for this net class. Other tools that combine GSPNs and queuing models are DyQN-Tool$^+$ [157] and SMART [68,69], the latter including solution techniques based on efficient storage. It implements SPN and QN models but offers a textual user interface only.

The DEDS toolbox [20] allows to specify Markovian models as queuing networks, GSPNs, and colored Petri nets. All model classes are transformed into a representation in an *abstract Petri net notation*, which is then used by evaluation algorithms.

The Möbius tool [72, 81, 82] is a multi-formalism, multi-solution software tool, in which model classes as well as analysis algorithms can be combined. This is done using an abstract model description similar to SDES of this text. Its implementation is consequently done as an abstract functional interface. The Möbius framework is able to integrate additional solvers that are applicable to some of its model classes in a modular way, and thus relieves the developers of prototype implementations from designing a complete tool from scratch. On the other hand, it is possible to use existing evaluation algorithms once a new model class has been described, with the abstract functional interface. A significant point of Möbius is the possibility to combine model parts from different classes. Möbius' understanding of models, however, does not allow for different action modes, which makes, e.g., colored Petri nets impossible to handle. Another difference is that enabling degrees can be captured with SDES. The graphical user interface of Möbius is not as easily adaptable to new model classes. The model-level abstract description of Möbius has later been extended by a state-level interface [86] using labeled transition systems to efficiently accommodate different solution algorithms.

Part III

Applications

13

Optimization of a Manufacturing System

The design of modern manufacturing systems is a complex task. High investments require that the planned system will fulfill the requirements. Moreover, complex interleaving of choices and synchronizations in manufacturing systems may lead to paradoxical behavior. For example, increasing the number of resources (i.e., tokens in a Petri net model) may result in a deadlocked system, and replacing a machine for a faster one can decrease global productivity. Methods and computer tools for the modeling and performance evaluation and optimization of manufacturing systems are therefore important. Manufacturing system design and operation is one of the fields where stochastic discrete event systems are widely used; see the bibliographical notes at the end of the chapter.

Direct optimization methods are not applicable for complex manufacturing systems if a level of detail is necessary in the model that goes beyond the first rough estimation steps in the design. Efficient methods based on a problem description as a linear programming problem are therefore unfortunately not applicable. Models of complex manufacturing systems are nonlinear in principle in addition to that. This is especially the case for selection problems where parameters determine different system layouts, strategies, or machine types. On the other hand, there are problems related to selecting the optimal speed of a transport system, number of transport pallets, and the like. It is not possible to guess the form of the function from a system model directly, which would make simple standard methods (e.g., a gradient Newton search) applicable. These restrictions lead to the application of heuristic search methods for an optimization.

This chapter reports on an application example modeled with a generalized stochastic Petri nets as described in Chap. 5, for which an optimal parameter set is searched with the optimization method introduced in Chap. 10. Types of manufacturing systems and typical design and optimization problems are explained in the first two sections. Elements of the profit function that are usually found in manufacturing systems are described with their specification in a Petri net model in Sect. 13.3. The manufacturing system application example and its GSPN model are covered in the subsequent two sections. Results for the example during the optimization algorithm of Chap. 10 are given finally, namely by reporting the performance measure approximation quality in Sect. 13.6 and the results and speedup of the two-phase optimization method in Sect. 13.7. The chapter concludes with some bibliographical remarks.

13.1 Types of Manufacturing Systems

Manufacturing systems can be characterized by the general type they belong to. Some important classes are explained later. More information on manufacturing system taxonomies can be found in [4, 32, 88, 89, 135, 270].

1. *Production lines* typically produce a low variety of products with a high production volume. The machines and production cells are organized in a sequence, which is guided by the flow of parts through the line. Issues to be investigated usually include impact of unreliable machines and buffer allocation.
2. *Assembly/disassembly* systems are sequentially organized for the different intermediate parts and final products like a production line. However, some resource sharing of machines is possible, such that competition takes place. Another similarity is that the production volume is usually high, while the number of different products is small. The main difference is that assembly (or disassembly) operations are the most important ones. In many cases only prefabricated workpieces that are delivered by a supplier are assembled without further manufacturing steps. Assembly and disassembly steps require the modeling of synchronizations (join) and splits (fork).
3. *Flow shops* are manufacturing systems where the main issue driving the structure and operation is the flow of parts. There are typically some different parts to be produced with some variations, but the overall set of production routes (i.e., the order of production steps and where they are executed) is fixed. However, the restriction of the sequential organization of the machines as in production lines is relaxed. The setup of machines and work places is a question of optimization depending on the flows of parts. The machines must have some degree of flexibility because of the different manufacturing steps that are executed on them. Setup times should be low with respect to the production times.
4. *Job shops* can be found in production environments with a high variety of products and a low production volume. Different products, often directly manufactured to the special needs of individual customers, can be produced at the same time. Because of the amount of variety, there are no sequential machines or fixed production routes. More important than the optimization of machine placement (to minimize transports) and production routes are flexible machines and production planning issues to schedule the shared resources. Adaptable machines e.g., with tool changing

possibilities and highly trained workers are necessary for the high degree of production variety. Resource utilization might be relatively low, especially if individual products need to be finished fast after they have been ordered.

5. *Flexible manufacturing systems* or short FMS aim at combining flexibility and productivity in one system. They represent an integrated computer controlled configuration of machines, production equipment, and material handling systems. Material and parts can be transported by automated guided vehicles (AGV), conveyors, or robots. The amount of manual work places is minimal. An automated computer system on different levels of hierarchy (such as plant level, cell level, machine level) is responsible for the control of the FMS. A wide variety of products can be produced in a medium production volume. The main advantage is the rapid adaptability to new products.

13.2 Typical Design and Optimization Issues

Some of the design problems that frequently occur in the design of a manufacturing system are identified later. They usually involve the selection of one out of several options, e.g., a size decision, machine, or material handling system selection. Numerical values can be discrete like the size of a buffer or continuous like a production mix. In practice, mixtures of these problems have to be considered in an optimization. Manufacturing system optimization characteristically deals with complex nonlinear evaluative models in high-dimensional search spaces. Typical design problems are:

– *Facility selection problems* cover examples like: should a faster machine with a higher initial investment and maintenance cost be bought? Is it preferable to select a machine with a better failure/repair behavior over a cheaper one? Very basic structural alternatives are for instance related to the selection of a material handling system. In a production line, conveyors might be the only alternative, but automated guided vehicles could be better in a more flexible setup.

There are usually advantages and disadvantages connected to either one of the alternatives. Every alternative has its own set of attributes (think of the speed of a conveyor, the price of a robot, or the setup time of a machine). All these values need to be included in the quantitative model to be evaluated. For every selection issue, we need one decision variable in the optimization problem. Such a variable is discrete and may assume values that correspond to the individual alternatives of the selection (like an enumeration). When the model has to be evaluated for a certain parameter set, the model attributes must be set according to the decision variable value during an optimization. This is easy for numerical values like a machine speed. Other alternatives require structural changes in the model.

In the worst case, there needs to be one prefabricated model for any global alternative, and thus for the cross product over all structural alternatives.

- *Facility duplication* might be useful in cases of bottlenecks where no faster machine is available (or is too expensive). Another similar question is the number of automated guided vehicles in a transport system setup. In any case, decision variables of this type are natural numbers, which model the number of resources that should be available in the manufacturing system. The corresponding models can typically be adapted to such a number by setting an initial number of tokens or selecting an appropriate degree of internal parallelism in an activity.

- *Placement problems* appear when the layout of a manufacturing system needs to be planned. It is related to both the physical placement (in the plant layout) as well as the logical layout e.g., of a production line. Line balancing for bottleneck avoidance is an issue for the latter. There are algorithms that calculate good layouts starting from production routes and volumes and corresponding transports. The overall impact of a layout with corresponding transport times, buffer, and conveyor sizes is not easy to calculate and requires an optimization of the global system. Placement of machines and other production facilities is modeled indirectly via different transport times and restrictions in a quantitative evaluation. One setup leads to one model (or one parameterization of a model). It thus needs to be handled like a selection problem.

- *Buffer allocation* deals with the question of how big intermediate buffers in the manufacturing system should be (buffer sizing). If a zero size means no buffer in the model, it contains the placement problem for buffers as a special case. This question is often found in a manufacturing system design. Additional buffer places might not be expensive, but can have a big impact on production capacity, especially when machine failures are significant. Buffering plays a major role also in the tradeoff between work in process and time between order and delivery (related key words are just-in-time supply vs. make-from-stock). Another design issue that is closely related is the number of pallets (or any type of work piece container). Both kinds of issues lead to decision variables that are natural numbers.

- *Production mix* is the question what amount of which products should be produced in a manufacturing system. The relative production rates need to be set such that the overall effect of different production and material costs combined with the profits is optimal. Resource restrictions of the manufacturing system as well as marketing constraints need to be obeyed. The decision variables can be represented by nonnegative real values, which model the production mix percentage or the individual production volume per part.

- *Production routes and work plans* can be chosen in manufacturing systems with some degree of flexibility such as job shops and FMS. This is a selection-type of design issue, because there is no meaningful way of assigning numbers to the different alternatives.

13.3 Profit Function Elements

A *profit function* has to be specified before an optimization can take place.[1] Typical profit functions consider the money earned from selling finished parts minus the costs arising from the production process. The price of raw parts, the money spent for work-in-process, machine and transport systems amortization, as well as utilization-dependent costs are examples. More complex functions can capture human factors and costs as well. All of them are determined by an appropriate performance analysis of a model, although it might not be obvious how reward variables can be specified for them. The complexity of the profit function depends on the needs of the modeler and has to include every significant influence. Selection of significant issues is similar to modeling itself, it requires human expertise and can hardly be automated.

To evaluate the mean performance of a system during normal processing, we restrict ourselves in the following to a *steady-state* evaluation.[2] It is thus clear that the absolute amount of profit gained depends linearly on the time interval under inspection. For every profit result, we need to know the corresponding amount of time. For the remainder of this section, let Δt be some fixed time interval for which profit and cost are determined. This value would for instance correspond to one week if we are about to evaluate and optimize the profit per week. The optimal set of parameters, i.e., the values of decision variables with the best possible profit result, would of course be the same if a different interval had been chosen – only the absolute result changes. It is further assumed that the manufacturing system produces N different parts.

A simple formula for the amount of money earned by selling finished parts is then

$$\text{Profit}_{sell} = \Delta t \sum_{i=1...N} \text{OutRate}_i * \text{SellPrice}_i$$

if OutRate_i is the output rate of parts of type i per time unit and SellPrice_i is the price paid for one part i. This is only valid if we assume that the market buys all finished parts and that the price is independent of the output. The value of OutRate_i is equal to the throughput of a transition in the Petri net modeling the removal or output of parts of type i, or an action in an SDES model that fires every time that such a product is finished.

An alternative formula can take into account that the price per part drops if more parts are produced and sold. Some kind of market saturation can thus be modeled. We assume that this behavior can be approximated by an exponential function. This function converges toward an upper limit MaxPrice_i, which is the maximum amount of money earned by selling any

---

[1] The optimization literature often uses a *cost function*, which has to be minimized. Both variants can obviously be transferred into each other, but profit maximization is more appropriate for our application area.

[2] Other transient issues could of course be optimized equally well with the method described Chap. 10.

number of parts of type i together. For the sake of simplicity, the different parts are assumed to be independent. The resulting function has the form

$$\text{Profit}_{sell} = \Delta t \sum_{i=1...N} \text{MaxPrice}_i \left(1 - e^{-a\, \text{OutRate}_i} \right)$$

The question is then how to adjust the factor a. If only one part is being produced and sold, its price is (almost) not decreased by selling a large number of products. On the other hand, it can not exceed its theoretical value SellPrice_i. Lets thus assume that the price gained per part is equal to its theoretical value at point zero of the above profit function. The derivative of the profit function in point zero should hence equal the theoretical price per part. For a fixed value of Δt, the following equations must then hold

$$\text{SellPrice}_i = \frac{d\, \text{Profit}_{sell}}{d\, \text{OutRate}_i}(0)$$

$$= \text{MaxPrice}_i \left(a\, e^{-a\, \text{OutRate}_i} \right)(0)$$

$$= \text{MaxPrice}_i\, a$$

And thus

$$a = \frac{\text{SellPrice}_i}{\text{MaxPrice}_i}$$

After examining the money gained by running a manufacturing system, different types of costs have to be subtracted for a total profit function. To be able to manufacture and assemble new products, raw parts have to be bought. For M different input parts, the amount of money spent is

$$\text{Profit}_{buy} = -\Delta t \sum_{i=1...M} \text{InRate}_i * \text{BuyPrice}_i$$

Where InRate_i is the input rate of supply parts of type i, which can be measured as the throughput of a corresponding transition in a stochastic Petri net model, and BuyPrice_i is the price to be paid for one raw part. It is assumed that this price is independent of the number of parts bought. This independent computation for buying and selling parts instead of a direct profit calculation takes into account waste parts.

Another cost that has attracted much attention in recent years (e.g., Kanban systems) is the money spent for work in process. This position calculates the money bound to intermediate parts in the system. The current value of the parts cannot be invested elsewhere during the time from buying raw parts until they are finished and sold. Even if it is not necessarily borrowed in reality, the cost is commonly calculated using some fictitious interest rate. Because this rate is usually given for the period of one year, the function relates this to the actual time interval Δt.

$$\text{Profit}_{WIP} = -\frac{\Delta t}{\text{one year}}\, \text{InterestRate} \sum_{i=1...L} \text{MeanNumber}_i * \text{Price}_i$$

MeanNumber$_i$ is the mean number of parts of type i in the manufacturing system in steady-state, which can be computed from the token distribution probabilities of the model. The simplified formula used here does not take into account the increasing worth of the different parts $1 \ldots L$ during processing, Price$_i$ should be some useful mean value. For an intermediate part of type i this will usually be in the range between the sum of BuyPrice$_j$ of the parts that are assembled up to now and the value of SellPrice$_i$.

In addition to the costs corresponding to the manufactured parts, the manufacturing system itself requires initial investment and maintenance. The investment costs Investment$_i$ of a resource i are distributed over its lifetime LifeTime$_i$ to calculate the amortization per time interval Δt for all K resources.

$$\text{Profit}_{amort} = -\Delta t \sum_{i=1\ldots K} \frac{\text{Investment}_i}{\text{LifeTime}_i}$$

For facility selection and duplication problems, quality and speed of machines as well as the number of AGVs and pallets influence this cost.

Some resources require more money when they are busy. Energy for machines, spare parts and tools, and others belong to this class of utilization-dependent working costs.

$$\text{Profit}_{util} = -\Delta t \sum_{i=1\ldots K} \text{Utilization}_i * \text{VarCosts}_i$$

It is assumed that the variable costs VarCosts$_i$ of a resource i linearly depend on its utilization Utilization$_i$. The degree of utilization can be computed from the token distribution probabilities of a Petri net model by analyzing the places that model the different states of resource i.

Constant working costs correspond to resources that need money no matter whether they are used or not. In general, there are utilization dependent *and* constant costs for each resource. Examples for the second are maintenance and repair, worker's wages, energy, and so on. General costs that do not belong to a specific manufacturing system of a company like rent, insurance, management, and others can be added here.

$$\text{Profit}_{const} = -\Delta t \sum_{i=1\ldots K} \text{ConstCosts}_i$$

ConstCosts$_i$ gives the amount of money per time unit for resource i.

After specifying the different profit and costs, the overall profit can be calculated as

$$\text{Profit} = \text{Profit}_{sell} - \text{Profit}_{buy} - \text{Profit}_{WIP} - \text{Profit}_{amort} - \text{Profit}_{util} - \text{Profit}_{const}$$

Whether or not all of these (and more specific) elements of a profit function are necessary is a decision of the modeler. For the sake of simplicity, one can neglect issues that are fixed and thus do not contribute to the optimal parameter set.

This spares the work necessary to estimate or measure costs and profits that are constant from the optimization standpoint. However, the absolute value of the resulting profit will be different from the real value in that case.

13.4 A Manufacturing System Example

This section introduces a manufacturing system, which is modeled by a generalized stochastic Petri net (GSPN) below and to which the two-phase optimization methodology presented in Chap. 10 is applied later on.

The example belongs to the class of flexible manufacturing systems (FMS). It comprises a robot, an AGV system, and four conveyors for material transport, two machines, one manual work place, and one assembly station. Figure 13.1 shows a sketch of the layout.

Parts to be processed arrive in the input buffer. Each part circulates through the system mounted on a pallet. The robot at the loading and unloading station takes raw parts from the input buffer and places them on an empty one. From the loading station, parts are taken to one of the two machines by a transport system with automated guided vehicles. After being processed in one of the machines, the work pieces are transported to the manual workplace on the conveyor that belongs to the machine. Another conveyor takes them to the assembly station, where additional parts can be assembled. The circle is closed by the fourth conveyor, which transports parts back to the unloading station. The robot can take a finished part from its pallet there and place it into the output buffer. However, pallets are not necessarily unloaded at the unloading station if the part mounted on them has to be processed further. In that case the pallet is moved on to one of the machines as described earlier.

Two types of products, named here A and B, have to be produced using the example FMS. Parts of type A can first be processed by any one of the two machines. A manual operation and an assembly of an additional part

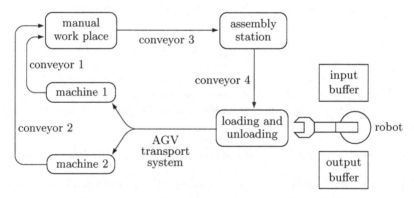

Fig. 13.1. Flexible manufacturing system example

have to follow, before the product is finished. B-type parts are first processed by machine 1. Afterwards, they are tested at the manual work place. Parts that have been correctly processed are transported to the assembly station. After an assembly operation the product is finished. However, five percent of the parts have to be reworked at machine 1. Those parts are detected at the manual work place, transported to machine 1 on the circle again, and can then be moved to the assembly station for the remaining processing steps. Further details of the example like the transport and manufacturing delays are not covered here.

We consider the following design issues for the example:

- The number of pallets P available in the system (pallet allocation). More pallets can lead to a higher throughput, but increase the work in process as well as the initial investment and running costs. The model is evaluated for pallet numbers from 2 to 30.
- The number of vehicles A of the AGV transport system in the range from one to four (facility duplication). Additional AGVs should decrease the waiting time of parts at the loading station and thus increase the throughput. The disadvantages are comparable with the ones described earlier for the number of pallets.
- Production route of part A is taken as a design issue. It is possible to process parts of type A on either one of the two machines, while B-parts are always processed by machine 2. The probability for a part of type A to be transported to machine 1 by the AGV system is, therefore, considered as a parameter.
- Finally, the production mix should be adjusted during the design. The percentage of parts of type A is assumed to be changeable in the range from 25% up to 80%.

The profit function is intended to compute the profit per day for the example and contains the following elements. The profit per sold part is set to 2 (part A) and 4 (part B) monetary units. Work in process costs are assumed to equal 10 monetary units per part in the system during one day. Amortization and constant costs over time are assumed to be 2 000 plus 250 per AGV vehicle and 20 per pallet for one day.

13.5 A Generalized Stochastic Petri Net Model of the Example

A Petri net model for the example from the previous section is shown in Fig. 13.2. Its basic layout follows the one shown in Fig. 13.1. The different processing tasks (modeled by transitions) and parts in buffers (modeled by tokens and places) had to be unfolded because of the use of simple Petri nets. Model elements that belong to one machine, work place, the robot, or the assembly station are highlighted with a grey background.

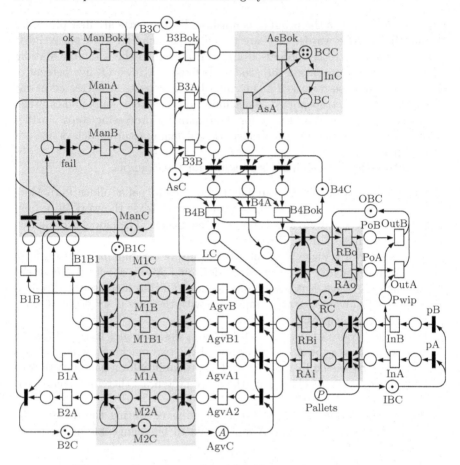

Fig. 13.2. Petri net model of flexible manufacturing system

The robot at the loading and unloading station is described with the transitions RAi (short for robot, part A input), RBi (part B input), RAo, and RBo as well as the surrounding places and immediate transitions. Immediate transitions are used to model the start of an activity and for decisions between different possible tasks of one resource. The firing of one of the immediate transition of the robot model takes away the token from place RC (short for robot capacity), which otherwise models an available robot. Places that are not modeling resource capacities are locations where parts can be inside the manufacturing system. Actual locations usually correspond to several places because of the mentioned unfolding.

The remaining parts of the shown model are similarly structured. Places whose names end with C (for capacity) ensure that the resource restrictions of buffers and machines is not exceeded. Elements with names starting by M1 and M2 model the two machines, Man the manual work place, and As the assembly station. The small model part at the assembly station model with places BCC and BC as well as transition InC models the arrival and storage of small parts (of type C) that are assembled to parts A and B. Whether a part of type B has been correctly processed is decided by the firing of the immediate transitions ok and fail. The probability to be processed correctly or not (95% and 5%) is defined with the firing probabilities of these transitions. Places and transitions with names starting with IB model the input buffer, OB the output buffer, and Agv the AGV transport system.

The four conveyors act as intermediate buffers between their connected stations. Hence their names begin with a B followed by the number. The number of AGV vehicles is set by the model parameter A, and P defines the number of pallets. In general, model elements with trailing A (B) refer to parts of type A (B). The production mix (one of the decision variables) can be set with the relative firing probabilities of immediate transitions pA and pB. These two immediate transitions are enabled when there is an empty space in the input buffer, and their respective firing corresponds to the arrival of a raw part of type A or B.

The individual transition delays are set according to the chosen example values but omitted in this text. One second is equivalent to one time instant of the model. The profit function for one day is set then up as a reward variable of the model as follows. The profit from earned parts was set to 2 (4) per part A (B) and can be specified using the throughput of finished parts through transitions OutA and OutB. A conversion factor of $24 * 60 * 60 = 86\,400$ (seconds per day) is used. Work in process is related to the overall number of parts in the system times the chosen factor of ten. This could be specified based on the sum of token numbers in numerous places of the model. For an easier specification, the place Pwip has been added to the model. The number of tokens in this place equals the number of parts in the system. Constant costs over time were $2\,000$ plus 250 per AGV vehicle (number is specified by parameter A in the model) and 20 per pallet (number is specified as parameter P in the model).

Finally, the complete profit function is defined as the following reward variable of the example GSPN:

$$\text{Profit} = 172800\text{TP}\{\#\texttt{OutA}\} + 345600\text{TP}\{\#\texttt{OutB}\}$$
$$-10\text{E}\{\#\texttt{Pwip}\} - 2000 - 250 \cdot A - 20 \cdot P$$

where TP$\{\#\texttt{OutA}\}$ denotes the throughput of transition OutA, and E$\{\#\texttt{Pwip}\}$ the expected number of tokens in place Pwip as explained in Sect. 5.1.

13.6 Profit Function Approximation Quality

The approximation techniques that are the basis of the optimization method of Chap. 10 are applied to the FMS example model shown in Fig. 13.2 to investigate the approximation quality, i.e., its deviation from the real values.

Hence, the model belongs to the class of FRT nets and can be analyzed approximately with the computation of bounds. The two basic types of approximations, namely the estimation of transition throughputs and mean number of markings, are considered first. An investigation of the approximation of the complete profit function for the example follows.

Figure 13.3 shows a plot of the throughput of parts A (at transition InA), comparing the different quantitative evaluation techniques considered during the optimization. Simulated results are computed with a setting of confidence interval and relative error to use them as the real values for the comparison. As expected, simulated values as well as the approximated results are between the computed lower and upper bounds. Both values are quite close, and the shape of the functions is very similar. A numerical comparison of absolute differences or relative error is omitted intentionally, because this would not make sense for such a small and random part of the overall solution space.

For an evaluation of the approximated values for average markings, Fig. 13.4 contains plots of the values computed by simulation and approximation for the mean number of tokens in place Pallets. The absolute values computed by approximation are not very close to the simulated ones. The experience was the same with other examples: throughput approximations are usually much better than the marking approximations. As already stated in Chap. 10, this is due to the fact that the marking bounds and approximations are computed using the throughput approximation, which already contains some error. An exact approximation is, however, not necessary for our optimization use, because only the shapes of the functions need to resemble each other. This is the case also for the marking approximation. One could also

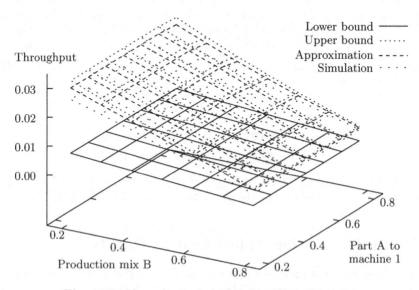

Fig. 13.3. Throughput approximation of transition InA

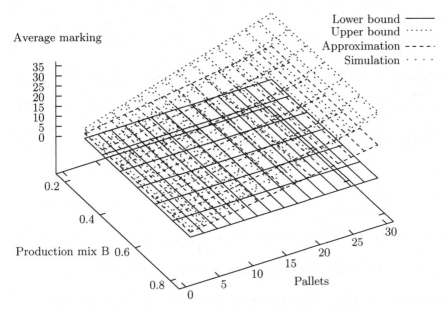

Fig. 13.4. Quality of marking approximation of place `Pallets`

Table 13.1. Improvement of marking approximation quality

	Throughput TP{#InA}	Marking E{#Pallets}	
		Approximation	Improved appr.
Lower bound	0.0041	0.0000	0.0000
Upper bound	0.0199	9.0668	5.9081
Appr. result	0.0183	4.5334	2.9540
Simulation	0.0152	0.1774	

say that the approximated values depend in a similar way on the changing parameters as the simulated ones.

The mentioned better approximation quality of throughputs lead to the idea to use the approximated throughputs in the marking approximation algorithms as if they were the real values. This way of *improved marking approximation* is assessed for the example. Table 13.1 compares simulated and approximated values of the example for production mix 50% parts B, two AGVs, 10 pallets, and a probability of sending parts of type A to machine 1 of 50%.

The approximated results are calculated from the bounds with $\alpha = 0.9$ and $\beta = 0.5$. The throughput is not very close to its upper bound. The choice of $\alpha = 0.9$ is still too big, as the actual throughput value is smaller than the approximated one. The error made in the throughput approximation is about 20% for this particular case. The mean marking approximation for place

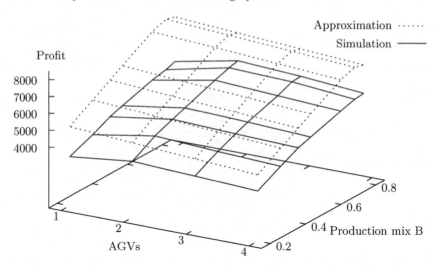

Fig. 13.5. Profit function approximation for the FMS example

Pallets is quite far from the actual value because upper and lower bounds are not close together, and the simulated value is near to the lower bound. The improved approximation method makes the result significantly better, but the error is still big. In the case of the FMS example, the only marking approximation needed in the profit function covers the work in process. The influence of it on the profit function (and thus the approximation error) is not very important, and therefore the approximate bounds computation does not change the results significantly.

To check the quality of the approximation method further, the profit function has been computed using simulation and approximation for a selected number of parameter sets that systematically cover the whole range of possible solutions. Because of the dimension of the parameter space, only selected parts can be shown in the plots.

Figure 13.5 shows a plot of the profit function vs. the number of AGV vehicles and the production mix in terms of parts B. Two different meshes are drawn for results obtained by simulation and approximation based on bounds. For the picture, the number of pallets is set to 12 and the probability of parts A to be sent to machine 1 is set to 30%. The actual values are in the range for which the optimal values are achieved with simulation. The shape of the functions is quite similar – the main difference is that the approximated profit is higher for one AGV, while simulation achieves the optimal result with two AGVs.

The most important question concerning approximation quality for our purposes is how far away the approximated parameter set is from the one that is computed by simulation. Table 13.2 shows a comparison of the best parameter sets for simulation and approximation using bounds for this example. Approximation results of the profit function are quite different from

Table 13.2. Results comparison for simulation and approximation

Parameter	Simulation		Approximation	
	Optimal	Best 5%	Optimal	Best 5%
AGV vehicles	2	2.13	1	1.26
Pallets	8	10.57	6	11.63
Part B prod. mix	80%	80%	80%	78%
Part A to machine 1	10%	23%	10%	11%
Profit	6 397	6 153	8 717	8 445

the simulated ones. More important, however, is the very good approximation of where the global maximum can be found. The best simulated profit of 6 397 for a systematic full search is achieved for two AGVs, eight pallets, 80% parts of type B, and a probability of 10% for parts of type A to be sent to machine 1. The optimal setting of the decision variables based on the approximated values is very close, the last two values are exactly the same. For a broader perspective not only the point of the optimum was calculated, but the mean parameter setting over all parameter sets for which the profit value was within a 5% range off the optimum. The results are shown in the table in the column "Best 5%", and are also very close for approximation and simulation.

As the approximately found parameter set is almost a neighboring solution to the one found by exhaustive simulation, the preoptimization should be able to find a very good starting point for the second optimization phase. However, the simulated cost function value of 5 464 for the best approximated parameter set reaches only 85% of the optimum of the simulated parameter sets. This underlines the importance of the second fine-grain optimization step, although the parameter sets are very close.

13.7 Results of the Two-Phase Optimization

After the general approximation quality investigation in the previous section, results for the application of the two-phase method are shown here. Emphasis is on the tradeoff between optimization result quality and speedup.

The application example was optimized using the standard simulated annealing algorithm and the two-phase version for a comparison of the results and run times. Each simulation run was executed with a quite high setting of 98% confidence level and 3% relative error. All computations were carried out on a Linux PC with a 266 MHz Intel Pentium II Mobile Celeron processor. The overall optimization needed 802 min. Standard optimization results in parameter values as two AGVs, nine pallets, 79% parts B, and probability 22% of parts A to machine 1. The profit value for this parameter set is calculated as 6 338. These results are in an acceptable range with the numbers calculated during the scan of the whole parameter space.

Afterwards, the two-phase optimization was run. The preoptimization phase only took 2 min to complete. One approximation typically needs between 1 and 2 s of computation time. The "optimal" parameter set is one AGV, five pallets, 78% parts B, and a probability of 10% for parts A to machine 1. This again is within a close range of the full scan of the approximated values as done in the previous section. The approximated profit value at the optimal point of the preoptimization is 8 841, while the "real" profit value at this point is 5 069.

The second optimization step is started with the final results of the preoptimization phase as initial values. Table 13.3 shows results (best found profit and corresponding parameter set, result value from simulation) and computation times for different selections of the optimization parameter TAnnealScale. TAnnealScale influences the cooling speed of the simulated annealing, cf. p. 227. The example shows that it is possible to make the cooling process much faster without loosing significant result quality. However, there is of course a limit on how fast the optimization can be made. It is interesting to see that for TAnnealScale = 10, the second phase finds a better result than for 20. This is due to the randomness in the underlying optimization and simulation algorithms. On the other hand, for TAnnealScale values of 5 and 1, the fine-grain optimization was not able to find the better solution with AGV vehicles equal to two. The simulated annealing process does not almost move away from the initial solution for TAnnealScale = 1, resulting in a nonoptimal parameter set. This shows that a significant speedup can be achieved, but the heuristic choice of the faster temperature scheme is important. Experiences from this and other examples show that a speedup by a factor around ten can easily be reached without a significant loss in the result quality.

It should be noted that the computation time values in Table 13.3 in all columns already contain the improvement of the algorithm by using a cache of simulation results. The effect of the cache thus adds to the one because of the

Table 13.3. Tradeoff between speedup and result quality

TAnnealScale	Standard	Phase I	Phase II			
	100	100	20	10	5	1
AGV vehicles	2	1	2	2	1	1
Pallets	9	5	8	9	10	6
Part B prod. mix	79%	78%	78%	78%	74%	78%
Part A to machine 1	22%	10%	13%	16%	10%	10%
Profit	6 338	5 069	6 267	6 326	5 575	5 388
Time (min)	802	2	135	89	38	14
Speedup (Ph. I+II)			5.8	8.8	20.0	50.0

Table 13.4. Reduction of computational complexity by using the cache

	Number of simulation calls
Exhaustive search	144 768
Standard simulated annealing	2 008
Simulated annealing with cache	274

two-phase optimization method. Table 13.4 compares the computational cost for an optimization using a exhaustive search of the parameter space with a simulated annealing algorithm with and without cache. The number of simulation calls necessary during a theoretical exhaustive search of all parameter sets is derived by multiplying the number of possible values for all decision variables (optimization parameters). The factor for continuous parameters is computed by dividing the search interval by the discretization distance that is also used for the cache parameter equality test.

For the example, the cost function was called 2 008 times. Only 274 of these calls needed a simulation evaluation, the remaining results could be taken from the cache, resulting in a cache hit rate of 86%. The cache thus speeds up the simulated annealing algorithm by almost one order of magnitude, which multiplies with the speedup because of the use of two-phase simulated annealing.

Notes

The example application in this chapter underlines that the optimization of complex systems is computationally expensive, even when iterative meta heuristics like simulated annealing are applied. This is due to the costly quantitative model evaluation e.g., by simulation. The chapter showed that by using the two-phase optimization method based on performance bounds together with a cache of intermediate results as introduced in Chap. 10, an overall speedup of about two orders of magnitude can be reached.

The results presented in the chapter are based on joint work presented in [351–353]. Additional heuristics for the further improvement of the two-phase optimization methodology have been reported in [281]. Related work on manufacturing system modeling and evaluation can, e.g., be found in [4, 7, 87, 88, 270, 276, 290, 292, 294, 295, 323, 336]. Optimization in automation is considered, e.g., in [288].

14

Communication System Performability Evaluation

The future **European Train Control System** (**ETCS**) will be based on mobile communication and it overcomes the standard operation with fixed blocks.[1] It is introduced to increase track utilization and interoperability throughout Europe while reducing trackside equipment cost. Data processing on board the train and in radio block centers (RBC) as well as a radio communication link are crucial factors for the safe and efficient operation. Their real-time behavior under inevitable link failures thus needs to be modeled and evaluated. This chapter presents a stochastic discrete event model of communication failure and recovery behavior. An additional model for the exchange of location and movement authority data packets between trains and RBC is presented and analyzed. Performance evaluation of the model shows the significant impact of packet delays and losses on the reliable operation of high-speed trains. It allows to obtain and compare the theoretical track utilization under different operation strategies.

Train control is an important part of a railway operations management system. It connects the fixed signaling infrastructure with the trains traditionally. With the European Union ERTMS/ETCS project (European Rail Traffic Management System/European Train Control System), a standardized European train control system is designed, which will gradually replace the great number of different train control systems in use today. It will allow trains to cross borders without the need to change locomotive or driver, as it is still necessary today. The system forms the cornerstone of a common system for train control and traffic management.

At the final stage of ETCS implementation throughout Europe, more or less all train control infrastructure will be either on-board the trains or distributed in control centers. There is no need for optical signals, wheel counters, or a fixed arrangement of track parts into blocks. Trains and control centers are connected by mobile communication links. The safety of passengers depends on the communication system reliability. Real-time communication and

---

[1] Sect. 14.2 explains the different types of train operation.

information processing thus play a major role for the implementation of ETCS. The application example presented here is thus a distributed real-time system. It is subject to hard safety requirements, but has to deal with inherent soft real-time aspects (communication delay jitter and packet losses).

The importance of quality of service parameters for the communication and specification of the real-time behavior of subsystems has been addressed in the specifications of ETCS (see e.g., [95, 96]). The requirements are however not very detailed – no distributions are considered, but only probabilities of meeting certain deadlines. Although it is important to specify subsystem characteristics, the real-time behavior of the system as a whole can only be assessed by evaluating their interaction. The work presented in this chapter goes a first step into that direction by evaluating one safety-critical communication structure together with its failure behavior.

In addition to offering interoperability between the different European railroad companies, another major goal is to increase track utilization with higher throughput of high-speed trains. It is obvious that dropping the standard block synchronization of trains and migrating to a virtual block system has the potential of allowing closer distances between trains. Transmission errors in the communication system influence the minimum possible distance between trains and thus the maximum track utilization. This dependency is addressed and evaluated for the first time. Communication system, failure behavior, and safety braking of trains are modeled and analyzed using different performance evaluation techniques in the following. The results show that the vision of "driving in brake distance" behind another train with ETCS would lead to a very unreliable train behavior.

The remainder of the chapter is organized as follows: After an overview of the ETCS communication architecture, classic block operation as well as the future virtual block mode of ETCS are explained. Communication system failures are modeled and analyzed in Sects. 14.3.1 and 14.3.2 with stochastic Petri nets and Statecharts alternatively. A condensed model is derived from the results in the sequel. Section 14.3.4 describes how a safety-critical part of the ETCS communication system is modeled and presents results of a real-time behavior evaluation. Rare-event simulation (cf. Sect. 9.2) is applied because of the small scale of the computed probability measures.

14.1 The Future European Train Control System ETCS

To facilitate fast and efficient train traffic across borders in Europe, a unified European Train Control System (ETCS) [95] is under development in several European countries. ETCS is the core part of the more general European Railway Traffic Management System (ERTMS). The normal fixed block operation with mechanical elements, interlockings, and optical signals will be substituted by a radio-based computerized train control system. It receives commands about the train routes that are to be set, and directs wayside objects along these routes. To simplify migration to the new standard, ETCS defines three levels of operation.

ETCS Level 1 uses spot transmission of information to the train via passive transponders. It is a supplement for the conventional, existing trackside signaling technology for lines with low to moderate train density. Block sections are defined by the existing signaling system. This level increases safety against passing signals at danger and in areas of speed restriction.

With the ETCS Level 2 system, radio communication replaces the traditional trackside signals, which allows considerable savings in infrastructure and maintenance costs. The system enhances safety considerably by monitoring train speed and, if necessary, intervening automatically. This allows higher speeds and shorter headways, increasing capacity. The traffic management system processes and sends information and instructions for the train driver directly onto a monitor in the driver's cab via radio communication. A RBC traces the location of each controlled train within its area. The RBC determines and transmits track description and movement authorities according to the underlying signaling system for each controlled train individually. The first ETCS Level 2 track has been installed between Olten and Luzern, Switzerland in April 2002 for Swiss Federal Railways (SBB).

ETCS Level 3 additionally takes over functions such as the spacing of trains. Radio communication replaces the traditional trackside signals. No trackside monitoring system is necessary as trains actively report their head and tail positions as well as train integrity to control centers. **Moving block** operation can be applied to increase line capacity. An essential advantage of level 3 is the reduction in life cycle costs through the abolition of the devices for track occupancy monitoring and trackside signals. The only trackside hardware necessary are so-called balises, small track-mounted spot devices, which communicate their identity and exact position to trains that drive over them. They are used to recalibrate the on-board position tracking system, which otherwise relies on wheel sensors and can thus be inaccurate during a longer trip.

Figure 14.1 depicts a simplified view of the communication architecture underlying ETCS. Each train features a train integrity control system and a computer that can control train speed. It communicates via GSM-R (see later) radio with base transceiver stations (BTS), which are connected to base station controllers (BSC) by cable. The BSCs are communicating with RBC via ISDN.

RBC are the trackside part for radio at ETCS levels 2 and 3. Their major functions include safe train separation based on allocation of routes by regulation and interlocking. Position and integrity reports are sent by the trains periodically or upon request. On the basis of this information and the train routes, safe track areas are assigned to trains. This is done with so-called *movement authority* messages.

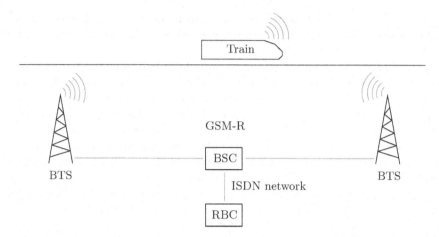

Fig. 14.1. Simplified ETCS communication architecture

The European Integrated Railway Radio Enhanced Network (EIRENE) project was started on behalf of the European Railways to define a new digital radio standard for application in the European High Speed Rail System. The EIRENE System Requirements Specification [98] defines the set of requirements that a railway radio system shall comply with to ensure interoperability between national railways. GSM (Global System for Mobile Communications) was chosen as the base technology because of availability and cost considerations. Additional functions that are tailored to the needs of railroad use (like area addressing, automatic international roaming, etc.) have been defined as *Railway GSM* (GSM-R [74]). For up-link and down-link there are different frequency bands reserved for GSM-R around 900 MHz.

The EURORADIO layer of the communication link specifies the Radio Communication System requirements to the air gap interface between train and trackside equipment [96,203]. The MORANE (Mobile Radio for Railway Networks in Europe [250]) project was set up to specify, develop, test, and validate prototypes of a new radio system. Trial sites exist in France, Italy, and Germany. Results of a quality of service test at one of these sites are presented in [287].

14.2 Train Operation with Moving Blocks Versus Fixed Blocks

This section explains how the future virtual "moving block" method and the standard fixed block train operation methods work. Position report message exchange and emergency braking due to communication problems under ETCS moving block operation is covered as well. This is used in Sect. 14.3 to derive and analyze a corresponding Petri net model. Standard block operation is covered to compare its track utilization with the results for ETCS later.

The aim of the application example analysis is to investigate the dependency between maximum throughput of trains and reliability of the communication system. ETCS is being introduced to maximize track utilization by high-speed trains. The maximum theoretical utilization will be achieved if trains are following each other with a minimum distance. The question is then: How close after each other can trains be operated theoretically under ETCS? We assume in the following a continuous track without stops, on which trains follow each other with a maximum speed v (current high-speed trains have a maximum speed of $300\,\mathrm{km\,h^{-1}}$) and a distance s. Moreover, for the following considerations we arbitrarily select w.l.o.g. two trains (*Train1* and *Train2*) that directly follow each other. To ensure safety of the system, worst-case assumptions are made for all timings, distances, etc. The final results will thus be upper bounds for the possible track utilization under worst-case assumptions of the available specifications. Practical values will be worse because trains have different speeds, need to follow their timetable, and accelerate or brake due to trackside conditions.

14.2.1 Fixed Blocks

The old type of block train operation ensures safety by fixed physical blocks, in which only one train may be located at any moment. Its operation is based on trackside equipment: Each block begins with a main signal and ends at the main signal of the following block. Trackside wheel counters check train integrity at the end of each block. The main signal at the beginning of a block may show green only if the number of wheels that have entered the block minus the number of wheels that left it at the end is zero.

A train must not enter the block if the corresponding main signal shows red. An approach signal is necessary because with bad weather conditions and high train speeds the train driver might not be able to stop the train before the main signal after seeing it. The approach signal can obviously not be located before the previous main signal. One of the problems with blocking operation is the "discretization" of track space, which leads to a waste of track utilization. From the throughput point of view, the blocks should therefore be as small as possible; this would, however, lead to a much bigger investment in trackside equipment. The minimum block size must be bigger than the distance from an approach signal to its main signal, which needs to allow for the maximum braking distance of a train.

The maximum theoretical train throughput is achieved when we assume an unlimited number of trains with identical speed v, which follow each other with a fixed head-to-head distance s. For a train to drive through a green approach signal of block i, the previous train must have left block i already. The minimum distance, therefore, includes twice the block length (assumed here to equal the braking distance for current high-speed trains $2\,800\,\mathrm{m}$), the train length (410 m), and a safety distance (50 m). From the railroad literature, it is known that the minimum travel distance between two trains in block operation also needs to obey a time to prepare the travel path (10 s), signal view time (12 s), and time to resolve the travel path (6 s). At a speed of 300 km h^{-1} of high-speed trains, the theoretical minimum distance between two trains is then calculated as 8393 m.

It should be noted that these considerations are only done to theoretically compare ETCS moving block operation with the old way of operating a block system. In reality, efficient operation of high-speed trains would not be possible that way. Real-life block sizes are between 2000 m and 4000 m, and approach signals are located 1000 m ahead. Different national systems for high-speed train operation exist in Europe. In Germany, an electronic system with trackside antenna cables ("Linienzugbeeinflussung," LZB) already allows much smaller block sizes and train driver information about the next train.

14.2.2 Moving Blocks

Continuous operation is introduced by ETCS with the notion of virtual *moving blocks*. Because there is no fixed block assigned to a train, and no physical block borders exist, the train movement is controlled by exchanging messages with the RBC. Each train periodically checks its integrity and sends this information together with the current position of the train head to the RBC. The time needed to check the train integrity is specified to be in the range between 2 and 5 s. Let Δt denote the time between two successive position reports of *Train1*. The requirements definition specifies $\Delta t \geq 5$ s. It is obvious that more frequent position reports will facilitate smaller train distances s, thus we choose $\Delta t = 5$ s in the following.

The integrity/position report is sent via GSM-R to the RBC and processed there, which takes 0.5 s typically. The resulting information is sent to the following *Train2*, telling it either that everything is fine to go on driving (by sending a new *movement authority* packet that extends the free track before it) or that an emergency braking is necessary immediately.

However, if a communication packet is delayed or lost on either the communication up-link (*Train1*→RBC) or down-link (RBC→*Train2*), *Train2* needs to decide on its own at what point of time emergency braking is inevitable out of safety reasons. There is obviously a deadline t after the last movement authority has been received, when the train needs to be stopped. The worst-case assumption is that after the last integrity check of *Train1* has been completed, a part of the train's carriages are lost from the main train and stop where they are or there is an accident. The movement authority, therefore, shall never exceed the "min safe rear end" of the preceding train [95] in moving block operation.

We would like to investigate the deadline and its dependency on the train head-to-head distance s (see Fig. 14.2 for an illustration). First of all

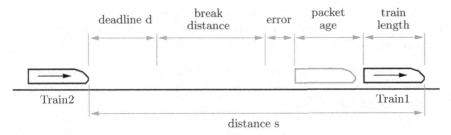

Fig. 14.2. Train distance and deadline

the train length (about 410 m for the German high-speed train "ICE") needs to be subtracted from the distance. Second, when the results of the position/integrity report of *Train1* arrive at *Train2*, the information is already some time old. Typical minimum (maximum) delays d can be estimated as follows: 2 (5) s to complete the integrity check, 0.5 (1) s end-to-end delay of position message to the RBC, 0.5 s to process the information there, again 0.5 (1) s for the downlink transfer to *Train2* plus assumed 1.5 s to process the information in the train and start braking if necessary. Packet ages d when arriving at *Train2* thus range typically between 5 and 9 s. Then there is a location error of not more than 20 m possible in the position report of *Train1*. The emergency braking distance needs to be subtracted as well, being between 2 300 and 2 800 m depending on the actual speed. For simplicity we assume in the following braking distance plus train length plus position error as $l = 3\,000$ m.

The deadline t is then given by $t = (s - l)/v - d$. The minimum theoretical distance for $v = 300\,\mathrm{km\,h^{-1}}$ is thus $s_{min} = 4\,000$ m. This simple consideration already shows that the common term of "driving in braking distance" with ETCS is misleading, because even if everything would run perfectly, trains cannot get closer than 4 km.

14.3 An ETCS Train Operation and Communication Model

The ability to exchange data packets with position and integrity reports as well as movement authority packets is crucial for the reliable operation of ETCS. In this section, a quantitative model of moving block operation and the necessary data exchange is built stepwise while taking into account the reliability of the communication channel. A communication link status model is presented first, constructed both as a stochastic Petri net and a Statechart model. The analysis results lead to a condensed failure model in Sect. 14.3.3, which is combined with the communication exchange and emergency stop behavior afterwards.

Model construction is based on the following sources of information about the qualitative and quantitative behavior of the communication system and its failures:

- A Quality of Service parameter specification (maximum connection establishment delay, etc.) is given in the *Euroradio form fit functional interface specification* (FFFIS) [96].
- Allowed parameter ranges for some system design variables like the minimum time between two subsequent position reports sent by a train are specified in the *ERTMS Performance Requirements for Interoperability* [97].
- Definitions of requirements of reliability, availability, maintainability and safety (RAMS) as well as acceptable numbers of failures per passenger-kilometer due to different reasons can be found in the *ERTMS RAMS Specification* [94].
- Some additional assumptions (mean time to complete the on-board train integrity check, etc.) are adopted from a description of simulation experiments carried out by the German railways company [257].
- Another detailed description of communication quality of service parameters is provided in [146], serving as an acceptance criteria for future measurements and tests of actual ETCS communication setups.
- Results of such a quality-of-service test at a railway trial site are presented in [287], thus facilitating a comparison with the original requirements. It turns out that the QoS parameters are in the required range, although often close to and even sometimes worse than the requirements.

In the following, we adopt worst-case assumptions based on the requirements, because otherwise there would be no guarantee of a working integrated system.

14.3.1 A Communication System Failure Model

The communication link between train and RBC is always connected in normal operation mode. In that situation the following failures may happen:

Transmission errors occur from time to time, possibly due to temporarily bad radio signal conditions. There is no action necessary, because after a short time the link is operable again.

Connection losses may happen e.g., because of longer radio signal problems in areas where the radio coverage is not complete. The train hardware detects this state after some timeout and tries to establish a new connection. There is a slight chance of failing to establish such a connection until a certain timeout has elapsed, after which the connection establishment procedure starts over again.

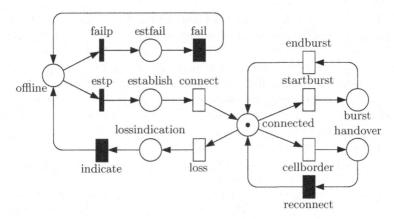

Fig. 14.3. Failure and recovery model for GSM-R communication channel

Handovers   take place every time the train crosses the border be-
                         tween the communication areas of two neighboring base
                         transceiver stations (BTS). The train connects to the
                         next BTS automatically, but this may take some time.

Figure 14.3 shows a stochastic Petri net (more specifically, a *deterministic and stochastic Petri net*, cf. Chap. 5) model of the described behavior. The firing delays and distributions have been chosen as follows. One unit of model time means one second in reality.

Transition **startburst** models the beginning of a transmission error. It has an exponentially distributed firing delay because of the stochastic nature of transmission errors. The corresponding firing time is comparable with a mean time to failure of the communication link because of transmission errors. The specification requires this value to be greater than or equal to 7 s for 95% of all cases. From the density and distribution functions of the exponential distribution

$$f(x) = \lambda e^{-\lambda x} \text{ and } F(x) = 1 - e^{-\lambda x}$$

we can calculate the necessary parameter λ value:

$$\lambda = -\frac{\ln p}{x} \approx 0.00733 \text{ with probability } p = 0.95 \text{ and } x = 7.$$

Transition **endburst** models the end of the transmission problem. The delay is assumed to be memoryless and the specification requires it to be smaller than 1 s in 95% of all cases. Thus, the transition's firing delay is assumed to be exponentially distributed with parameter $\lambda \approx 3$ ($F(1) = 1 - e^{-\lambda} = 0.95$).

The crossing of a cell border and connection setup with a new BTS is modeled by transitions **cellborder** and **reconnect**, respectively. The BTS are situated a few meters away from the track normally and have a typical density of $0.1 \ldots 0.3$ BTS per kilometer. Another source specifies 7 km as the mean BTS distance, which is adopted here. Unlike for personal use of a mobile phone, handovers happen quite often due to the speed of the train. ETCS is required to work for speeds up to $500 \, \mathrm{km \, h^{-1}}$ ($139 \, \mathrm{m \, s^{-1}}$). Thus the worst-case mean time between two handovers is $50.4 \, \mathrm{s}$. The firing delay of `cellborder` is thus exponentially distributed with parameter 0.0198 (the mean delay being equal to $1/\lambda$). From the specification we know that a reconnection is required to take at most $300 \, \mathrm{ms}$, which is taken as a worst case with a deterministic transition `reconnect`.

Following the specification, a complete connection loss takes place only rarely, namely 10^{-4} times per hour or $2.77 * 10^{-8}$ per second. The parameter of the exponential transition `loss` is set accordingly. There is a certain amount of time needed to detect the communication loss, which is required to be not greater than $1 \, \mathrm{s}$. This is modeled by the deterministic transition `indicate` with one as the fixed delay.

After being offline, the train communication system tries to reestablish the link at once. The requirements specify that a connection attempt must be successful with 99.9% probability, while in the remaining cases the establishment is canceled after $7.5 \, \mathrm{s}$ and retried. This behavior is modeled with immediate transitions carrying the success/fail probabilities `estp` and `failp`, and the deterministic transition `fail` with delay of 7.5. Connection establishment times are random, but required to be less than $5 \, \mathrm{s}$ for 95% of the cases. The corresponding firing distribution of transition `connect` is thus exponential with parameter 0.6.

The model shown in Fig. 14.3 depicts states and state transitions of the communication link. The initial state is `connected`. It is obvious that there will always be exactly one token in the model, letting the Petri net behave like a *state machine*, and the reachability graph is isomorphic to the net structure.

Because in every marking there is at most one transition with non-exponentially distributed firing delay enabled, the model can be numerically analyzed with standard solution algorithms for non-Markovian stochastic Petri nets as described in Sect. 7.3.3. Because of the state machine structure, it would also be possible to exchange all deterministic transitions (delay τ) with their exponential "counterpart" (with firing rate $\lambda = 1/\tau$), without changing the resulting steady-state probability vector. It could then be analyzed as a simple generalized stochastic Petri net.

Numerical analysis of the example is computationally inexpensive because of its small state space. Despite the "stiffness" of the problem (e.g., firing rates of transitions `endburst` and `loss` differ by eight orders of magnitude), the exact solution is a matter of seconds. A simulation with proper confidence interval control would take quite some time because of the mentioned rare events.

Table 14.1 shows the results of the numerical analysis. The connection is working with a probability of 99.166%, being worse than the required availability of 99.95% as specified in EEIG ERTMS User Group [96]. This requirement is commented to be a coverage requirement, although we see

Table 14.1. Performance results of the communication failure model

Place/state	Probability
Connected	0.916
Burst	$2.4305 * 10^{-3}$
Handover	$5.9027 * 10^{-3}$
Loss indication	$2.7546 * 10^{-8}$
Establish	$4.5910 * 10^{-8}$
Estfail	$2.0680 * 10^{-10}$

from the model evaluation that it is already violated by the allowed handover downtimes.

In fact, handovers account for more than 70% of the overall unavailability. To avoid their impact on the communication link, there are discussions about installing two independent GSM-R devices in each train. For instance in the Rome-Naples ETCS installation, all electronic units have been duplicated for a higher reliability and availability. The connection to the next BTS can then be established when the train gets close to the cell border already, thus avoiding any offline state due to handovers. Bursts are responsible for another 29% of communication outage, while the other failures have only a small influence.

Another direction of current research is to predict necessary handovers to speed them up. This is possible for train control especially because train routes and locations are well known in the system [242]. When a handover is imminent, it is clear what the next base station will be, and the handover process can be significantly sped up by a preparation. GSM management data are loaded at the future base station for this reason.

14.3.2 Alternative UML Statechart Model and its Transformation

UML Statecharts as extended by the UML profile for schedulability, performance, and time [255] (cf. Sect. 3.4) may be used for the communication link modeling instead of stochastic Petri nets. Figure 14.4 depicts such a Statechart describing the ETCS radio communication link state following the descriptions mentioned earlier.

The radio link operates in `Normal Mode` initially, which is specified by the destination state of the black dot. All times are given in seconds. Transmission errors are modeled by a Statechart transition leading from state `Normal Mode` to state `Transmission Error` with an `<<RTdelay>>` of {RTduration = (percentile, 5, 7)} (less than 7 s in 5% of all cases). It takes the radio link less than 1 s in 95% of all cases to operate in `Normal Mode` again, which is modeled by the Statechart transition with an `<<RTdelay>>` of {RTduration = (percentile, 95, 1)}.

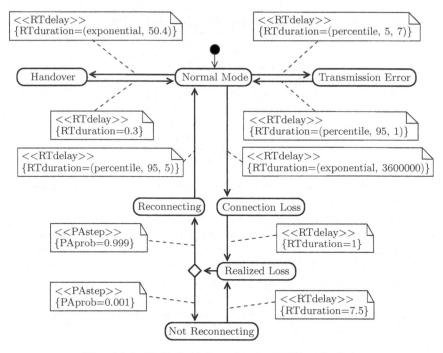

Fig. 14.4. UML Statechart for the ETCS radio link

Time between two handovers is assumed to be exponentially distributed with a mean of 50.4 s. The inscription of the corresponding transition from state `Normal Mode` to state `Handover` is {RTduration = (exponential, 50.4)}. The connection to the next BTS is modeled by a Statechart transition with a fixed delay of 0.3 s: <<RTdelay>> {RTduration = 0.3}.

Total connection losses result in a transition from state `Normal Mode` to `Connection Loss` with an exponentially distributed delay with mean $3.6 * 10^6$ s, specified by {RTduration = (exponential, 3600000)}. Connection loss detection is modeled by a transition with a fixed delay of 1 s <<RTdelay>> RTduration = 1. The behavior during a reconnection attempt is modeled in the Statechart using a choice state with two outgoing transitions: One with a probability of 99.9% (<<PAstep>> {PAprob = 0.999}, Success) and the other with a probability of 0.1% (<<PAstep>> {PAprob = 0.001}, Failed). The cancellation after 7.5 s is represented by a SM-transition with the fixed delay of 7.5 s {RTduration = 7.5}. In the case of a successful immediate reconnection it takes not more than 5 s in 95% of all cases until the radio link operates in `Normal Mode` again. This is modeled by the transition from state `Reconnecting` to state `Normal Mode` with the inscription <<RTdelay>> {RTduration = (percentile, 95, 5)}.

The UML Statechart model can now be translated into a stochastic Petri net following the method described in Sect. 3.5. The result is an analyzable

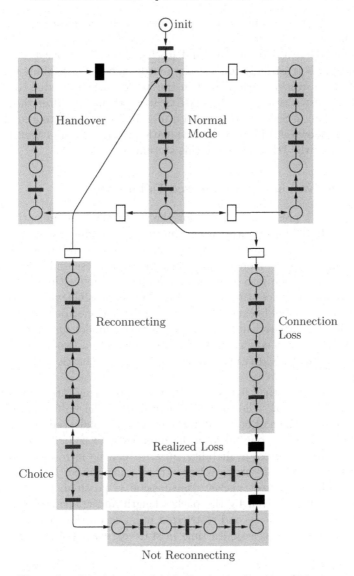

Fig. 14.5. Resulting stochastic Petri net after translation

Petri net model for which performance measures can be derived by a software tool. For the example, the resulting stochastic Petri net model is shown in Fig. 14.5. To avoid cluttering the picture, names of model elements are omitted. Section 3.5 uses a systematic way to attach names to the Petri net model elements derived from the Statechart state names. Petri net places and transitions resulting from one state of the UML Statechart are shaded together and marked accordingly. The probabilistic choice between failure and success of a reconnection is modeled with the immediate transitions in conflict.

The model is a deterministic and stochastic Petri net because of the transition types. It is strongly connected and safe (1-bounded). Many unnecessary immediate transitions are present in the model, which are due to the empty *entry* and *exit* activities in the UML Statechart model. After the simplification described in Sect. 3.5 the model in Fig. 14.5 is reduced to a model similar to the one shown in Fig. 14.3. The evaluation results given in Sect. 14.3.1 thus also apply to the UML model of this section.

14.3.3 Derivation of a Condensed Failure Model

Section 14.3.4 presents a model for the real-time communication between trains and RBC. Its performance evaluation is, however, computationally expensive, which is in part due to the combination of the failure model with the normal operation model. The failure model as presented in Sect. 14.3.1 is, therefore, condensed into a smaller model here, to make the later evaluation of the combination less time consuming. This is possible without considering the operation model because the failure model does not depend on it.

By doing so, there will be a tradeoff between model *complexity* and *accuracy*. We decided to condense the failure model into a two-state system with the basic states ok and failed. A corresponding stochastic Petri net is shown in Fig. 14.6.

The question is then how to specify transition firing rates to minimize the approximation error. The main characteristic of the failure model is mean availability, which shall be equal in the exact and condensed model. Thus the probability of having one token in place ok needs to be 0.99166.

Even with a correct availability, an error can be introduced by selecting a wrong speed of state changes between ok and failed. If the speed would be too high, no correct packet transmission is possible, because a certain undisturbed time (given by the packet length and transmission bit rate) is always necessary. The second restriction imposed on the condensed model is thus to keep the mean sojourn time in state ok exactly as it was in the full

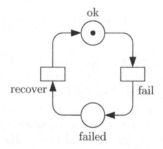

Fig. 14.6. Condensed failure model

model. This time is the reciprocal value of the sum of all transition firing rates going out from the state, in our case $1/(\lambda_{startburst}+\lambda_{cellborder}+\lambda_{loss}) \approx 36.77$.

With these two restrictions, the transition rates can be easily calculated. Let λ denote the transition rates and π the state probability vector in steady-state. Then

$$\pi_{ok} = 0.99166, \ \pi_{ok} + \pi_{failed} = 1 \quad \text{(probabilities)}$$
$$\pi_{ok}\,\lambda_{fail} = \pi_{fail}\,\lambda_{recover} \quad \text{(balance equations)}$$
$$\lambda_{fail} = \tfrac{1}{36.77} \quad \text{(sojourn time)}$$

and thus $\qquad \lambda_{recover} = 3.236$

The model is then completely defined and will be used as a simplified failure model in the subsequent section.

14.3.4 A Moving Block Operation Model

A model of the position report message exchange and emergency braking due to communication problems is developed below. The goal is to analyze the dependency between maximum throughput of trains and reliability measures of the communication system.

Figure 14.7 shows a Petri net model for the ETCS movement authority data exchange as explained in Sect. 14.2. The upper part models the generation of the position/integrity report and its transmission to the following train via the RCB. Transition GenMsg models the generation of a new message that assures train integrity and contains the current position. The send process to the RBC is modeled by place sendingUp and transition TransmitUp (delay specification see below). Processing at the RBC corresponds to the

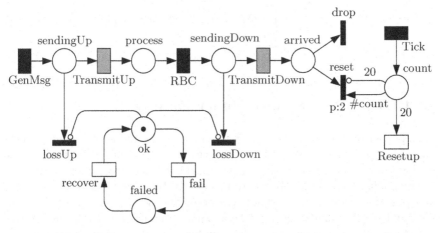

Fig. 14.7. Model of communication during moving block operation

deterministic transition RCB, and the transmission to the following train analogously to the uplink.

The failure behavior of the communication link is given by the condensed model as derived in the previous section. It is connected to the main model in a way that all messages are lost (tokens are removed from places sendingUp and sendingDown) as long as the link is failed.

For the application of the chosen simulation method, a Petri net place was needed in which the number of tokens can be used as the thresholds and final event. For this reason, the time until deadline t is "counted" with tokens in place count. For a reasonable number of possible thresholds, we define 20 as the number of tokens modeling the violation of the deadline. The deterministic firing time τ_{Tick} of transition Tick then needs to be set accordingly depending on the deadline under evaluation: $\tau_{\text{Tick}} = d/20$.

Every time a new movement authority message arrives at the second train (place arrived), the current elapsed time is set to zero: transition reset fires and removes all tokens from place count. The train stops after an exceeded deadline and we assume a Resetup time of 15 min before the train can move on. Movement authority packets arriving during that time are dropped (transition drop has lower priority than reset).

The end-to-end transmission delay for messages is specified in the requirements as being between 0.4 and 0.5 s on the average, but being less than 0.5 for 95%, less than 1.2 s for 99%, and less than 2.4 s in 99.99% of all cases. For a realistic mapping of this timing behavior into the stochastic Petri net model, we used two *generalized transitions* with expolynomial firing delays, how they are allowed in the class of extended and deterministic stochastic Petri nets (see Sect. 5.1). The actual data transmission times (0.13 s for a packet) have to be incorporated as well. The transition firing delay of TransmitUp and TransmitDown is defined by the following distribution.

$$f_X(x) = \begin{cases} 9.5 & \text{for } 0.53 \leq x < 0.63 \\ 0.057 & \text{for } 0.63 \leq x < 1.33 \\ 0.00842 & \text{for } 1.33 \leq x < 2.53 \\ 0.0 & \text{otherwise} \end{cases} \tag{14.1}$$

14.4 ETCS Performance Under Failures

For the performance evaluation of the model, the numeric analysis methods of Chap. 7 cannot be used, because the restriction of not more than one enabled nonexponential transition per marking is violated when transitions TransmitUp and TransmitDown are both activated. For a discrete time scale analysis, the resulting state space size would be too big because of the complex and highly differing firing delays. Thus simulation was the only choice, but standard methods could only be used for a very limited number of evalua-

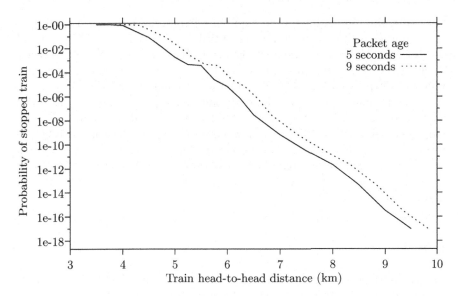

Fig. 14.8. Train stop probability vs. train distance

tions. Acceptable stop probabilities are naturally very small, and thus a rare-event simulation technique is necessary to successfully derive the measure of interest.

The presented Petri net model of the moving block operation has been evaluated using the RESTART implementation in TimeNET [198] (cf. Sect. 9.2). Computations were made on a cluster of Sun workstations under SunOS. Thresholds are defined based on the number of tokens in place count, whereas the rare event A is defined by having 20 tokens in it. The tool selects suitable thresholds based on a prior standard simulation run with limited computation time. Near-optimal thresholds are set for numbers of tokens where the probability has been computed, while the rest are set linearly depending on a manual choice of the overall number of thresholds.

Figure 14.8 shows the resulting probability for a train to be stopped due to an exceeded deadline vs. the train distance. To get a notion of the impact of the movement authority packet age at arrival, two curves are shown representing the minimum (5 s) and maximum (9 s) typical age. Beginning at a point where the train distance is big enough for the trains to be not almost always stopped (4 km in the 5 s case), the curves exhibit a logarithmic dependency between probability and distance.

The analysis shows a significant impact of communication delays and packet losses on the possible track utilization under ETCS level 3/moving block operation. As it has been already pointed out in a previous paper [346], the vision of "driving in braking distance" is unrealistic. However, the refined model presented here leads to more realistic values for the probability of a stopped train. Depending on the required reliability of high-speed train operation (and thus the accepted number of train brakings due to ETCS communication problems), it might be necessary to select a relatively long distance between trains. Another possibility will be to revise the current ETCS communication specifications. Compared with a theoretical setup of fixed blocks with the minimum block size, the performance measures show that an ETCS controlled train will be stopped about $1.3 * 10^{-6}$ seconds per year, or a train stops once every $7.1 * 10^8$ years. However, the real competitors to ETCS in terms of performance are the current national high-speed train operation systems. As they are usually based on standard electronic trackside equipment, their main problem is not communication system reliability, but installation and maintenance cost.

Notes

Model-based performance evaluation is helpful during the design of fault-tolerant distributed real-time systems. The chapter investigated safety-critical communication inside the future European Train Control System. Stochastic Petri nets and Statecharts are used to model and evaluate the failure and recovery behavior of the communication link as well as its combination with the exchange of vital train information between trains and radio block centers. Quantitative results are presented that put into perspective quality of service specifications and theoretical high-speed track utilization. The model evaluations show the significant influence of communication system reliability on efficient train operation. It will be crucial for the success of the final ETCS implementation level to analyze the real-time behavior of train operation and communication system under failures with more details.

The contents of the chapter are an extended presentation of the work published in Zimmermann and Hommel [346, 347]. Statechart model and translation in Sect. 14.3.2 were introduced in Trowitzsch and Zimmermann [308]. Design, analysis, and simulation of the presented models has been done using the tool TimeNET (cf. Sect. 12.1) and prototype extensions of it.

The first model presented in Zimmermann and Hommel [346] has been adopted as a case study for the modeling and evaluation with StoCharts in Jansen and Hermanns [180], a stochastic extension of state chart models.

Petri nets and their stochastic timed extensions have proven to be a useful formalism for real-time systems. They are considered to describe discrete event systems in a concise and appropriate way. An additional advantage is the availability of many different analysis and simulation techniques as well as software tools. Petri nets have been used in the context of real-time systems many times, see e.g., [34, 136, 236].

Most of the work in the area of train control systems deals with *qualitative* aspects like validation of correctness, absence of forbidden safety-critical states, etc. Yet in a real-time system like a distributed communication-based train control system, critical safety questions can only be answered when also *quantitative* aspects are considered and evaluated. Failures and other external influences on the model require stochastic model values, but fixed values for deadlines or known processing times are equally important. Modeling and evaluation techniques need to support both to be applicable in this area.

In references [182, 243], the ETCS communication structure is modeled with colored Petri nets. The model is used for a division of the system into modules, visualization of the functional behavior, and a check of different scenarios.

A verification of the radio-based signaling system together with a case study of a rail/street crossing is carried out in reference [77]. Live sequence charts are used to model the system, which is analyzed with the STATEMATE software tool. The same example is specified and validated by the authors of [10] with a formalism called "Co-Nets." The model is based on timed Petri-Nets and contains object-oriented features as well as rewriting logic.

The ETCS radio block center is formally modeled and validated in reference [47]. Message sequence charts are used to model and check different scenarios.

ETCS train traffic is compared with today's standard train control operations in Germany in a simulation study of Deutsche Bahn (German railways company) [257]. Using a proprietary simulation program, the movement of a set of trains through an example line is simulated. The results say that ETCS operation in its final stage will increase track utilization by about 30% for the example. However, the communication is not modeled, and failures are not taken into account.

15

Supply Chain Performance Evaluation and Design

Supply chains and logistic networks play a major role in today's businesses because of the growing importance of external suppliers to final products and partition of work between distributed plants. Material buffer levels are kept small to decrease the bound investments. Timely deliveries of intermediate parts are thus necessary to avoid shipment delays while keeping the amount of work in process small.

Model-based quantitative evaluation of logistic networks is a vital tool to aid in the decision-making at various stages of planning, design, and operation of supply chain operation. A robust control and management of supply chain loops can be achieved by analyzing performance measures such as the throughput rate, average resource utilization, expected number of parts in a buffer, setup costs, work-in-process inventory, mean order queue time, etc. All of these measures are indicators of how well the supply chain is operated.

Typical decisions during the planning and design stages include the number of containers or transport facilities, buffer storage capacity in a logistic center, scheduling of material and parts, number of links in inbound and outbound logistics, and location of distribution centers. During the operational phase, performance modeling and analysis can help in making decisions related to predicting the probability of a material shortage.

An adequately complex and flexible model class is necessary to capture the detailed behavior of a logistic network and the individual nodes, such as manufacturing plants (assembly, fabrication), logistics centers, suppliers, and dealerships. Colored stochastic Petri nets as described in Chap. 6 are applied to an example in the following.

The remainder of the chapter is organized as follows. The subsequent section introduces the supply chain example that is considered throughout the chapter. Section 15.2 presents a colored stochastic Petri net model of the example, which is finally analyzed in Sect. 15.3. The modeling method applied in the chapter is covered in detail in Chap. 6.5, while the used quantitative evaluation technique (discrete event simulation) has been explained in Sect. 9.

15.1 A Supply Chain Logistics Example

A supply chain system from the area of vehicle production is taken as an example here. It represents an adapted version of a real-life industrial problem that has been considered in a project involving a major U.S. car maker, and groups of Stanford University and Technische Universität Berlin. Details of system and model have been altered in order to keep the original information confidential. Model and problems are, however, typical. The main issue considered here (thus driving the modeling and evaluation process) is the question how long customers have to wait for their vehicle from the day of the purchase decision. This delay between order and delivery (or order-to-delivery time) is denoted by *OTD time* in the sequel.

The background is the way how the U.S. vehicle business (in contrast to European car makers) works today. Vehicle dealerships keep a high number of cars in their lot to have a large selection available for prospective customers. Buyers usually select a vehicle directly from the lot and thus do not need to worry about the time it would take to get an individually ordered vehicle. There are, however, big drawbacks for the car making companies and dealers. The high amount of available cars binds money and makes fast reactions to changing customer needs impossible. In fact, the selection of vehicles and the installed options are guessed based on past sales patterns. What thus often happens is that many of the vehicles can only be sold by giving large rebates, decreasing the earnings substantially. Another issue is that individually manufactured vehicles could be sold with a higher price considering the different options. A navigation system would for instance result in high earnings when sold to an interested customer, but might only be sold to a certain percentage of customers.

The current way of operating the supply chains, plants, and logistics implies a very long OTD time for an individually ordered vehicle. Car makers have realized that an agile supply chain and a short OTD time can be a substantial marketing factor and lead to higher earnings. Even in Europe, where individually ordered vehicles are sold by tradition, OTD times have become an issue to attract more customers lately. The example considered in this chapter shows how OTD times of a certain operation style can be evaluated and how possible improvements are quantifiable to aid in strategic design decisions.

Figure 15.1 shows a rough overview of the covered entities. Customer, dealership, and (assembly) plant as well as the logistic network that transports vehicles from the plant to the dealership are considered. Internal and external suppliers which deliver intermediate parts to the assembly plant (*inbound logistic*) are not considered here. We assume that enough material is available at the plant to avoid downtimes. How this can be efficiently achieved has been considered in the project as well.

A pure on-demand production setup would work as follows. When a customer comes to a dealership, he has a certain vehicle in mind that he or she wants to buy. This specification of a desired vehicle is ordered by the dealer

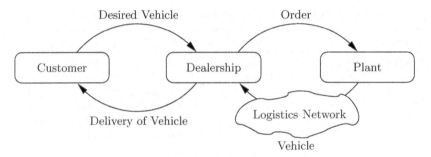

Fig. 15.1. Sketch of supply chain entities

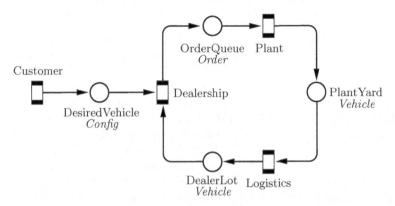

Fig. 15.2. Main colored Petri net model of the example

from the plant. The order is stored in an order queue, which is after some time processed by the plant, resulting in a new vehicle in the plant yard. From there it is transported by the logistics network to the dealer where the customer can pick it up.

15.2 Colored Petri Net Model of the Supply Chain Example

The following sections present the supply chain model and its hierarchical decomposition into submodels. Each model describes the actions and decisions of one typical entity, for example, a dealer or customer. Communication takes place by exchanging tokens via interface places. Behavior and decisions based on rules are specified using transition firing attributes.

Figure 15.2 shows the topmost level of hierarchy of the supply chain model. It describes the main entities and their interaction. Transitions with thick bars depict substitution transitions, which are refined with a submodel (see the following subsections). Types of tokens in places are depicted in italics (e.g., *Config* or *Vehicle*). Transition Customer models the customer

Table 15.1. Token types (colors) of the supply chain model

Color	Element	Element Type	Remarks
Config	model	string	"ModelA" or "ModelB"
	drive	string	"TwoWd" or "FourWd"
	interior	string	"'Leather," "Vinyl," or "Cloth"
	color	string	"Black," "Blue," "Grey," "Green" "Silver," "Tan," "Red," "Orange" "Yellow" or "White"
Order	Conf	Config	Vehicle specification
	OrderTime	DateTime	Time of ordering
	Origin	string	"Customer" or "Dealer"
Vehicle	Order	Order	Order for the production
	ProdTime	DateTime	Time when production finished

behavior; see Sect. 15.2.1 for more details. The selected vehicle type and configuration information is transferred to the dealership model (transition Dealership, Sect. 15.2.2) through place DesiredVehicle. The queue of waiting orders at the plant is modeled by place OrderQueue. After production in the Plant (explained in Sect. 15.2.3), new vehicles arrive in the PlantYard and are transported by Logistics (see Sect. 15.2.3) to the DealerLot. The model considered here only takes into account one dealership and one plant in contrast to the original detailed model.

Tokens model complex entities and thus have a set of corresponding attributes. The set of token types used in the supply chain model are listed in Table 15.1. String and DateTime are the only necessary base types. The latter is a convenient way of handling a point in time by specifying day and time. Configuration of a vehicle and the type of car that a customer wants to buy are described by the type Config. It specifies the model, type of drive, interior, and color. This is a simplification of the actual configuration attributes for the sake of readability. An order for a vehicle production is described by color Order, describing the order time as well as the ordered configuration. An actual vehicle has the attributes Order – the order that initiated the production of the vehicle – and the time when its production was finished.

15.2.1 Customer Model

For our purposes it is important to specify *when* a customer comes to a dealership to purchase a car and *what configuration* he wants to buy. This customer behavior is captured in the customer submodel shown in Fig. 15.3. The timed transition NewCustomer models a new customer purchase. Its firing delay is exponentially distributed with a mean interfiring time of 5 000 s, the assumed time between two successive customers at a dealership.

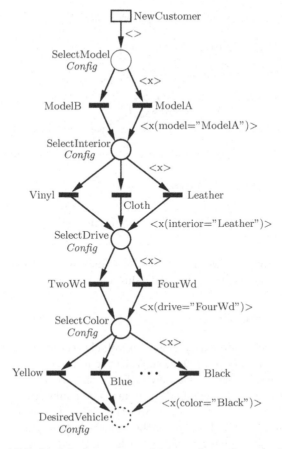

Fig. 15.3. Model of customer vehicle configuration selection

The configuration of the customer's desired vehicle is generated in a step-wise fashion. A newly created token of type *Config* without set attributes is generated by the firing of NewCustomer in place SelectModel. The token then follows through a series of places (SelectModel, SelectInterior, SelectDrive, and SelectColor), in which one attribute of the configuration is set one after another. Firing probabilities that correspond to known customer choices are associated to the immediate transitions like ModelB and ModelA in the model selection case. The different choices are obviously not independent in reality, but a previous analysis showed that this simplification does not lead to a significant error. The firing of any one of the immediate transitions sets one configuration attribute to the corresponding value. Firing ModelA for instance sets model to "ModelA." To simplify the figure, only one arc inscription pair is shown for every decision, and some transitions are neglected in the color selection case. The fully specified configuration token

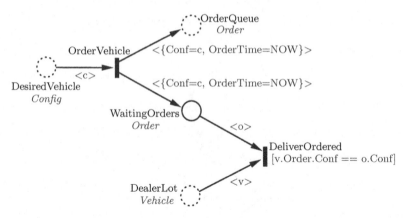

Fig. 15.4. Dealership model

finally arrives in place `DesiredVehicle`, which acts as an interface to the upper model level and thus the dealership model.

15.2.2 Dealership Model

When a customer wants to purchase a vehicle, the corresponding token of type *Config* arrives in place `DesiredVehicle` in the dealership model shown in Fig. 15.4. The dealer sends a new order to the order queue (by creating a token of type *Order* in place `OrderQueue`). The configuration of the order is copied from the customer's choice, and the order time is set to the current model time (`NOW`). A similar copy of the order is kept in place `WaitingOrders` at the dealership to identify waiting orders later on when new vehicles are available in the dealer lot.

If a vehicle arrives at the dealer lot (place `DealerLot`) that matches a waiting order, transition `DeliverOrdered` fires. The guard function of that transition ensures that only vehicles that match exactly are delivered.

15.2.3 Plant Model

The plant submodel is depicted in Fig. 15.5 and describes how orders in the order queue are processed. The model contains two parts: the bottom model specifies at what times new orders are scheduled for production, while the upper part models the actual scheduling and production process.

Material and production planning takes place in the presented model once a week. The firing delay of transition `StartSchdule` is thus $7 * 24 * 60 * 60$ s. Its firing generates a token without a specified type in place `Scheduling`, enabling transition `Schedule` due to its firing guard `#Scheduling>0`. Transition `Schedule` is immediate, and thus moves all currently waiting orders from the `OrderQueue` to place `Waiting` by a series of subsequent firings. The scheduling

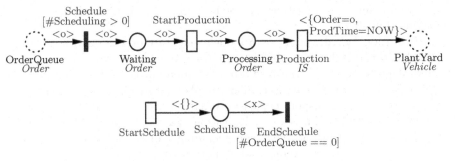

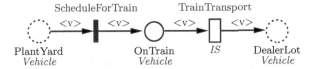

Fig. 15.5. Plant model

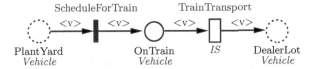

Fig. 15.6. Vehicle logistics: Train transport

phase is finished when the order queue is empty, because that state enables transition EndSchedule. Its guard function is #OrderQueue==0, and the firing removes the token from place Scheduling.

Order tokens wait in place Waiting for the corresponding start of vehicle production. The sequencing of orders is done by transition StartProduction, which has *single server* semantics (the default) for this reason. It may fire every 2 600 s, which is an estimation of the time between two successive vehicle production starts that are available for one dealership. Vehicles in production are modeled by order tokens in place Processing, and the production time (about one week) is specified at the *infinite server* timed transition Production. As mentioned earlier, material buffering and supply chain issues are not considered in this simplified model.

Finished vehicles are modeled as new tokens that are created in place PlantYard. Attributes *Order* and *ProdTime* are set accordingly: The order containing the configuration information is copied from the initiating *Order* token, while the production time is set to the current model time.

15.2.4 Vehicle Logistics Model

New vehicles need to be transported from the plant to the dealership, which is done by the logistics network. Vehicles are picked up from the plant yard and transported to the dealer lot. Figure 15.6 shows a first simple version of how this can be done, considering only train transport. Vehicle transport by train has the advantage of being very cost-efficient. The downside is its long delay and high variance. Moreover, train transport is externally managed by the train companies, and thus cannot be influenced significantly by the car making company.

Transition `ScheduleForTrain` models the start of a train transport for a vehicle token. All vehicle tokens are moved immediately to place `OnTrain`, which depicts the actual transport. To avoid unnecessary model complexity, no railcar loading and scheduling is modeled. We realistically assume that there is always enough transport capacity available on trains; therefore, there is no resource restriction necessary. Transition `TrainTransport` models the actual delay of the transport operation, including waiting and scheduling times, which is assumed to be exponentially distributed with a mean of one million seconds (about 11 days). The firing semantics of `TrainTransport` is *infinite server*, because all vehicle tokens are delayed concurrently. A more complex logistics model is developed and analyzed in Sect. 15.3.3.

15.3 Order-To-Delivery Time Evaluation and Improvement

The order-to-delivery time for customers is evaluated based on the model that has been explained in the previous sections. We assume that the initially planned setup of the supply chain (as described earlier) is developed further to achieve better OTD times. Results of the quantitative model evaluation lead to ideas how the design and operation of the supply chain can be changed in order to decrease the OTD time. The subsequent sections present the corresponding updated models as well as the resulting OTD values. It serves as a demonstration of how a series of model changes and quantitative evaluations can be successfully exploited to improve the performance of a supply chain significantly.

The following result measures are defined:

– The most important value we are interested in is the OTD time in days as a mean over all customers, defined as measure `OTD`. This is a nontrivial measure, because we need to average over customers and not over time as it is offered by default. The waiting time of a customer is given directly by the time that the corresponding *Order* token resides in place `WaitingOrders` of the dealership model. However, waiting times cannot be directly measured.

The OTD time can be derived indirectly using Little's Law (compare p. 70). The waiting time of a token in a place is thus equal to the mean number of tokens in that place divided by the mean interarrival rate (all in steady-state). The mean time between the arrival of new customers is known from the mean firing delay of transition `NewCustomer` in the customer submodel, which was set to 5 000 s (17.28 firings per day). What remains to be computed is the mean number of waiting tokens in place `WaitingOrders`, which is simply specified by `#WaitingOrders`. The OTD time in days can thus be computed in steady-state, averaged over an infinitely long time interval as

$$\texttt{OTDtime} = \frac{\texttt{\#WaitingOrders}}{17.28}$$

- In later model variants some vehicles are stored in the dealer lot. It makes sense then to analyze the number of vehicles available in the lot, which is defined as

$$\texttt{NDealerLot} = \texttt{\#DealerLot}$$

Measure `NDealerLot` is analyzed in steady-state as well.

Discrete event simulation is used for the quantitative evaluation of the model. Numerical analysis techniques are not applicable because of the transitions with nonexponentially distributed firing delay distributions. Moreover, the storage of model time values in tokens as necessary for the OTD time evaluation leads to a state space that is infinite. Performance measures of the supply chain example model have been analyzed with the software tool TimeNET (cf. Sect. 12.1).

All evaluations have been carried out on a PC with Intel Pentium III Mobile processor running at 1 GHz under Windows XP. A simulation run of 6 years of model time typically took 45 s. Measure samples for the first year of model time are discarded to avoid influences of the initial transient phase. Statistical analysis shows that the remaining simulation length leads to sufficiently accurate results for our purposes. A typical evaluation like the one with two trucks in Sect. 15.3.3 shows that considered 26 553 samples result in a confidence interval for the mean `OTDtime` of $(7.9617, 8.2873)$ (i.e., a maximum relative error of only 2%) for a confidence level of 99%. Because of the high percentage of vehicles that are immediately delivered in that case, the deviation of the samples is quite high, which means that this example represents one of the worse confidence settings.

The initial setup of the model as specified in Sect. 15.2 is evaluated first. There are obviously no vehicles available in the dealer lot, because every one is produced on demand and immediately delivered; thus `NDealerLot` $= 0$. The mean OTD time is computed as `OTDtime` $= 25.08$ days.

15.3.1 Popular Configuration Storage at the Dealership

The first change in the model includes the details of how vehicles can be produced and stored without prior customer order. A rough estimate of the probability of vehicle configurations of customer purchase decisions is known from past sales numbers. The dealership can thus order popular configurations, keep them available at the dealer lot, and sell them to customers with matching vehicle desire immediately. The dealership can check the influence

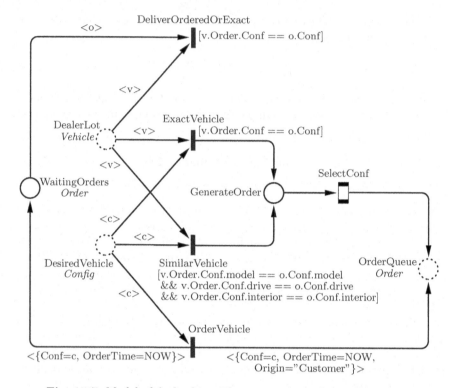

Fig. 15.7. Model of dealership with storage of popular configurations

of a certain selection of popular configurations and the number of available vehicles at the lot using the performance evaluation of the model.

Figure 15.7 shows a more detailed model of a dealership with appropriate changes. A new customer order arrives as a corresponding token in place DesiredVehicle. If there is a completely matching vehicle available in the dealer lot (place DealerLot), transition ExactVehicle fires and delivers that vehicle to the customer. In the case that a vehicle is available that is almost exactly as wanted (only the color may differ), that vehicle is sold to the customer by firing transition SimilarVehicle with an assumed probability of 50%. The other half of almost matching configurations as well as all nonmatching ones are handled by transition OrderVehicle. Transitions OrderVehicle and SimilarVehicle have priority one to be in conflict, while transition ExactVehicle has priority two to ensure that an exactly matching vehicle is always sold. The matching of transitions DeliverOrderedOrExact, ExactVehicle and SimilarVehicle is achieved by the comparison of the token attributes in the shown guard function.

Firing of OrderVehicle works like in the first simplified model. An *Order* token is placed in the OrderQueue with appropriate configuration setting and current order time. Furthermore, it is marked in the token that the order originated from a customer. A similar token is placed in `WaitingOrders`, from which waiting orders are removed by firing `DeliverOrderedOrExact` when a matching new vehicle arrives in the dealer lot. Waiting orders should be fulfilled first, thus `DeliverOrderedOrExact` has priority three, superseding all other delivery transitions. It should be noted that the vehicle that is delivered in response to a waiting order has not necessarily been produced to that order. It could have arrived by chance in the dealer lot after the order was placed.

A new vehicle is ordered by the dealership whenever a vehicle is sold directly from the lot. Firing `SimilarVehicle` or `ExactVehicle` creates one token in place `GenerateOrder`, which initiates the selection of an appropriate configuration in subnet `SelectConf` (see below), which is placed in the order queue just like customer orders. This keeps the number of available vehicles at the dealer lot at a steady level. The initial marking of place `DealerLot` is chosen appropriately with a number and selection of vehicles that should be available.

Selection of a vehicle model and configuration for a dealer order is done in submodel `SelectConf`, which is shown in Fig. 15.8. There is simply one immediate transition `PopConf`i for every vehicle configuration that is considered for being available at the dealer lot. The output arc inscription sets the appropriate order information. Relative firing probabilities of the conflicting immediate transitions are chosen for simplicity according to the previous

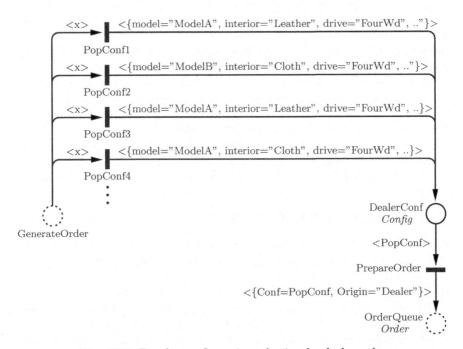

Fig. 15.8. Popular configuration selection for dealer orders

Table 15.2. Influence of vehicle storage on the OTD time in days

Number of configurations	NDealerLot	Immediate delivery	OTDtime	
			All	Waiting
2	153.3	34%	16.07	24.36
9	43.8	53%	10.29	21.87
9	168.3	55%	9.99	22.33

purchase decisions at the dealership. In the actual project an optimization was carried out to find a small number of configurations that matches the set of popular configurations with small deviations. One of the decisions to be made at the dealership is how many of the most popular configurations should be ordered and stored. The presented model contains at most nine configurations, which together account for almost half of all vehicle sales. Transition PrepareOrder creates an order token in the order queue with the generated configuration and a marker that this order was placed by a dealership.

This model change will not result in a significantly smaller OTD time for plant-ordered vehicles. There will, however, be a certain probability of a zero OTD time, leading to a smaller average mean OTD time for all customers. Different setups have been evaluated after the model change. Results are shown in Table 15.2.

The table shows the number of popular configurations that are stored in the dealer lot and the mean number of available vehicles. Both numbers mainly influence the probability with which a customer buys a vehicle from the lot, which is listed as "Immediate delivery." The results show that the number of dealer-ordered configurations is more important than a very high number of available vehicles at the lot. The two OTD time values represent mean numbers, taking into account all customers or only the ones that do not purchase an available vehicle. We choose to order nine different configuration types and to keep the amount of available vehicles in the range between 60 and 120 as a consequence of the results. The actual number of available vehicles indirectly depends on other model parameters and can thus not be chosen directly. The new OTD time of 9.99 days represents a 60% improvement.

15.3.2 Order Scheduling at the Plant

At the plant there are two obvious details that can be changed for a smaller OTD time. First, customer-ordered vehicles should be processed with priority over the ones that will be stored in the dealer lot. In addition to that it is possible to change the order scheduling scheme from once-a-week to daily.

Figure 15.9 shows the updated plant model. The delay between two successive firings of transition StartSchedule is changed to 86 400 s. The only visible difference is that scheduling and production start is now individually done for customer orders (trailing C) and dealer orders (trailing D in names).

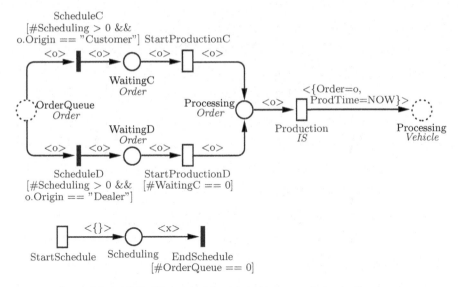

Fig. 15.9. Plant model with customer order priority

The two scheduling transitions can only be activated for the appropriate order tokens due to their guard functions and thus sort the order tokens into the two places WaitingC and WaitingD. Production of dealer-ordered vehicles (transition StartProductionD) may only start if there is no waiting customer order.

This change has the following influence on the performance measures. The mean number of available vehicles is 124.4 and 55% of customers directly buy a vehicle from the lot. OTD time of all customers drops to 8.37 and the time for waiting customers to 18.62, an improvement of 16%.

15.3.3 Truck Transport of Customer-Ordered Vehicles

To reduce the average OTD time we need to concentrate on the OTD time of waiting customers, because the percentage of immediate deliveries could only be increased by storing more vehicles with additional configurations in the dealer lot. Other means of vehicle transport are considered for this reason. Specialized trucks for vehicles will be responsible for the transport between plant yard and dealer lot. However, because of the higher cost, this will only apply to customer-ordered vehicles. Train transport is chosen for dealer-ordered vehicles and in cases where there is no truck available for transport.

An additional type (color) *Truck* is introduced into the model. It has six attributes: V1, V2, V3, V4, and V5 denote the vehicles that are stored in the truck token and are of type *Vehicle*. Attribute Num contains the number of vehicles that are loaded on the truck currently.

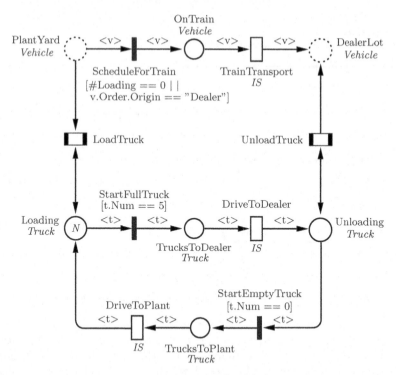

Fig. 15.10. Vehicle logistics with truck and train transport

Figure 15.10 shows the updated logistics submodel including truck transports. The upper row of transitions and places models train transport as in the previous submodel version shown in Fig. 15.6. However, vehicles are scheduled for train transport only if they are either dealer-ordered or if there is currently no truck available for transport.

The lower part of the figure shows the truck logistics behavior. *N* trucks are available in place `Loading` initially. When fully loaded with five vehicles, transition `StartFullTruck` fires and puts the truck token into place `TrucksToDealer`. The driving time is modeled by transition `DriveToDealer` with *infinite server* firing semantics and a delay that is 250 000 s plus an exponentially distributed part with mean 50 000. This models the shorter and less variating transport time with respect to train transports. Unloading of trucks happens in place `Unloading`, from where empty trucks drive back to the plant as shown in the model.

Loading and unloading of vehicles onto trucks is specified in the submodels `LoadTruck` and `UnloadTruck`. Both are quite similar; therefore, only the loading case is shown in Fig. 15.11. There are five transitions that model the loading of one vehicle each, because of the five vehicle slots on the truck. Any one of the transitions takes a vehicle token from the plant yard and copies

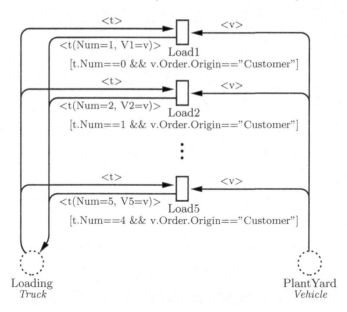

Fig. 15.11. Vehicle logistics: Loading of trucks

its information into the space that the transition is responsible for. The truck token is removed from place **Loading** during the firing and put back with the additional vehicle information later. The number of vehicles is updated as well. Only customer-ordered vehicles are taken into consideration, which is specified in the guard functions. This is not strictly necessary here, because other tokens are removed from **PlantYard** immediately by **ScheduleForTrain**.

An important issue is to have almost always a truck available for loading of vehicles, because they otherwise need to be transported by train. The probability that a truck is waiting in place **Loading** has been evaluated for different numbers of trucks in the logistics network. Its limiting value for bigger numbers of trucks is obviously one. Figure 15.12 shows the results graphically.

A sensible number of trucks for the example can be chosen based on the resulting OTD times. Figure 15.13 depicts mean OTD times for all and for waiting customers depending on the overall number of trucks. About 15 trucks are necessary to achieve the minimum OTD time of 5.92 days, which is an improvement of 29%. The mean OTD time of waiting customers is 12.9 days.

Another interesting evaluation shows how the OTD times are distributed for different numbers of trucks. Figure 15.14 shows probability density functions for selected truck numbers, ignoring all zero OTD times. The curves start at the left with the remaining probabilities that a waiting order is fulfilled by an incoming dealer-ordered vehicle, which has not been available before. The peak value at 12.5 is due to the mean delays of production and transport, while the stochastic influences lead to the distribution around this value. For smaller number of trucks the transport time is heavily influenced by the long

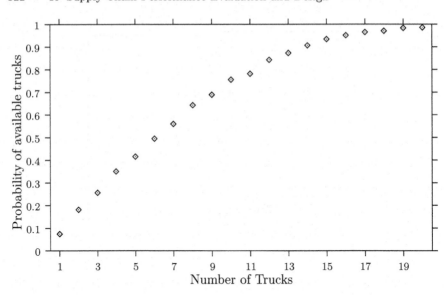

Fig. 15.12. Truck availability at plant yard for loading

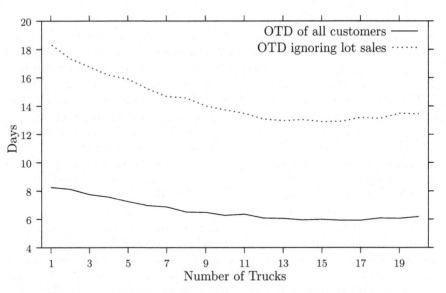

Fig. 15.13. Order-to-delivery times versus overall truck number

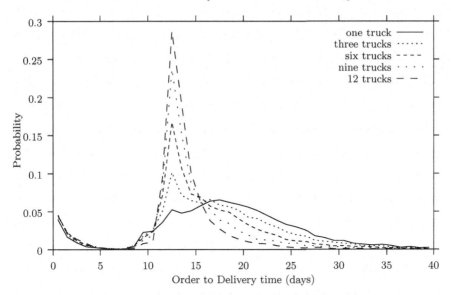

Fig. 15.14. Order-to-delivery time distributions of waiting customers

and variating train transport, resulting in a much flatter curve without the
characteristic peak.

Notes

The application example has shown that strategic decisions during the design
of a supply chain can be efficiently aided using a model and performance
evaluation. The mean OTD time for customers as the significant measure for
our experiment could be improved from about 25 to less than 6 days in the
model.

The chapter is based on results of an industrial project, in which the
specific class of colored stochastic Petri nets of Sect. 6 has been developed
and implemented in the software tool TimeNET (Sect. 12.1). The goal of the
project was to model and analyze supply chain issues of a major U.S. vehicle
manufacturer. Background material has been presented in [172, 331, 363].

Generalized stochastic Petri nets are used for modeling and analyzing supply chain networks in reference [272]. Make-to-stock and assemble-to-order
systems are compared in terms of total cost.

A toolset for modeling and analysis of logistic networks is presented in
Bause [19]. Another software tool that can be used to model and analyze
logistic systems is ExSpect [312], which uses hierarchical colored Petri nets.

The discrete behavior of logistic systems is modeled by timed Petri-Nets
with individual tokens in Lemmer and Schnieder [221]. The application of
timed colored Petri nets to logistics is also covered in van der Aalst [1].

Model-Based Design and Control
of a Production Cell

Support for the efficient design and operation of manufacturing systems requires an integrated modeling, analysis, and control methodology as well as its implementation in a software tool. Colored Petri nets are able to capture the characteristic features of manufacturing systems in a concise form. Stochastic as well as deterministic and more general distributions are necessary and thus adopted for the firing times of transitions.

Different evaluation techniques are available for an efficient performance and dependability prediction: direct numerical analysis, approximate analysis, and simulation. Finally, the model can be used to control the manufacturing system directly. There is no need to change the modeling methodology, thus avoiding additional effort e.g., for model conversion.

The chapter demonstrates the modeling, performance evaluation, and control of a production cell using variable-free Petri nets (cf. Sect. 6.5). The considered application example is described in Sect. 16.1. Its vfSCPN model is presented in Sect. 16.2.

The throughput of the example is derived using the iterative approximation technique introduced in Sect. 8, and results are compared with those obtained by standard methods. Online control of the production cell using the model is presented in Sect. 16.4, before some final notes are given. The software tool implementation of the described methods is discussed in Sect. 12.1.

16.1 A Production Cell Application Example

This section describes the application example that is used in the remainder of the chapter. It is a manufacturing cell built of parts from the "Fischertechnik" construction kit for education and research purposes. Figure 16.1 shows the system and Fig. 16.2 the layout. The application example has been chosen to demonstrate the benefits of model-based evaluation and control.

In the considered production cell, new work pieces are initially stored in the high bay racking on pallets. The rack conveyor can fetch one of them and

Fig. 16.1. The considered production cell

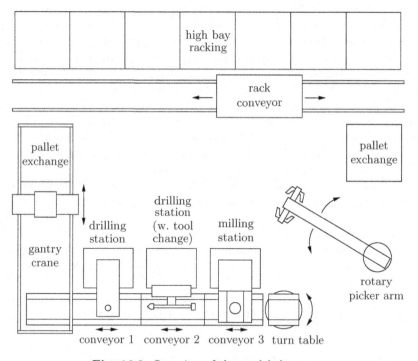

Fig. 16.2. Overview of the modeled system

deliver it to one of the pallet exchange places. A gantry crane may take it to the first conveyor belt. Three of these belts move work pieces from one processing station to another. There are two drilling stations, the second one having three different interchangeable drilling tools. The last station is a milling machine. Work pieces stay on the conveyor during processing. After leaving the machines, they arrive at a turn table. This table puts them into position for the rotary picker arm, who takes the work piece to the right pallet exchange place. From there it is brought back to a slot in the high bay racking by the rack conveyor.

Exchange of unprocessed and finished work pieces with the world outside the production cell takes place via the rack storage. In the following it is assumed that work pieces have to be machined by the two drilling machines and the milling machine, in this order. Work pieces move counterclockwise through the system.

The production cell example comprises 22 motors and 84 sensing devices altogether. For the online control they can be accessed through a standard RS 232 serial communication interface.

16.2 Production Cell Modeling with vfSCPN

The application example is modeled with variable-free colored Petri nets, which have been covered in Sect. 6.5. Simple Petri nets would be too restrictive because of the different production steps that need to be modeled, while the full expressional power of colored Petri nets is not necessary.

Two color types are defined for the application area of manufacturing systems: **Object tokens** model work pieces inside the manufacturing system, and have the product name (*name*) and the current state (*step*) as attributes (both have the base type *string*). **Elementary tokens** as they are known from simple Petri nets are used to model states of resources. Places can contain only tokens of one type or color; a place marking is thus a multiset of the sort of the type. Places are drawn thicker to mark them as corresponding to object tokens. Types of places are thus not written into the model figures. The model is hierarchically structured using substitution transitions. One of the refining submodels is described after the main model explanation.

The used net class and restriction to two place (and token) types strongly encourages the modeler to describe the system as it is structurally. This means that e.g., a buffer should be modeled by exactly one place with the appropriate capacity. One location of parts (one active resource) should not be modeled with more than one place (transition). However, if a complex manufacturing system is modeled, one can also start at a higher level of abstraction, e.g., one production cell is modeled by a transition without further details. Later on, this transition could then be changed into a substitution transition, which is then hierarchically refined by submodels at lower levels of hierarchy. Besides that, the degree of abstraction is left to the modeler.

16.2.1 Main Hierarchical Model

The example is modeled as a vfSCPN in this section. Structural information of the production cell is considered first, while the integration of work plan information (i.e., the processing steps and their sequence) is covered in

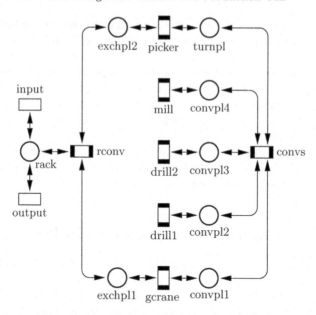

Fig. 16.3. Highest level of the hierarchical colored model

Sect. 16.2.3. This means that for the moment no transition details about the color of removed and added tokens is required in the models, as it will be added later on.

Figure 16.3 shows the top level (the prime page) of the model. Its composition follows the layout of the modeled system depicted in Fig. 16.2. Please note that this and all subsequent model figures show the system after a counterclockwise rotation with respect to the layout sketch in Fig. 16.2.

Places model buffers and other possible locations of work pieces. Place rack corresponds to the rack storage, places exchpl1 and exchpl2 to the pallet exchange places, and place turnpl to the turn table. The remaining four places represent locations of work pieces on the conveyors, which are directly in front of the machines or the gantry crane. As described earlier, input and output of work pieces takes place through the rack storage and is modeled with transitions input and output.

In principle, there are two different operations that can be performed: transport and processing of work pieces. The former corresponds to moving a token to another place, while the latter is modeled by a change in the color of the work piece token, specifically of the *step* attribute. Transitions modeling machines specify processing steps that only change the token color. This is emulated by removing the former token from the place and instantly adding a token with the new color by firing the transition. Therefore, many transitions and places are connected by arcs in both directions (loops), which are conveniently drawn on top of each other and thus look like arcs in both directions. The shown structural model contains all possible actions of the resources, even if they are not used for the processing. The gantry crane could e.g., move work pieces from the conveyor to the exchange place as well.

Transitions with thick bars depict substitution transitions, which are refined by a submodel on a lower level of hierarchy. The next subsection shows such a refining submodel for the rotary picker arm (transition picker). Substitution transitions are e.g., used to describe the behavior of a machine with more detail during a top–down design. Submodels from a library of standardized building blocks (templates) can be parameterized and instantiated while refining the model [344]. This alleviates the creation of complex manufacturing system models, where many structurally similar parts can be found.

Transition rconv contains the model of the high bay rack conveyor, while transitions gcrane and picker correspond to the gantry crane and rotary picker arm, respectively. For the transport of a work piece from one machine to the next, two of the three conveyor belts have to operate simultaneously. All three conveyors are, therefore, treated together as one transport facility and are modeled by transition convs. Thus, their synchronization is hidden at a lower level and can be specified together. The meaning of the remaining model elements should be clear from the layout figure.

16.2.2 Refined Model of the Rotary Picker Arm

The Petri net model shown in Fig. 16.4 is a hierarchical refinement of the substitution transition picker in Fig. 16.3. It specifies the inner behavior of the rotary picker arm as well as the correlated control of the turn table. The system states of the rotary picker arm are modeled by elementary places and arcs (drawn thin in Fig. 16.4). Possible locations of work pieces are modeled by the object places TurnTable and PalletExch (drawn thick). Because these places are interface places to the upper model level, they are depicted with dashed circles. They refer to the places turnpl and exchpl2 on the upper level, respectively.

Transitions having names beginning with G describe actions of the picker arm gripper (lower, close, raise, open). The ArmTurn transitions model the turning of the picker arm, and the turn table is described with transitions named TTurn.

The rotary picker arm can execute two useful actions: take a work piece from the turn table to the upper pallet exchange and the reverse. The current state of the picker arm (and of the turn table) corresponds to the location of the elementary token in the model. Figure 16.4 shows the state after initialization, where the token is in place Idle.

Either one of the two immediate transitions StartF and StartB can fire if the resources are idle, thus starting one of the two possible transport actions. The decision is made by firing guards (marking dependent boolean expressions) of the two transitions, which are added later depending on the work plan information. This ensures that the picker arm is only activated for useful

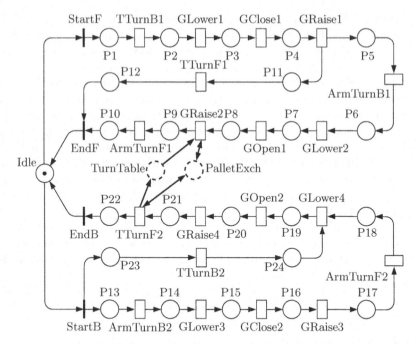

Fig. 16.4. Refined model of the rotary picker arm

transport activities. After the firing of one of the two starting transitions, the corresponding transport action begins. The different transitions become enabled and fire in succession. They describe individual steps of each transport. Actions that may be performed concurrently are modeled that way, e.g., TTurnB2 in parallel with ArmTurnB2 and its successors after StartB has fired.

16.2.3 Integration of Work Plan Information

The model presented so far only describes structural information of the production cell, i.e., properties of resources and transport connections. The full production cell model needs to capture work plan specific information as well. This can be done using the *transition modes* of vfSCPN (cf. Sect. 6.5). Every mode corresponds to one individual way of how the transition behaves, including all necessary information such as input and output tokens and delay.

Work plan information could be added to the model manually, which is a rather tedious task for complex applications. A method to model work plans with variable-free colored Petri nets in a way similar to the manufacturing system structure has been introduced in references [339, 354, 361]. The structural model describes the abilities and work plan independent properties of the manufacturing system resources, such as machines, buffer capacities, and transport connections.

Work plan models specify the work piece dependent features of the manufacturing system. The different model parts are automatically merged resulting in a *complete model* with the algorithm of [339, 354, 361]. The final model then includes both the resource constraints of the system and the synchronization of the production steps.

Figure 16.5 shows the first part of the work plan model for the running example. This model describes the sequence of operations and transports for a work piece A at the highest level of hierarchy. Each step can only be carried out by a resource that is available in the manufacturing system layout. Therefore, only transitions, places, and their connecting arcs from the structural model can be used here. Arc inscriptions show the name (A) and processing state (unpr or drilled) of the work piece, separated by a dot. These two inscription parts refer to the *name* and *step* attributes of the product tokens. It is always possible to use this simplified arc inscription notation because every transition occurrence in a work plan model has exactly one transition mode. It should be noted, however, that the same (structural) transition may appear several times in such a model, possibly resulting in different transition modes in the complete model.

A work plan model usually consists of a simple succession of transitions and places. An exception is the modeling of alternative routes, assembly, and disassembly operations. More than one input or output arc is connected to a transition then. Although it cannot be seen immediately in Fig. 16.5, an assembly operation is also needed for the example work plan. Each work piece is transported and processed while being fixed to one pallet. For an input of a new work piece into the rack storage, there has to be an empty pallet in it (place rack contains a token of color P.empty). The input operation (transition input fires) removes this token and puts back a token with color A.unpr (i.e., unprocessed). The inverse operation is carried out by transition output.

A pallet without a mounted work piece has no different states. The transport strategy of empty pallets is described in an additional work plan model, which is not shown here.

After the structure and work plans have been modeled with individual vfSCPNs as described earlier, a complete model is generated automatically. This is done by adding the information contained in the work plan models

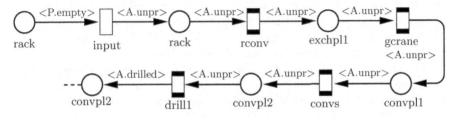

Fig. 16.5. Part of the work plan model

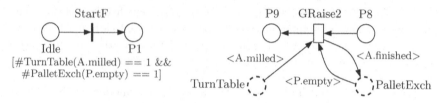

Fig. 16.6. Example transition modes

to the structural model. Each transition of the structural model is extended with descriptions of its different modes. For every transition mode, the input/output behavior, firing time distribution, and guard function may be specified. Every time a transition appears in a work plan model, a new transition mode is created. Thus, the resource constraints that are imposed by the structure and the synchronization of the work plans are compiled into one model. The resulting model contains all necessary information from the structural and work plan models; details are covered in Zimmermann et al. [339].

Figure 16.6 shows the resulting inscriptions for two transitions of the model in Fig. 16.4 as an example. Firing of `StartF` initiates a transport of a product after the last processing step (`A.milled`) from the turn table on a pallet (`P.empty`) at the exchange place of the high-bay racking. The guard of the immediate transition ensures that this operation is only started if the right product and pallet are available. Transition `GRaise2` is the step of the picker arm to which the actual state change of product tokens is attached in the model. The ready product (`A.milled`) is taken from the `TurnTable` and put on an empty pallet in place `PalletExch`. The empty pallet token there is removed and substituted by a `A.finished` token, which models a finished product on a pallet.

16.3 Performance Evaluation

Performance and dependability of the manufacturing system example is evaluated based on the model to answer design questions. Throughput, utilization, work in process, and other properties of a certain layout can for instance be derived. Different variations of the system and their resulting performance and dependability measures can thus be computed and compared. The aim of this investigation is to obtain a better understanding of the correlations between details of the manufacturing system (e.g., the buffer capacities) and the main performance measures (like the throughput).

For the application example, the aim is to evaluate the throughput of work pieces. A performance measure is defined in the model, which gives the throughput of all finished work pieces per hour. This can be done using the throughput of a transition that each work piece passes exactly once like output. The mean number of transition firings per time unit is measured for this reason with a performance variable #output in steady-state.

Direct numerical analysis, approximate analysis, or discrete event simulation can be used to obtain the desired measures in steady-state from the model. Advantages and disadvantages of the three algorithms depend heavily on the actual model and the performance measures of interest (state space size, numerical stiffness, rare events, etc.). Unfortunately, thus no automatic decision can be made of which method will be the best one. Already the definition of "best" at least necessitates the specification of a balance between required speed and result exactness. Therefore, the modeler decides which method is applied, or can compare the results of different ones, as it is done for the application example in the following.

The throughput of the modeled manufacturing system is computed with different algorithms that were explained in Part II, facilitating a comparison of results and computational efforts. The software tool TimeNET (cf. Sect. 12.1) is used for all evaluations. The example is used here especially to demonstrate the iterative approximation technique of Sect. 8. The intermediate steps are thus shown in the following. For the standard numerical analysis and simulation techniques only the results are shown for a comparison further below.

16.3.1 Partition and Aggregation

Following the steps of the approximate performance evaluation method described in Chap. 8, the original model is partitioned into subsystems SS_i first. These subsystems are aggregated and used to build low-level systems LS_i and a basic skeleton BS as the basis for the iterative approximation technique. This section shows the results of this step for the running example.

The partition step divides the original main model (Fig. 16.3) into submodels, which can be selected according to the substitution transitions. However, the submodels of the drilling and milling machines are considered together with the conveyor system and are not used as individual subsystems because of their simple structure. In fact, because of the path-preserving aggregation, their internal behavior can later be completely aggregated (cf. the basic skeleton shown in Fig. 16.8). In our example, thus all remaining four transitions (besides input and output) are subsystems. Decisions about a partition can also be done based on an estimation of the submodel state space size [171].

The partition thus results in four subsystems for the example, namely for rconv, gcrane, picker, and convs (the latter together with the three machine-modeling transitions). Figure 16.7 sketches the chosen partitioning. The places that are neighbors of subsystems constitute the set of buffer places, in our example rack, exchpl1, exchpl2, turnpl, and convplpl.

The next step is the aggregation of all subsystems following the rules described in Sect. 8.2.2. Every subsystem SS_i is aggregated individually. It leads to four aggregated subsystem models SS_i^*, which are substantially less complex and thus have a smaller potential state space. The low-level systems

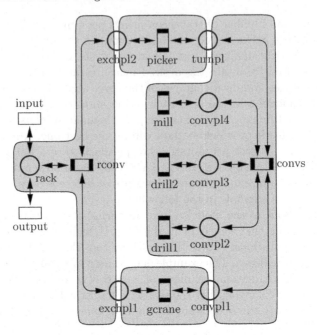

Fig. 16.7. Partition of the example model

LS_i are derived in the following step. For a subsystem SS_i, the corresponding LS_i consists of the original SS_i and all other subsystems $j \in \{1, \ldots, n\}, j \neq i$ in their aggregated versions SS_j^*.

Figure 16.8 shows the basic skeleton BS of the example, where all subsystems SS_i are aggregated. Only transitions input and output are kept from the original model in all low-level subsystems as well as the basic skeleton.

The resulting transition modes are not shown in the figure for simplification. The model part corresponding to the rack conveyor is covered in more detail (rc is short for rconv in the model). In the original model, there are five path combinations between the buffers exchpl1, exchpl2, and rack. New work pieces (A.unpr) are moved from rack to exchpl1, while empty pallets are transported both from exchpl1 and rack to exchpl2. For empty pallets, there is an additional way from exchpl1 to rack. If the work pieces are finished, they are taken from place exchpl2 to rack. Thus, the transportation of both unprocessed work pieces and empty pallets between rack and exchpl1 is possible. Therefore, it is important to distinguish between different colors in the rconv subsystem (see Table 16.2) during the throughput computation.

The five aggregated paths can be found in the aggregated version: (rc_in1, rc_out2) represents the path of unprocessed work pieces and empty pallets between rack and exchpl1 (using different colors), while (rc_in2, rc_out3) and (rc_in2, rc_out1) represent the transport of empty pallets to exchpl2

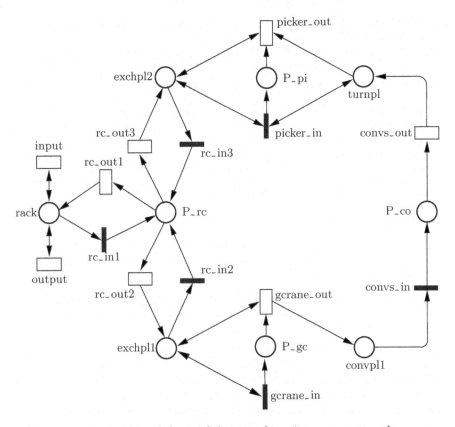

Fig. 16.8. Basic skeleton of the manufacturing system example

and **rack**, respectively. The path (rc_in3, rc_out1) corresponds to the transport of finished work pieces to **rack**.

16.3.2 Performance Evaluation Results

The iterative approximation technique of Chap. 8 has been applied using the derived low-level systems and basic skeleton of the example. Table 16.1 compares computational effort and achieved accuracy with numerical analysis and simulation. The CPU times were measured on a Sun Ultra 5 workstation running at 333 MHz.

For the simulation (see Chap. 9), three runs with different accuracies have been carried out. Table 16.1 shows confidence level and relative error in percent. It is obvious that the computational effort increases dramatically with the desired accuracy. The computation took about 70 min for a relative error of 1% .

The numerical analysis method (see Chap. 7.3.2) yields the most exact results. For the modeled system, it calculates a throughput of 19.6692 work

Table 16.1. Accuracy and computational effort of evaluation techniques

Technique	Numerical analysis	Iterative approx.	Simulation 99%, 1%	98%, 2%	90%, 10%
Throughput	19.67	19.12	19.59	19.48	20.75
Error	0%	2.8%	0.4%	1.0%	5.5%
CPU time (s)	1625	186	4 176	1 104	56

Table 16.2. Results after the final iteration of the approximation algorithm

Subsystem SS_k	Place p (Color)	Sojourn time for p in LS	$\Lambda(t)$ in BS $t \in OT_k$ in LS	Sojourn time for p in BS
rconv	P_rc			
	(A.finished)	0.095063	0.0227272	0.095605
	(A.unpr)	0.117001	0.0555555	0.116851
	(Pallet.empty)	0.095063	0.0243902	0.095605
gcrane	P_gc	0.183017	0.0294117	0.183017
convs	P_co	0.252994	0.0217666	0.252995
picker	P_pi	0.139954	0.0384615	0.139954

pieces per hour. The computation took about 27 min and generated a Markov chain with 41 279 states for the original model. The approximation algorithm only took 3 min to finish despite its more complex algorithm, because the sizes of the reachability graphs and thus CTMCs it had to cope with were about one magnitude smaller than for the original model. The state space sizes of the low-level systems ranged between 306 and 1 128 states, while the basic skeleton only has 15 states. For the initialization of the algorithm, service rates of all output transitions of the aggregated parts in the low-level systems and the basic skeleton were set to 1. Convergence was, nevertheless, reached after only three iterations. Table 16.2 shows the computed sojourn time of tokens in the preset places of the output transitions. With an error of less than 3%, the approximation algorithm achieves a good tradeoff between result quality and computational effort.

In addition to that the influence of the rack conveyor speed on the overall throughput was evaluated. All rack-related delays were multiplied by a factor in the range from 0.4 up to 5.0. Figure 16.9 shows the resulting throughput values.

It follows from the results that the rack conveyor is the main bottleneck of the example system, which is in accordance with observations of the real system. For a planned real-life production system, this would surely be not acceptable, because the usually expensive processing stations should be utilized more than a transport facility. From the resulting throughput values and the associated profit of work pieces as well as additional costs for a faster rack conveyor, the right rack conveyor can be selected.

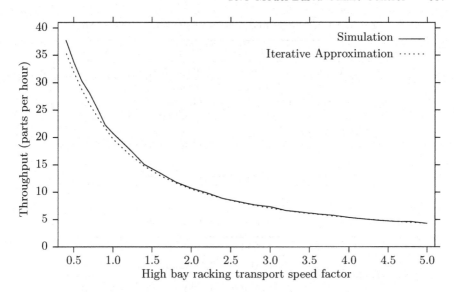

Fig. 16.9. Throughput vs. rack conveyor speed

16.4 Model-Based Online Control

Motivated by the use of separate models for structure and work plans of a production system, the task of designing the control system can be divided in two steps. The specification of the work plans ensures that the manufacturing system produces (in the model) the desired products. However, the afterwards automatically generated main model does not necessarily have to be free of deadlocks or optimal with respect to some performance measures like the throughput. The model can be analyzed and the problems detected, leading to control strategies that improve the system behavior.

A controller can only forbid activities in any state of the production system that would otherwise be possible. This corresponds to disabling transitions inside the model. Using marking dependent guard functions, this can be done easily with the used class of Petri nets. The guard function of a transition (or of one firing possibility) is a boolean function of the net marking, which has to be true to allow the enabling. The second step of the control design thus corrects and optimizes the production process. Finally, the model-based online control ensures the execution of the specified behavior.

The control technique explained in Chap. 11 has been used to actually control the application example. It is briefly explained here using the submodel of the rotary picker arm as shown in Fig. 16.4.

Table 16.3 lists the motors and sensors that are present in the rotary picker arm and the turn table (compare the layout in Fig. 16.2), which must be controlled together. In the case of the Fischertechnik application model, every motor movement has an associated sensor for the final position. These

Table 16.3. Motors and sensors of the rotary picker arm

Motor	Direction	Final position sensor
Turntable	Forward	TTurnIsFront
	Backward	TTurnIsBack
GripperPosition	Open	GripperOpen
	Close	GripperClosed
GripperHeight	Raise	GripperIsUp
	Lower	GripperIsDown
ArmTurn	Forward	ArmIsFront
	Backward	ArmIsBack

Table 16.4. Picker arm transitions and their associated control signals

Transitions	Control output		Sensor input
	Motor	Direction	
TTurnB1,2	Turntable	Backward	TTurnIsBack
TTurnF1,2	Turntable	Forward	TTurnIsFront
GOpen1,2	GripperPosition	Open	GripperOpen
GClose1,2	GripperPosition	Close	GripperClosed
GRaise1..4	GripperHeight	Raise	GripperIsUp
GLower1..4	GripperHeight	Lower	GripperIsDown
ArmTurnF1,2	ArmTurn	Forward	ArmIsFront
ArmTurnB1,2	ArmTurn	Backward	ArmIsBack

sensors are hard-wired such that no movement is possible beyond the physical constraints. It is, therefore, not necessary to switch off a motor explicitly from the model-based control.

The different steps for the two possible transport actions (forward and backward movement of a part) are described by sequences of elementary transitions and places. Each timed transition corresponds to one controllable activity. After the inclusion of the work plan information in the transition modes of the model, transitions StartF and StartB can only fire if tokens that correspond to work pieces to be transported are located in places TurnTable and PalletExch. Firing guards (marking-dependent boolean expressions) of the transitions are used for this. It ensures that if the work plan models are being specified correctly, the rotary picker arm is only activated for useful transport orders. Modeling errors or wrong transport strategies leading e.g., to deadlocks or bottlenecks can be detected with a prior analysis.

The corresponding transport action begins after the firing of one of the two starting transitions. Each transition describes the individual steps of each transport, and have input and output signals associated to them. Table 16.4 lists the transitions in the model that have input or output signals attached.

There are always two or three transitions with identical input and output signals, because the possible elementary movements are necessary in both transport movement sequences. As an example, transition TTurnB1 models the movement of the turn table into the backward position. When the transition becomes enabled, the turn table motor is switched on by the associated output signal. When the final position is reached, the corresponding sensor TTurnIs-Back is activated and the input signal leads to the firing of the transition.

Notes

Manufacturing and automation systems are a classic application area of Petri nets; see [88,270,276,286,292,294,336] for surveys. Two popular extensions including stochastic timing are stochastic Petri nets (SPNs, [5]) and generalized stochastic Petri nets (GSPNs, [53]). Both have been used for manufacturing system modeling and evaluation [7,295]. However, the use of these simple Petri net classes requires that if more than one product is processed by a machine, the machine's model has to be replicated because of the lack of distinguishable tokens. In general, using nets without distinguishable tokens leads to models that do not reflect the actual structure, making the models less understandable. Simple Petri nets, therefore, do not appear to be adequate for the modeling of more complex production systems [339].

Therefore, colored Petri nets (CPNs, [188]) have been applied to manufacturing systems. Viswanadham and Narahari [322] used colored Petri nets for the modeling of automated manufacturing systems. On the basis of these models, deadlocks can be found by analyzing the invariants. A colored Petri net model of a manufacturing cell controller is described by Kasturia, DiCesare and Desrochers [192]. After obtaining its invariants, the liveness of the model is checked.

Ezpeleta and Colom [100] generate a model that is used for deadlock prevention control policies. Martìnez, Muro and Silva [233] show how the coordination subsystem of a flexible manufacturing system can be described by a colored Petri net. The obtained model is embedded into the surrounding levels of control (local controllers and scheduling subsystem), and a terminology based upon the Petri net colors is used for the interaction. Analyzing the model detects deadlocks, decision problems, and gives performance measures that depend on variations in the system being modeled.

The independent modeling of structural and functional system parts is of high importance for the modeling of complex production systems. Only then it is possible to change parts of the work plans without having to redesign the whole model. Villaroel, Martìnez and Silva [316] proposed to model work plans with colored Petri nets, while for the structural model predefined building blocks are used.

Simple Petri net models of complex production systems tend to be large as well as hard to understand and maintain. Colored Petri nets overcome this problem with their more sophisticated modeling capabilities. A drawback is the need to define textual elements like color types and variables. The class of variable-free stochastic colored Petri nets used throughout this chapter has been introduced especially for the modeling of manufacturing systems [339, 354, 361], aiming at resolving this problem for a specific application domain. An industrial case study can be found in reference [340]. Later work showed the applicability to workflow systems [83, 85].

17

Summary and Outlook

This text takes a look at stochastic discrete event systems from various viewpoints. It intends to give an overview of models classes, algorithms taking a model as input, and what benefits we may finally get from a model. Several application examples are presented in Part III to motivate advantages of model-based design especially for complex systems. Examples and case studies moreover aim at covering a broad range of model classes as well as engineering issues, which can be supported based on a formal model. The importance of stochastic discrete event systems as a family of models and their applications should be evident from the literature available today and the number of projects based on it.

A variety of model classes for their description have been developed in the past. Probabilities of random actions and randomly distributed times have to be incorporated in the models, which were in part originally developed as qualitative models only. The text covers some of the most prominent model classes allowing a derivation of measures related to performance, reliability, or timeliness. Part I introduces automata, queuing models, and simple as well as colored Petri nets. The selection of "the" right model class, a good abstraction level, as well as how to finally construct a specific model is, however, impossible with a formal algorithm. It is preferably learned from experience or with intuitive examples.

Despite the wealth of model classes and their variants, we believe that the general understanding of stochastic discrete event systems allows a unified treatment. An abstract model class SDES has been defined to capture models that are expressed as automata, Petri nets, or other families of models. The goal was not a definition allowing any imaginable model; in fact, details of existing model classes were intentionally ignored to simplify the presentation. This is, however, not a general restriction, as the SDES model class could be extended accordingly. A similar approach has been taken in the development of the Möbius tool [81,82]. The introduced SDES model class extends this work by allowing variants of actions and enabling degrees among others. Colored

Petri nets and other complex models can thus be captured. The corresponding translations of all presented model classes is formally given in the text.

Several algorithms are then presented in the second part of the text, based on the abstract SDES model class. The approach of using one model class as a kind of abstract interface between modeling and evaluation allows to extend the framework by new model classes and algorithms without touching the other side. A serious drawback is, however, that specialized analysis algorithms for selected model classes can not be used. An example are product-form solutions of queuing networks.

The behavior of each model class is specified indirectly by the definition of the stochastic process resulting from the corresponding SDES model. Algorithms for the standard numerical analysis as well as discrete event simulation were given in the context of SDES. Further contributions include an iterative approximation technique, a distributed simulation method, as well as an efficient indirect optimization technique.

The software tool TimeNET has been used for the case studies. It is developed in the group of the author, and implements efficient evaluation methods for several classes of stochastic discrete event systems.

Although several fields of research in stochastic discrete event systems have reached a certain maturity, there are numerous open problems to be addressed in the future. From the modeling point of view, new model classes or variants of existing ones are required to capture the special needs of restricted application areas. Colored Petri nets seem to be general enough to capture very complex systems. Specialized subclasses may pave the way for an improved industrial acceptance, just like the evolution of general simulation software shows.

Another important aspect regards the modularity of models. With the number of aspects to be considered in a system, and the shared work in a design team, we cannot expect that a single formalism will be sufficient to describe a complex system. Modular multiformalism approaches aim at combining model parts from different model classes. In the future, it appears to be more important to develop interfaces and analysis algorithms that cope with heterogeneous models than to develop additional model classes. A vision of design engineering is that the supplier of a system part delivers a behavioral model together with it, which can then be "plugged into" the model of the appliance it is assembled to.

Another main area of future efforts comprises algorithms for the evaluation of models. It is comparably simple to develop a new model class; its analytical tractability is, however, often an open issue. Simulation and numerical analysis based on the full reachability graph are two extremes that are available today. Both have their advantages in certain environments. In the future, it is expected that numerical analysis will go beyond its current limits by the use of structured representations of the reachability graph and transition matrix, and even further with condensed representations of the solution vector itself. The boundaries of numerical analysis techniques with respect to more general delay distributions are well understood today, leaving only little hope for the development of efficient general solution algorithms. Approximation techniques with controlled accuracy are required, which differ between important and insignificant parts of the reachability graph. We expect the clear boundaries between exact analysis and single-path simulation to fade with combination techniques.

Theoretical developments will require an efficient implementation in software tools. Industrial acceptance is, however, only achieved with a graphical user interface as an intuitive modeling environment as well as robust and efficient evaluation algorithms. Important steps for this development include international standardization efforts of Petri nets, the definition of an exchange format for software tools, and the advent of software tools supporting multiformalism models.

Symbols and Abbreviations

$\lvert A \rvert$	Number of elements of a multiset A
A/B	Restriction of a set A to elements of B
$\parallel$	Concurrency relation between events
$x^{\bullet}$	Postset of x
$^{\bullet}x$	Preset of x
$\to$	Relation between events, which is true if the first one may causally affect the second one
$\longrightarrow$	It follows that
$\sigma \xrightarrow{V} \sigma'$	Firing sequence V leading from a state σ to σ'
$\mathbf{1}$	vector with all entries equal to one, with appropriate size
2^A	Power set of a set A, i.e. the set containing all subsets of A
A	A stochastic automaton
a	An action of a SDES model
$A^{\star}$	Set of actions of a SDES
Activity	An enabled action variant together with its scheduled execution time or remaining activity delay
Arr	SDES actions that correspond to customer arrivals in a queuing network
as	Action state: a set of currently active action variants together with their remaining activity delays
AS	Set of all possible action states AS
au	One atomic unit of a distributed simulation
AU	Set of atomic units of a distributed simulation
$AU_{affected}$	Set of atomic units that are influenced by an action in a distributed simulation
AV	Set of all possible action variants v of a SDES
AV^{exp}	Action variants with an exponentially distributed delay
AV^{gen}	Action variants with an associated generally distributed delay
AV^{im}	Action variants with zero delay
$\mathbb{B}$	Set of $\{\text{True}, \text{False}\}$

B	Set of states in a RESTART simulation
β^*	Set of possible bindings of a colored Petri net transition
β	A binding of a transition in a colored Petri net, one possible setting of values for transition variables
BS	Basic skeleton
$\mathbf{C}$	Matrix of conversion factors (numerical analysis)
C	A cut in a distributed simulation
C	Color (type) of a place, variable or expression in a colored Petri net
$\mathbf{C}$	Incidence matrix of a Petri net
Cap	Capacity of a place; maximum allowed number of tokens in it
ce	An event of the complete stochastic process $CProc$; marks the execution of an action variant or activity
$\chi+$	Upper bound on the throughput of a Petri net transition
$\chi-$	Lower bound on the throughput of a Petri net transition
$\chi\simeq$	Approximation of the throughput of a Petri net transition
χ	Mean throughput (number of firings) of a Petri net transition
$Cond^\star$	Condition function of an SDES; allows or forbids to enter states
cost	Optimization function or cost function
CPN	Colored Petri net
$CProc$	Complete stochastic process underlying a SDES
cs	Complete state of the stochastic process $CProc$; contains state variable values as well as the set of ongoing activities (the action state as)
CS	Set of all theoretically possible complete states cs of a SDES
cst	An individual compound simulation time of a distributed simulation
CST	All possible compound simulation times in a distributed simulation
CTMC	Continuous-time Markov chain
$\overline{\mathbf{D}}^{(1)}$	Average service demands of transitions in a Petri net
Deg	Degree of concurrency of a Petri net transition, either SS or IS
$Deg^\star$	Enabling degree of an SDES action
$Delay^\star$	Delay of an SDES action variant; described by a probability distribution function
Δ	Dirac impulse
DTMC	Discrete-time Markov chain
e	A specific event of a stochastic automaton A
$\mathcal{E}$	Event set of a stochastic automaton A
$\mathrm{E}\{x\}$	Expected value of x
E	Event (in probability theory)

$\mathbb{E}$	Event space in probability theory
E	Additional boolean SDES state variables that model the enabling of transfer actions *Trans*
e	An event in a distributed simulation
E	Set of possible events in a distributed simulation
EC	Equal conflict relation between two Petri net transitions
EMC	Embedded Markov chain
$Ena^\star$	Enabling function of a SDES variant
Enabled	Enabling function that returns the enabled action variants of an action for a specific state σ
ETCS	European Train Control System
EvList	Event list in a distributed simulation
$Exec^\star$	Execution function of a SDES action variant; describes state change
Executable	Set of action variants that are actually executable in a state σ. Differs from *Enabled* by considering priorities and immediate versus timed delays
$Expr_{Var}$	Set of all expressions built from variables out of *Var*
$\mathcal{F}^+$	Set of non-negative distribution functions
f	State transition function of a stochastic automaton
F	(Cumulative) distribution function
$\mathcal{F}^{det}$	Deterministic probability "distribution" function set
$\mathcal{F}^{exp}$	Exponential probability distribution function set
$\mathcal{F}^{gen}$	General probability distribution function set
$\mathcal{F}^{im}$	Immediate probability "distribution" function
f_I	Importance function that guides a RESTART simulation
$Future_{cst}^{EvList}$	Future part of the event list *EvList* with respect to a simulation time *cst*
G	Distribution function defining interevent times in a stochastic automaton
G	Boolean guard function that specifies which bindings of a transition may be enabled for a state in a colored Petri net
Γ	Set of feasible (or enabled) events in a state of a stochastic automaton
Γ	Average interfiring time of a Petri net transition
GSPN	Generalized stochastic Petri net
$\mathbf{I}$	Identity matrix
I	Initial state of a queuing network model
Inh	Inhibitor arc multiplicities in a Petri net
IS	Infinite server semantics of a Petri net transition; degree of concurrency
K	Capacity of a queuing node
Λ	Delay of a transition in a Petri net; specified by a probability distribution function

Λ	Interarrival time of customers to a queuing network; specified by a probability distribution function
$LastBefore_{cst}^{E}$	Last event time in the event list E before a simulation time cst
$Level$	Level of a state in a RESTART simulation
lp	A logical process of a distributed simulation
LP	Set of logical processes of a distributed simulation
LS	Low-level system
$\mathbf{m}'$	Subsequent marking in a Petri net; e.g. after firing a transitions
$\mathbf{m}_+$	Upper bound on the mean number of tokens in a Petri net place
$\mathbf{m}_-$	Lower bound on the mean number of tokens in a Petri net place
$\mathbf{m}_0$	Initial marking of a Petri net
M	Number of levels in a RESTART simulation
$\overline{\mathbf{m}}$	Mean number of tokens in a Petri net place
$\mathcal{M}_A$	Set of all multisets over A
$\mathbf{m}_{\simeq}$	Approximated value for the mean number of tokens in a Petri net place
$\mathbf{m}$	A marking of a Petri net
M	Set of theoretically possible markings of a Petri net
M	Number of servers of the queues in a queuing network model
$PV_{\min}^{e_1,e_2}$	Minimal path priority in a distributed simulation
$mode$	Action mode, one of the possible sub-actions of an action
$Modes^{\star}$	Set of action modes (of an action) for a SDES model
μ	Service time of a queue in a queuing network model; specified by a probability distribution function
mv_t	SDES action variable corresponding to the transition mode selection in a vfSCPN model
$\mathbb{N}^+$	Set of positive natural numbers
n	Index variable of the complete stochastic process $CProc$
$\mathbb{N}$	Set of natural numbers
$Node$	Computing node on which an atomic unit is executed in a distributed simulation
ω	Weight of a path in a RESTART simulation
OT	Output transitions of a subsystem
$P\{\sigma \xrightarrow{V} \sigma'\}$	Path probability of taking the firing sequence V from state σ to σ'
p	A single place of a Petri net
P	Set of places of a Petri net
$\mathrm{P}\{x\}$	Probability of x
$\mathbf{P}$	Matrix of state-transition probabilities
$Past_{cst}^{EvList}$	Past part of the event list $EvList$ with respect to a simulation time cst

$\pi_j(q)$	Probability that there are j customers in queue q
Π	Transition priority of a Petri net
$\boldsymbol{\pi}$	Vector of state probabilities
Post	Forward incidence matrix of a Petri net, also used in function form
Pre	Backward incidence matrix of a Petri net, also used in function form
$Pri^\star$	Priority of an SDES action
$Pri^\star_{Global}$	Global priority of an action in a distributed simulation
PV	Path priority vector in a distributed simulation
$\mathbf{Q}$	Matrix with state-transition rates
q	One node in a queuing network model
Q	Set of simple queuing systems (or nodes) in a queuing network model
QN	Queuing network
QS	Queuing system
$\mathbb{R}^+$	Set of positive real numbers $0 < x < \infty$
$\mathbb{R}^{0+}$	Set of non-negative real numbers $0 \leq x < \infty$
$\bar{r}$	Average token waiting time in a Petri net place
r	Routing probability for a specific pair of queues in a queuing network; equals one entry in the routing matrix R
R	Routing matrix of a queuing network model
$\mathbb{R}$	Set of real numbers $-\infty < x < \infty$
R	Number of retrials in a RESTART simulation
$\mathbf{R}$	Routing matrix of a Petri net
RAD	Remaining activity delay
$ravg^\star$	Specifies whether a SDES reward variable should be averaged over the observation interval or not
RE	Edges of the reachability graph RG
$Region$	Model region of an atomic unit in a distributed simulation
RESTART	Repetetive simulation trials after reaching thresholds
$rexpr$	Set of considered events or states for an automaton-based reward variable
$rexpr$	Filter expression of a reward variable in a colored Petri net
$rexpr$	State-dependent numerical expression or considered queue for a reward variable of a queuing model
RFT	Remaining firing time
RG	Reachability graph of a SDES
$rimp^\star$	Impulse reward part of a SDES reward variable
$R_{inst}{}^\star$	Instantaneous reward at a certain point of time
$rint^\star$	Observation interval of a SDES reward variable
$robj$	Reward variable object in a colored Petri net: either a place or a transition
$rrate^\star$	Rate reward part of a SDES reward variable
RRE	Edge set of the reduced reachability graph RRG

RRG	Reduced reachability graph
RRS	Reduced reachability set
RRS^{exp}	Subset of the reduced reachability set RRS, containing the states where only activities with exponentially distributed delays are active
RRS^{gen}	Subset of the reduced reachability set RRS, containing the states where only activities with non-exponentially distributed delays are active
RS	Reachability set (of states) for a SDES
RS^{tan}	Subset of the reachability set RS with tangible states
RS^{van}	Subset of the reachability set RS with vanishing states
$rtype$	Boolean value denoting the type of a reward variable for a stochastic automaton; corresponds to a rate or impulse reward
$rtype$	Type of a reward variable in a Petri net
$rtype$	Boolean value denoting the type of a reward variable for a queuing model; corresponds to a rate or impulse reward
RV	Set of reward variables
$RV^\star$	Set of SDES reward variables
$rvar$	One reward variable
$rvar^\star$	An individual reward variable of a SDES
$s(.)$	Step function; used for probability distribution functions
$S^\star$	Set of all sorts
$S^\star$	Sort function of an SDES
S	Sample space in probability theory
$S(q)$	System (or response) time of a customer at queue q
$\overline{s}$	Mean service time (delay) of a Petri net transition
SC	Structural conflict: Relation between SDES actions marking the ones with intersecting input variable sets
SCPN	Stochastic colored Petri net
SDES	Stochastic discrete event system
se	One event of the simplified process $SProc$
SE	Set of events that are executed at a time point in the simplified process $SProc$
σ	A state of a SDES as captured in the state variable values
$\boldsymbol{\sigma}$	Firing count vector of transitions in Petri net
Σ	Set of all possible states of a SDES model
$Signals_{In}$	Set of input signals of a control-interpreted SDES model
$Signals_{Out}$	Set of output signals of a control-interpreted SDES model
sink	Imaginary sink node in a queuing network, used to model leaving customers
EMC	Subordinated Markov chain
SPN	Stochastic Petri net
$SProc$	Simplified stochastic process of a SDES
SS^*	Aggregated version of a subsystem

SS	Single server semantics of a Petri net transition; degree of concurrency
SS	Subsystem of a model
SSC	Symmetrical structural conflict: Relation between SDES actions
ST	Simulation time
$States_{In}$	State inputs of a control-interpeted SDES model
$States_{Out}$	State outputs of a control-interpeted SDES model
$StList$	State list in a distributed simulation
ST^{safe}	Safe simulation time in a distributed simulation
sv	A state variable of a SDES
$SV^\star$	State variable set of a SDES
$SV^\star_{affected}$	Set of state variables that may be affected by an action execution in a distributed simulation
$SV^\star_{input}$	Input state variables of an SDES action
$SV^\star_{local}$	Set of state variables that are locally known in an atomic unit of a distributed simulation
$SV^\star_{required}$	Set of state variables that are related to an action in a distributed simulation
$\mathcal{T}$	Set of all types (i.e. colors) allowed in a specific colored Petri net
t	A single transition of a Petri net
T	Set of transitions of a Petri net
$T(q)$	Throughput of queue q (rate of served customers)
t	Index variable of the simplified stochastic process $SProc$ (continuous time)
θ	State sojourn time of the stochastic process $CProc$
Thr	Threshold of the importance function in a RESTART simulation
T^{im}	Set of immediate transitions of a Petri net; i.e. with a zero delay
TP	Throughput of output transitions for a subsystem
$Trans$	SDES actions for a queuing network model that correspond to the transfer of a customer to a different queue
$Trigger$	Associates triggered output signals to action variants as well as the triggered action variants to input signals in a control-interpeted SDES model
T^{tim}	Set of timed transitions of a Petri net, i.e. with a delay other than zero
$U(q)$	Utilization of a server of queue q (fraction of busy time or number of busy servers)
UML-SC	UML Statechart
UML	Unified modeling language
$\mathbf{v}^{(1)}$	Visit ratios of transitions in a Petri net
v	One of the variables in a colored Petri net

v	Action variant — a pair of action and corresponding action mode $(a, mode)$
$Val_0{}^\star$	Initial value of a SDES state variable
$Value$	Value of a state output in a certain state
Var	Set of all variables in a colored Petri net
$var^\star$	An individual action variable of a SDES
$Vars^\star$	Action variable set of a SDES
$VDeg^\star$	Enabling degree of an SDES action variant
vfSCPN	Variable-free stochastic colored Petri net
VT	Vector time in a distributed simulation
W	Relative firing weight for transitions in a Petri net
$W(q)$	Waiting time of a customer at queue q
$Weight^\star$	Weight of an SDES action variant; relative probability of execution
X^*	Set of all theoretically possible states of a queuing network
x_0	Initial state of a stochastic automaton
x	An individual state of a stochastic automaton A
$\mathcal{X}$	State space of a stochastic automaton
$X(q)$	Queue length at queue q (number of customers)
X	State of a queuing network
X	Random variable
$\mathbf{x}$	T-semiflow of a Petri net
XML	Extensible markup language
$\mathbf{y}$	P-semiflow of a Petri net

List of Figures

Figures 14.2, 14.3, 14.6, 14.7, and 14.8 have been reprinted from the Journal of Systems and Software, Volume 77, Issue 1, A. Zimmermann and G. Hommel: Towards modeling and evaluation of ETCS real-time communication and operation. Pages 47–54, Copyright 2004, with permissions from Elsevier.

Figure 13.1, 13.2, and 13.5 have been reprinted from the Journal of Intelligent Manufacturing, Volume 12, Issue 5/6, A. Zimmermann, D. Rodriguez, and M. Silva: A Two Phase Optimisation Method for Petri Net Models of Manufacturing Systems. Pages 409–420, Copyright 2001, with permissions from Springer.

List of Algorithms

List of Tables

References

1. W. van der Aalst, "Timed coloured Petri nets and their application to logistics," PhD Thesis, Eindhoven University of Technology, 1992.
2. E. Aarts and J. Korst, *Simulated Annealing and Bolzmann Machines*. Wiley, 1989.
3. M. Ajmone Marsan, G. Balbo, G. Chiola, G. Conte, S. Donatelli, and G. Francheschinis, "An introduction to generalized stochastic Petri nets," *Microelectronics and Reliability, Special Issue on Petri Nets*, pp. 1–36, 1989.
4. M. Ajmone Marsan, G. Balbo, G. Conte, S. Donatelli, and G. Franceschinis, *Modelling with Generalized Stochastic Petri Nets*, Series in parallel computing. John Wiley and Sons, 1995.
5. M. Ajmone Marsan, "Stochastic Petri nets: An elementary introduction," in *Advances in Petri Nets 1989*, Lecture Notes in Computer Science, G. Rozenberg, Ed. Springer Verlag, 1990, vol. 424, pp. 1–29.
6. M. Ajmone Marsan and G. Chiola, "On Petri nets with deterministic and exponentially distributed firing times," in *Advances in Petri Nets 1987*, Lecture Notes in Computer Science, G. Rozenberg, Ed. Springer Verlag, 1987, vol. 266, pp. 132–145.
7. R. Y. Al-Jaar and A. A. Desrochers, "Petri nets in automation and manufacturing," in *Advances in Automation and Robotics*, G. N. Saridis, Ed. JAI Press, 1990, vol. 2.
8. R. Alur and D. L. Dill, "A theory of timed automata," *Theoretical Computer Science*, vol. 126, pp. 183–235, 1994.
9. R. Alur, "Timed automata," in *Proc. 11th Int. Conf. on Computer-Aided Verification*, Lecture Notes in Computer Science. Springer Verlag, 1999, vol. 1633, pp. 8–22.
10. N. Aoumeur and G. Saake, "Towards an adequate framework for specifying and validating runtime evolving complex discrete-event systems," in *Proc. 1st Workshop on Modeling of Objects, Components, and Agents*, 2001, pp. 1–20.
11. Y. Atamna, "Definition of the model "stochastic timed well formed coloured nets"," in *Proc. 5th Int. Workshop on Petri Nets and Performance Models*, Toulouse, 1993, pp. 24–33.
12. O. Babaoğlu and K. Marzullo, "Consistent global states of distributed systems: Fundamental concepts and mechanisms," in *Distributed systems (2nd Ed.)*, S. Mullender, Ed. New York, NY, USA: Addison-Wesley, 1993, pp. 55–96.

13. G. Balbo, G. Chiola, S. C. Bruell, and P. Z. Chen, "An example of modeling and evaluation of a concurrent program using colored stochastic Petri nets – Lamport's fast mutual exclusion algorithm," *IEEE Transactions on Parallel and Distributed Systems*, vol. 3, no. 2, pp. 221–240, 1992.

14. G. Balbo, G. Chiola, G. Franceschinis, and G. Molinar Roet, "On the efficient construction of the tangible reachability graph of generalized stochastic Petri net models," in *Proc. 2nd Int. Workshop on Petri Nets and Performance Models*, Madison, Wisconsin, 1987, pp. 136–145.

15. G. Balbo and M. Silva, Eds., *Performance Models for Discrete Event Systems with Synchronisations: Formalisms and Analysis Techniques*. Universidad de Zaragoza, Spain, 1998, MATCH Advanced School.

16. J. Banks, J. Carson, B. Nelson, and D. Nicol, *Discrete-Event System Simulation*, 4th ed. Prentice Hall, 2004.

17. J. Banks (Ed.), *Handbook of Simulation: Principles, Methodology, Advances, Applications, and Practice*. Wiley-Interscience, 1998.

18. F. Baskett, K. M. Chandy, R. R. Muntz, and F. G. Palacios, "Open, closed and mixed networks of queues with different classes of customers," *J. Assoc. Comp. Mach.*, vol. 22, pp. 248–260, 1975.

19. F. Bause, H. Beilner, M. Fischer, P. Kemper, and M. Völker, "The ProC/B toolset for the modelling and analysis of process chains," in *12th Int. Conf. Computer Performance Evaluation, Modelling Techniques and Tools (TOOLS 2002)*, Lecture Notes in Computer Science, T. Field, P. Harrison, J. Bradley, and U. Harder, Eds., no. 2324. London, UK: Springer Verlag, Apr. 2002, pp. 1–51.

20. F. Bause, P. Buchholz, and P. Kemper, "A toolbox for functional and quantitative analysis of DEDS," in *Proc. 10th Int. Conf. on Modeling Techniques and Tools for Computer Performance Evaluation*, Lecture Notes in Computer Science, vol. 1469, Palma de Mallorca, Spain, 1998, pp. 356–360.

21. ——, "QPN-tool for the specification and analysis of hierarchically combined queueing Petri nets," in *Quantitative Evaluation of Computing and Communication Systems*, H. Beilner and F. Bause, Eds. Springer Verlag, 1995, vol. 977, pp. 224–238.

22. F. Bause and P. Kemper, "QPN-tool for the qualitative and quantitative analysis of queueing Petri nets," in *Quantitative Evaluation of Computing and Communication Systems*. Springer Verlag, 1994, vol. 794, pp. 224–238.

23. A. J. Bayes, "Statistical techniques for simulation models," *The Australian Computer Journal*, vol. 2(4), pp. 180–184, 1970.

24. M. Beaudouin-Lafon, W. E. Mackay, P. Andersen, P. Janecek, M. Jensen, H. M. Lassen, K. Lund, K. H. Mortensen, S. Munck, A. V. Ratzer, K. Ravn, S. Christensen, and K. Jensen, "CPN/Tools: A post-WIMP interface for editing and simulating coloured petri nets." in *Proc. 22nd Int. Conf. Application and Theory of Petri Nets 2001 (ICATPN 2001)*, Lecture Notes in Computer Science, J. M. Colom and M. Koutny, Eds., vol. 2075. Newcastle upon Tyne, UK: Springer, 2001, pp. 71–80.

25. V. E. Beneš, *Mathematical Theory of Connecting Networks and Telephone Traffic*. New York: Academic Press, 1965.

26. S. Bernardi, S. Donatelli, and J. Merseguer, "From UML Sequence Diagrams and Statecharts to analysable Petri Net models," in *Proc. 3rd Int. Workshop on Software and Performance (WOSP)*, Rome, Italy, July 2002, pp. 35–45.

27. J. Billington, S. Christensen, K. M. van Hee, E. Kindler, O. Kummer, L. Petrucci, R. Post, C. Stehno, and M. Weber, "The Petri net markup language: Concepts, technology, and tools," in *24th Int. Conf. Applications and Theory of Petri Nets 2003 (ICATPN 2003)*, Lecture Notes in Computer Science, W. M. P. van der Aalst and E. Best, Eds., vol. 2679. Eindhoven, The Netherlands: Springer, 2003, pp. 483–505.

28. A. Blakemore, "The cost of eliminating vanishing markings from generalized stochastic Petri nets," in *Proc. 3rd Int. Workshop on Petri Nets and Performance Models*, Kyoto, Japan, 1989, pp. 85–92.

29. A. Bobbio, A. Puliafito, M. Scarpa, and M. Telek, "WebSPN: Non markovian stochastic Petri net tool," 1997, in 18th Int. Conf. on Application and Theory of Petri Nets, (Toulouse, France), June 23-27, 1997; tool description.

30. G. Bolch and M. Kirschnick, "PEPSY-QNS - performance evaluation and prediction system for queueing networks," Universität Erlangen-Nürnberg; Institut für Mathematische Maschinen und Datenverarbeitung IV, Tech. Rep. TR-I4-92-21, Oct. 1992.

31. G. Bolch, S. Greiner, H. de Meer, and K. S. Trivedi, *Queueing Networks and Markov Chains*, 2nd ed. Wiley Interscience, 2006.

32. J. Browne, C. Heavey, and H. T. Papadopoulos, *Queueing theory of Manufacturing Systems. Analysis and Design*. Chapman-Hall, 1993.

33. R. E. Bryant, "A switch-level model and simulator for MOS digital systems," *IEEE Transactions on Computers*, vol. 33, no. 2, pp. 160–177, 1984.

34. G. Bucci and E. Vicario, "Compositional validation of time-critical systems using communicating time Petri nets," *IEEE Transactions on Software Engineering*, vol. 21, no. 12, pp. 969–992, 1995.

35. P. Buchholz, G. Ciardo, S. Donatelli, and P. Kemper, "Complexity of memory-efficient Kronecker operations with applications to the solution of Markov models," *INFORMS J. Comp.*, vol. 12, no. 3, pp. 203–222, 2000.

36. P. Buchholz, *Die strukturierte Analyse Markovscher Modelle*, Informatik-Fachberichte. Springer Verlag, 1991, vol. 282.

37. J. Campos, B. Sanchez, and M. Silva, "Throughput lower bounds for Markovian Petri nets: Transformation techniques," in *Proc. 4th Int. Workshop on Petri Nets and Performance Models*, Melbourne, Australia, 1991, pp. 322–331.

38. J. Campos and M. Silva, "Structural techniques and performance bounds of stochastic Petri net models," in *Advances in Petri Nets 1992*, Lecture Notes in Computer Science, G. Rozenberg, Ed. Springer Verlag, 1992, vol. 609, pp. 352–391.

39. J. Campos, G. Chiola, and M. Silva, "Properties and performance bounds for closed free choice synchronized monoclass networks," *IEEE Transactions on Automatic Control*, vol. 36, no. 12, pp. 1368–1382, 1991.

40. J. Campos, J. M. Colom, and M. Silva, "Approximate throughput computation of stochastic marked graphs," *IEEE Transactions on Software Engineering*, vol. 20, no. 7, pp. 526–535, July 1994.

41. J. A. Carrasco, "Automated construction of compound Markov chains from generalized stochastic high-level Petri nets," in *Proc. 3rd Int. Workshop on Petri Nets and Performance Models*, Kyoto, Japan, 1989, pp. 93–102.

42. C. G. Cassandras and S. Lafortune, *Introduction to Discrete Event Systems*. Kluwer, 1999.

43. K. M. Chandy and J. Misra, "Distributed simulation: A case study in design and verification of distributed programs," *IEEE Transactions on Software Engineering*, vol. 5, no. 5, pp. 440–452, Sept. 1979.

44. K. M. Chandy and L. Lamport, "Distributed snapshots: Determining global states of distributed systems," *ACM Transactions on Computer Systems*, vol. 3, no. 1, pp. 63–75, Feb. 1985.

45. K. C. Chang, R. F. Gordon, P. G. Loewner, and E. A. MacNair, "The research queuing package modeling environment (RESQME)," in *Proc. 25th Winter Simulation Conference*, Los Angeles, United States, 1993, pp. 294–302.

46. M. Chetlur and P. A. Wilsey, "Causality representation and cancellation mechanism in time warp simulations," in *PADS '01: Proc. 15th Workshop on Parallel and Distributed Simulation*. Lake Arrowhead, CA, USA: IEEE Computer Society, May 2001, pp. 165–172.

47. A. Chiappini, A. Cimatti, C. Porzia, G. Rotondo, R. Sebastiani, P. Traverso, and A. Villafiorita, "Formal specification and development of a safety-critical train management system," in *SAFECOMP*, 1999, pp. 410–419.

48. G. Chiola, "A software package for the analysis of generalized stochastic Petri nets," in *Proc. Int. Conf. on Timed Petri Nets*, Torino, Italy, 1985, pp. 136–143.

49. G. Chiola, C. Anglano, J. Campos, J. M. Colom, and M. Silva, "Operational analysis of timed Petri nets and application to the computation of performance bounds," in *Proc. 5th Int. Workshop on Petri Nets and Performance Models*, Toulouse, 1993, pp. 128–137.

50. G. Chiola and A. Ferscha, "Distributed simulation of timed Petri nets: Exploiting the net structure to obtain efficiency," in *Proc. 14th Int. Conf. on Application and Theory of Petri Nets*, Chicago, Illinois, USA, 1990, pp. 146–165.

51. G. Chiola, G. Franceschinis, R. Gaeta, and M. Ribaudo, "GreatSPN 1.7: Graphical editor and analyzer for timed and stochastic Petri nets," *Performance Evaluation*, vol. 24, pp. 47–68, 1995.

52. G. Chiola, "Compiling techniques for the analysis of stochastic Petri nets," in *Proc. 4th Int. Conf. on Modeling Techniques and Tools for Performance Evaluation*, Palma de Mallorca, Spain, 1988, pp. 11–24.

53. G. Chiola, M. Ajmone Marsan, G. Balbo, and G. Conte, "Generalized stochastic Petri nets: A definition at the net level and its implications," *IEEE Transactions on Software Engineering*, vol. 19, no. 2, pp. 89–107, 1993.

54. G. Chiola, C. Dutheillet, G. Franceschinis, and S. Haddad, "On well formed coloured Petri nets and their symbolic reachability graph," in *Proc. 11th Int. Conf. on Application and Theory of Petri Nets*, Paris, 1990.

55. ——, "Stochastic well-formed colored nets and symmetric modeling applications," *IEEE Transactions on Computers*, vol. 42, no. 11, pp. 1343–1360, 1993.

56. G. Chiola and A. Ferscha, "Distributed simulation of Petri nets," *IEEE Parallel & Distributed Technology: Systems & Technology*, vol. 1, no. 3, pp. 33–50, 1993.

57. H. Choi, V. G. Kulkarni, and K. S. Trivedi, "Markov regenerative stochastic Petri nets," *Performance Evaluation*, vol. 20, pp. 337–357, 1994.

58. G. Ciardo, "Discrete-time Markovian stochastic Petri nets," in *Numerical Solution of Markov Chains*, Raleigh, NC, USA, Jan. 1995, pp. 339–358.

59. G. Ciardo, A. Blakemore, P. F. J. Chimento, J. K. Muppala, and K. Trivedi, "Automated generation and analysis of Markov reward models using stochastic reward nets," in *Linear Algebra, Markov Chains, and Queueing Models*, IMA Volumes in Mathematics and its Applications, C. Meyer and R. J. Plemmons, Eds. Springer Verlag, 1992, vol. 48.

60. G. Ciardo, M. Forno, P. Grieco, and A. Miner, "Comparing implicit representations of large CTMCs," in *Numerical Solution of Markov Chains*, Urbana, IL, USA, Sept. 2003, pp. 323–327.

61. G. Ciardo and C. Lindemann, "Analysis of deterministic and stochastic Petri nets," *Performance Evaluation*, vol. 18, no. 8, 1993.

62. G. Ciardo and A. Miner, "A data structure for the efficient Kronecker solution of GSPNs," in *Proc. Petri Nets and Performance Models (PNPM)*. IEEE CS press, 1999, pp. 22–31.

63. ——, "Implicit data structures for logic and stochastic systems analysis," *Performance Evaluation Review*, vol. 32, no. 4, pp. 4–9, 2005.

64. G. Ciardo, J. Muppala, and K. S. Trivedi, "SPNP: Stochastic Petri net package," in *Proc. 3rd Int. Workshop on Petri Nets and Performance Models*, Kyoto, Japan, 1989, pp. 142–151.

65. ——, "On the solution of GSPN reward models," *Performance Evaluation*, vol. 12, no. 4, pp. 237–253, 1991.

66. G. Ciardo, "Analysis of large stochastic Petri net models," Ph.D. dissertation, Duke University, Durham, NC., 1989.

67. G. Ciardo, R. German, and C. Lindemann, "A characterization of the stochastic process underlying a stochastic Petri net," *IEEE Transactions on Software Engineering*, vol. 20, pp. 506–515, 1994.

68. G. Ciardo, R. L. Jones III, A. S. Miner, and R. Siminiceanu, "Logical and stochastic modeling with SMART," in *13th Int. Conf. Computer Performance Evaluations, Modelling Techniques and Tools (TOOLS 2003)*, Lecture Notes in Computer Science, P. Kemper and W. H. Sanders, Eds., vol. 2794. Urbana, IL, USA: Springer, 2003, pp. 78–97.

69. G. Ciardo and A. S. Miner, "SMART: Simulation and Markovian analyzer for reliability and timing." in *Proc. IEEE Int. Computer Performance and Dependability Symp. (IPDS'96)*. Urbana-Champaign, IL, USA: IEEE Comp. Soc. Press, Sept. 1996, p. 60.

70. G. Ciardo and R. Zijal, "Well-defined stochastic Petri nets," in *Proc. 4th Int. Workshop on Modeling, Analysis and Simulation of Computer and Telecommunication Systems (MASCOTS'96)*. San Jose, CA, USA: IEEE Computer Society Press, Apr. 1996, pp. 278–284.

71. E. Cinlar, *Introduction to Stochastic Processes*. Englewood Cliffs: Prentice-Hall, 1975.

72. G. Clark, T. Courtney, D. Daly, D. Deavours, S. Derisavi, J. M. Doyle, W. H. Sanders, and P. Webster, "The Möbius modeling tool," in *Proc. 9th Int. Workshop on Petri Nets and Performance Models (PNPM)*, Aachen, 2001, pp. 241–250.

73. J. M. Colom and M. Silva, "Convex geometry and semiflows in P/T nets. A comparative study of algorithms for computation of minimal P-semiflows," in *Advances in Petri Nets 1990*, Lecture Notes in Computer Science, G. Rozenberg, Ed. Springer Verlag, 1991, vol. 483, pp. 79–112.

74. A. Coraiola and M. Antscher, "GSM-R network for the high-speed line Rome-Naples," *Signal und Draht*, vol. 92, no. 5, pp. 42–45, 2000.

75. J. Couvillion, R. Freire, R. Johnson, W. D. Obal, M. A. Qureshi, M. Rai, and W. H. Sanders, "Performability modeling with UltraSAN," *IEEE Software*, vol. 8, pp. 69–80, 1991.

76. J. S. Dahmann, R. Fujimoto, and R. M. Weatherly, "The department of defense High Level Architecture," in *Winter Simulation Conference*, 1997, pp. 142–149.

77. W. Damm and J. Klose, "Verification of a radio-based signaling system using the STATEMATE verification environment," *Formal Methods in System Design*, vol. 19, no. 2, pp. 121–141, 2001.

78. P. D'Argenio, J. Katoen, and E. Brinksma, "A stochastic automata model and its algebraic approach," in *Proc. PAPM'97*. University of Twente, Enschede, 1995, pp. 1–16, CTIT Technical Report 97-14.

79. R. David, "Grafcet – a powerful tool for specification of logic controllers," *IEEE Trans. on Control Systems Technology*, vol. 3, no. 3, pp. 253–268, 1995.

80. R. David and H. Alla, *Petri Nets and Grafcet (Tools for modelling discrete event systems)*. Prentice Hall, 1992.

81. D. Deavours, G. Clark, T. Courtney, D. Daly, S. Derisavi, J. Doyle, W. Sanders, and P. Webster, "The Möbius framework and its implementation," *IEEE Transactions on Software Engineering*, vol. 28, no. 10, pp. 956–959, 2002.

82. D. D. Deavours, "Formal specification of the Möbius modeling framework," Ph.D. dissertation, University of Illinois at Urbana-Champaign, 2001.

83. J. Dehnert, J. Freiheit, and A. Zimmermann, "Modeling and performance evaluation of workflow systems," in *Proc. 4th World Multiconference on Systemics, Cybernetics and Informatics (SCI'2000)*, vol. VIII, Orlando, 2000, pp. 632–637.

84. ——, "Workflow modeling and performance evaluation with colored stochastic Petri nets," in *AAAI Spring Symposium - Bringing Knowledge to Business Processes*, Stanford, USA, Mar. 2000, pp. 139–141.

85. ——, "Modelling and evaluation of time aspects in business processes," *Journal of the Operational Research Society*, vol. 53, pp. 1038–1047, 2002.

86. S. Derisavi, P. Kemper, W. H. Sanders, and T. Courtney, "The Möbius state-level abstract functional interface," *Performance Evaluation*, vol. 54, no. 2, pp. 105–128, 2003.

87. A. A. Desrochers, *Modeling and Control of Automated Manufacturing Systems*. Washington, DC, USA: IEEE Computer Society Press, 1990.

88. F. DiCesare, G. Harhalakis, J. M. Proth, M. Silva, and F. B. Vernadat, *Practice of Petri Nets in Manufacturing*. London: Chapman and Hall, 1993.

89. B. L. Dietrich, "A taxonomy of discrete manufacturing systems," *Operations Research*, vol. 39, no. 6, pp. 886–902, Nov. 1991.

90. C. Dimitrovici, U. Hummert, and L. Petrucci, "Semantics, composition and net properties of algebraic high-level nets," in *Advances in Petri Nets 1991*, Lecture Notes in Computer Science, G. Rozenberg, Ed. Springer Verlag, 1991, vol. 524, pp. 93–117.

91. S. Donatelli, "Superposed generalized stochastic Petri nets: definition and efficient solution," in *Application and Theory of Petri Nets 1994*, Lecture Notes in Computer Science, R. Valette, Ed. Springer Verlag, 1994, vol. 815, pp. 258–277.

92. S. Donatelli and P. Kemper, "Integrating synchonization with priority into a Kronecker representation," *Performance Evaluation*, vol. 44, pp. 1–4, 2001.

93. B. P. Douglass, *Real-Time UML*, Object Technology Series. Addison-Wesley, 2004, 3rd Edition.

94. EEIG ERTMS User Group, *ERTMS/ETCS RAMS Requirements Specification*, UIC, Brussels, 1998.

95. ——, *ERTMS/ETCS System Requirements Specification*, UIC, Brussels, 1999.

96. ——, *Euroradio FFFIS*, UIC, Brussels, 2000.

97. ——, *Performance Requirements for Interoperability*, UIC, Brussels, 2000.

98. EIRENE Project Team, *EIRENE System Requirements Specification*, UIC, Brussels, 1999.

99. A. K. Erlang, "The theory of probabilities and telephone conversations," *Nyt Tidsskrift Mat. B*, vol. 20, pp. 33–39, 1909.

100. J. Ezpeleta and J. M. Colom, "Automatic synthesis of colored Petri nets for the control of FMS," *IEEE Transactions on Robotics and Automation*, vol. 13, no. 3, pp. 327–337, June 1997.

101. K. Feldmann, W. Colombo, and C. Schnur, "An approach for modeling, analysis and real-time control of flexible manufacturing systems using Petri nets," in *Proc. European Simulation Symposium*, Erlangen-Nürnberg, 1995, pp. 661–665.

102. A. Ferscha and S. K. Tripathi, "Parallel and distributed simulation of discrete event systems," University of Maryland, College Park, MD, USA, Tech. Rep., August 1994.

103. A. Ferscha, "Simulation," in *Performance Models for Discrete Event Systems with Synchronisations: Formalisms and Analysis Techniques*. Universidad de Zaragoza, Spain, 1998, pp. 819–924, MATCH Advanced School.

104. C. Fidge, "Logical time in distributed computing systems," *Distributed Computing*, vol. 24, no. 8, pp. 28–33, 1991.

105. G. S. Fishman, *Discrete-Event Simulation*, Springer Series in Operations Research. Springer, 2001.

106. M. P. I. Forum, "MPI: A message-passing interface standard," *Int. Journal of Supercomputer Applications and High Performance Computing*, vol. 8, no. 3/4, pp. 159–416, 1994.

107. B. L. Fox and P. W. Glynn, "Computing poisson probabilities," *Communications of the ACM*, vol. 31, no. 4, pp. 440–445, 1988.

108. J. Freiheit and A. Zimmermann, "Extending a response time approximation technique to colored stochastic Petri nets," in *Proc. 4th Int. Workshop on Performability Modeling of Computer and Communication Systems (PMCCS)*, College of William and Mary, Williamsburg, VA, USA, Sept. 1998, pp. 67–71.

109. ——, "A divide and conquer approach for the performance evaluation of large stochastic Petri nets," in *Proc. 9th Int. Workshop on Petri Nets and Performance Models (PNPM)*, Aachen, 2001, pp. 91–100.

110. J. Freiheit and J. Billington, "Closed-form token distribution computation of GSPNs without synchronisation," in *Proc. 11th Int. Conf. on Analytical and Stochastic Modelling Techniques and Applications (ASMTA 2004)*. Magdeburg, Germany: SCS Publishing House, 2004, pp. 199–206.

111. J. Freiheit and A. Heindl, "Novel formulae for GSPN aggregation," in *Symp. on Modeling, Analysis and Simulation of Computer and Telecommunication Systems (MASCOTS 2002)*, Fort Worth, Texas, 2002, pp. 209–216.

112. J. Freiheit, "Matrizen- und zustandsraumreduzierende Verfahren zur Leistungsbewertung großer stochastischer Petrinetze," Dissertation, Technische Universität Berlin, July 2002.

113. R. Fricks, S. Hunter, S. Garg, and K. Trivedi, "IDEAS: an integrated design environment for assessment of computer systems and communication networks," in *Proc. 2nd IEEE Int. Conf. on Engineering of Complex Computer Systems*, Montreal, Canada, 1996, pp. 27–34.

114. R. M. Fujimoto, "Parallel discrete event simulation," *Communications of the ACM*, vol. 33, no. 10, pp. 30–53, Oct. 1990.

115. R. Fujimoto, *Parallel and Distributed Simulation Systems*, Wiley Series on Parallel and Distributed Computing. Wiley-Interscience, 1999.

116. A. Furfaro, L. Nigro, and F. Pupo, "Distributed simulation of timed coloured Petri nets," in *Proceedings. Sixth IEEE Int. Workshop on Distributed Simulation and Real-Time Applications (DS-RT'02)*. IEEE Computer Society, Oct. 2002, pp. 159–166.

117. R. Gaeta, "Efficient discrete-event simulation of colored Petri nets," *IEEE Transactions on Software Engineering*, vol. 22, no. 9, pp. 629–639, 1996.

118. M. J. Garvels and D. P. Kroese, "A comparison of RESTART implementations," in *Proc. 1998 Winter Simulation Conference*, 1998.

119. M. Gen and R. Cheng, *Genetic Algorithms and Engineering Design*. Wiley, 1997.

120. H. J. Genrich, "Predicate/Transition nets," in *Advances in Petri Nets 1986*, Lecture Notes in Computer Science, W. Brauer, W. Reisig, and G. Rozenberg, Eds. Springer Verlag, 1987, vol. 254, pp. 207–247.

121. H. J. Genrich and K. Lautenbach, "System modelling with high-level Petri nets," *Theoretical Computer Science*, vol. 13, pp. 109–136, 1981.

122. ——, "The analysis of distributed systems by means of Predicate / Transition nets," in *Semantics of Concurrent Computation*, Lecture Notes in Computer Science, G. Kahn, Ed. Springer Verlag, 1979, vol. 70, pp. 123–146.

123. R. German, "New results for the analysis of deterministic and stochastic Petri nets," in *Proc. IEEE Int. Performance and Dependability Symp.*, Erlangen, 1996, pp. 114–123.

124. R. German, C. Kelling, A. Zimmermann, and G. Hommel, "TimeNET – a toolkit for evaluating non-Markovian stochastic Petri nets," in *Proc. 6th Int. Workshop on Petri Nets and Performance Models*, Durham, North Carolina, 1995, pp. 210–211.

125. R. German, D. Logothetis, and K. S. Trivedi, "Transient analysis of Markov regenerative stochastic Petri nets: A comparison of approaches," in *Proc. 6th Int. Workshop on Petri Nets and Performance Models*, Durham, North Carolina, 1995, pp. 103–112.

126. R. German and J. Mitzlaff, "Transient analysis of deterministic and stochastic Petri nets with TimeNET," in *Proc. Joint Conf. 8th Int. Conf. on Modelling Techniques and Tools for Performance Evaluation*, Lecture Notes in Computer Science. Springer Verlag, 1995, vol. 977, pp. 209–223.

127. R. German, "Analysis of stochastic Petri nets with non-exponentially distributed firing times," Dissertation, Technische Universität Berlin, 1994.

128. ——, "SPNL: Processes as language-oriented building blocks of stochastic Petri nets," in *Proc. 9th Conf. Computer Performance Evaluation, Modelling Techniques and Tools*, St. Malo, France, 1997, pp. 123–134.

129. ——, "Cascaded deterministic and stochastic Petri nets," in *Proc. 3rd Int. Meeting on the Numerical Solution of Markov Chains*, Zaragoza, Sept. 1999, pp. 111–130.

130. ——, *Performance Analysis of Communication Systems, Modeling with Non-Markovian Stochastic Petri Nets*. John Wiley and Sons, 2000.

131. R. German, C. Kelling, A. Zimmermann, and G. Hommel, "TimeNET – a toolkit for evaluating non-Markovian stochastic Petri nets," *Performance Evaluation*, vol. 24, pp. 69–87, 1995.

132. R. German and C. Lindemann, "Analysis of stochastic Petri nets by the method of supplementary variables," *Performance Evaluation*, vol. 20, pp. 317–335, 1994.

133. R. German, A. P. A. van Moorsel, M. A. Qureshi, and W. H. Sanders, "Expected impulse rewards in Markov regenerative stochastic Petri nets," in *Application and Theory of Petri Nets 1996*, Lecture Notes in Computer Science, J. Billington and W. Reisig, Eds. Springer Verlag, 1996, vol. 1091.

134. R. German, A. Zimmermann, C. Kelling, and G. Hommel, "Modellierung und bewertung von flexiblen fertigungssystemen mit TimeNET," *CIM Management*, no. 3, pp. 24–27, June 1994.

135. S. B. Gershwin, *Manufacturing Systems Engineering*. Prentice Hall, 1994.

136. C. Ghezzi, D. Mandrioli, S. Morasca, and M. Pezze, "A unified high-level Petri net formalism for time-critical systems," *IEEE Transactions on Software Engineering*, vol. 17, no. 2, pp. 160–172, Feb. 1991.

137. S. Gilmore and J. Hillston, "The PEPA Workbench: A Tool to Support a Process Algebra-based Approach to Performance Modelling," in *Proc. 7th Int. Conf. on Modelling Techniques and Tools for Computer Performance Evaluation*, Lecture Notes in Computer Science, no. 794. Vienna: Springer-Verlag, May 1994, pp. 353–368.

138. S. Gilmore, J. Hillston, and M. Ribaudo, "An efficient algorithm for aggregating PEPA models," *IEEE Transactions on Software Engineering*, vol. 27, no. 5, pp. 449–464, 2001.

139. C. Girault and R. Valk, *Petri Nets for System Engineering*. Springer, 2003.

140. P. Glasserman, P. Heidelberger, P. Shahabuddin, and T. Zajic, "Splitting for rare event simulation: Analysis of simple cases," in *Proc. Winter Simulation Conference*, 1996, pp. 302–308.

141. ——, "Multilevel splitting for estimating rare event probabilities," *Operations Research*, vol. 47, pp. 585–600, 1999.

142. F. Glover, J. P. Kelly, and M. Laguna, "Genetic algorithms and tabu search: Hybrids for optimization," *Computers Ops. Res.*, vol. 22, no. 1, pp. 111–134, 1995.

143. P. W. Glynn, "A GSMP formalism for discrete event systems," *Proceedings of the IEEE*, vol. 77, no. 1, pp. 14–23, Jan. 1989.

144. P. W. Glynn and D. L. Iglehart, "Importance sampling for stochstic simulations," *Management Science*, vol. 35, no. 11, pp. 1367–1392, Nov. 1989.

145. D. E. Goldberg, *Genetic Algorithms in Search, Optimization and Machine Learning*. Addison-Wesley, 1989.

146. M. Göller and L. Lengemann, "Measurement and evaluation of the quality of service parameters of the communication system for ERTMS," *Signal und Draht*, vol. 94, no. 1+2, pp. 19–26, 2002.

147. C. Görg, E. Lamers, O. Fuß, and P. Heegaard, "Rare event simulation," Computer Systems and Telematics, Norwegian Institute of Technology, Tech. Rep. COST 257, 2001.

148. A. Goyal, W. C. Carter, E. de Souza e Silva, S. S. Lavenberg, and K. S. Trivedi, "The system availability estimator," in *Proc. 16th Symp. Fault-Tolerant Computing*, Vienna, Austria, 1986, pp. 84–89.

149. D. Gross and C. Harris, *Fundamentals of Queueing Theory*, 3rd ed. Wiley, 1998.

150. P. J. Haas, *Stochastic Petri Nets: Modelling, Stability, Simulation*, Springer Series in Operations Research. Springer Verlag, 2002.

151. J. Hamilton, D. Nash, and U. P. (Eds), *Distributed Simulation*. CRC Press, 1997.

152. H.-M. Hanisch, A. Lüder, and J. Thieme, "A modular plant modeling technique and related controller synthesis problems," in *Proc. Int. Conf. on Systems, Man, and Cybernetics (SMC '98)*. IEEE, 1998, pp. 686–691.

153. D. Harel and M. Politi, *Modeling Reactive Systems with Statecharts: The Statemate Approach*. New York: Wiley, 1998.

154. D. Harel, "Statecharts: A Visual Formalism for Complex Systems," *Science of Computer Programming*, vol. 8, no. 3, pp. 231–274, June 1987.

155. D. Harel and A. Naamad, "The STATEMATE semantics of statecharts," *ACM Transactions on Software Engineering Methods*, vol. 5, no. 4, 1996.

156. B. Haverkort and K. Trivedi, "Specification techniques for Markov reward models," *Discrete Event Dynamic Systems: Theory Appl.*, vol. 3, pp. 219–247, 1993.

157. B. R. Haverkort, "Performability evaluation of fault-tolerant computer systems using DyQN-Tool$^+$," *Int. Journal of Reliability, Quality, and Safety Engineering*, vol. 2, no. 4, pp. 383–404, 1995.

158. B. R. Haverkort and I. G. Niemegeers, "Performability modelling tools and techniques," *Performance Evaluation*, vol. 25, no. 1, pp. 17–40, 1996.

159. P. Heegard, "Speed-up techniques for simulation," *Telektronikk*, vol. 91, no. 2, 1995.

160. P. Heidelberger, "Fast simulation of rare events in queueing and reliability models," *ACM Transactions on Modeling and Computer Simulation (TOMACS)*, vol. 5, no. 1, pp. 43–85, 1995.

161. A. Heindl and R. German, "A fourth order algorithm with automatic stepsize control for the transient analysis of DSPNs," *IEEE Transactions on Software Engineering*, vol. 25, pp. 194–206, 1999.

162. J. Hillston, *A Compositional Approach to Performance Modelling*, Distinguished Dissertations Series. Cambridge University Press, 1996.

163. C. Hirel, R. A. Sahner, X. Zang, and K. S. Trivedi, "Reliability and performability modeling using SHARPE 2000," in *11th Int. Conf. Computer Performance Evaluation: Modelling Techniques and Tools (TOOLS 2000)*, Lecture Notes in Computer Science, B. R. Haverkort, H. C. Bohnenkamp, and C. U. Smith, Eds., vol. 1786. Schaumburg, IL, USA: Springer, 2000, pp. 345–349.

164. C. Hirel, B. Tuffin, and K. S. Trivedi, "SPNP: Stochastic Petri nets. version 6.0." in *Computer Performance Evaluation, Modelling Techniques and Tools — 11th Int. Conf., TOOLS 2000*, Lecture Notes in Computer Science, vol. 1786. Schaumburg, IL, USA: Springer Verlag, 2000, pp. 354–357.

165. Y. C. Ho, R. Sreenivas, and P. Valiki, "Ordinal optimisation of DEDS," *Journal of Discrete Event Dynamic Systems*, vol. 2, pp. 61–88, 1992.

166. L. E. Holloway, B. H. Krogh, and A. Giua, "A survey of Petri net methods for controlled discrete event systems," *Discrete Event Dynamic Systems: Theory and Applications*, vol. 7, pp. 151–190, 1997.

167. L. E. Holloway and B. H. Krogh, "Controlled Petri nets: A tutorial overview," in *11th Int. Conf. on Analysis and Optimization of Systems*, Lecture Notes in Control and Information Sciences, G. Cohen and J.-P. Quadrat, Eds., vol. 199. Sophia-Antipolis: Springer-Verlag, 1994, pp. 158–168.

168. J. E. Hopcroft and J. D. Ullman, *Introduction to Automata Theory, Languages, and Computation*. Reading, MA: Addison-Wesley, 1979.

169. R. P. Hopkins, M. J. Smith, and P. King, "Two approaches to integrating UML and performance models," in *Proc. 3rd Int. Workshop on Software and Performance*, July 2002, pp. 91–92.
170. R. A. Howard, *Dynamic Probabilistic Systems*. New York: Wiley, 1971, vol. 2: Semi-Markov and Decision Processes.
171. A. Huck, J. Freiheit, and A. Zimmermann, "Convex geometry applied to Petri nets: State space size estimation and calculation of traps, siphons, and invariants," Technische Universität Berlin, Tech. Rep. 2000-6, 2000.
172. A. Huck, M. Knoke, and G. Hommel, "A simulation framework for supply chain management in an e-business environment," in *Proc. Workshop The Internet Challenge: Technology and Applications*, G. Hommel and S. Huanye, Eds. Berlin, Germany: Kluwer Academic Publishers, 2002, pp. 73–82.
173. L. Ingber, "Very fast simulated re-annealing," *Journal of Mathematical Computer Modelling*, vol. 12, no. 8, pp. 967–973, 1989.
174. ——, "Adaptive simulated annealing (ASA): Lessons learned," *Journal of Control and Cybernetics*, vol. 25, no. 1, pp. 33–54, 1996.
175. ——, *ASA - Adaptive Simulated Annealing for Nonlinear Systems*, 1998, http://www.ingber.com.
176. N. N. Ivanov, "Semi-Markov processes in timed stochastic Petri nets," *Automation and Remote Control*, vol. 55, no. 3, pp. 400–408, 1994.
177. J. R. Jackson, "Networks of waiting lines," *Operations Research*, vol. 5, pp. 518–521, 1957.
178. F. Jakop, "Entwurf und Implementierung einer generischen, betriebssystemunabhängigen Benutzungsoberfläche für grafische Modelle (PENG – Plattformunabhängiger Editor für Netz-Graphen)," Master's thesis, Technische Universität Berlin, Aug. 2003.
179. V. Janousek and R. Koci, "Towards an open implementation of the PNtalk system," in *Proc. 5th EUROSIM Congress on Modeling and Simulation*, Paris, France, 2004.
180. D. N. Jansen and H. Hermanns, "Dependability checking with StoCharts: Is train radio reliable enough for trains?" in *Proc. of the 1st Int. Conf. on the Quantitative Evaluation of Systems (QEST)*, Enschede, Netherlands, 2004, pp. 250–250.
181. D. N. Jansen, H. Hermanns, and J.-P. Katoen, "A probabilistic extension of UML statecharts: Specification and verification," in *Proc. 7th Int. Symp. Formal Techniques in Real-Time and Fault-Tolerant Systems (FTRTFT)*, Lecture Notes in Computer Science, W. Damm and E.-R. Olderoog, Eds., no. 2469. Oldenburg, Germany: Springer Verlag, 2002, pp. 355–374.
182. L. Jansen, M. Meyer zu Hörste, and H. Schnieder, "Technical issues in modelling the European train control system," in *Proc. 1st CPN Workshop, DAIMI PB 532*, Aarhus University, 1998, pp. 103–115.
183. D. Jefferson, "Virtual time," *ACM Transactions on Programming Languages and Systems*, vol. 7, no. 3, pp. 405–425, 1985.
184. D. Jefferson and H. Sowizral, "Fast concurrent simulation using the time warp mechanism," in *Distributed Simulation '85*, P. Reynolds, Ed. La Jolla, California: SCS, 1985, pp. 63–69.
185. A. Jensen, "Markov chains as an aid in the study of Markov processes," *Scand. Aktuarietidskrift*, vol. 3, pp. 87–91, 1953.
186. K. Jensen and G. Rozenberg, Eds., *High-Level Petri Nets: Theory and Applications*. Springer Verlag, 1991.

187. K. Jensen, "Coloured Petri nets and the invariant-method," *Theoretical Computer Science*, vol. 14, pp. 317–336, 1981.

188. ——, *Coloured Petri Nets: Basic Concepts, Analysis Methods and Practical Use*, EATCS Monographs on Theoretical Computer Science. Springer Verlag, 1992.

189. A. M. Johnson Jr. and M. Malek, "Survey of software tools for evaluating reliability, availability, and serviceability," *ACM Computing Surveys*, vol. 20, no. 4, pp. 227 – 269, 1988.

190. H. Jungnitz, B. Sanchez, and M. Silva, "Approximate throughput computation of stochastic marked graphs," *J. Parallel and Distributed Computing*, vol. 2, pp. 282–295, 1992.

191. H. Kahn and T. E. Harris, "Estimation of particle transmission by random sampling," *National Bureau of Standards Applied Mathematics Series*, vol. 12, pp. 27–30, 1951.

192. E. Kasturia, F. DiCesare, and A. A. Desrochers, "Real time control of multi-level manufacturing systems using colored Petri nets," in *Proc. Int. Conf. on Robotics and Automation*, 1988, pp. 1114–1119.

193. C. Kelling, "Control variates selection strategies for timed Petri nets," in *Proc. European Simulation Symposium*, Istanbul, 1994, pp. 73–77.

194. C. Kelling and G. Hommel, "Rare event simulation with an adaptive "RESTART" method in a Petri net modeling environment," in *Proc. 4th Int. Workshop on Parallel and Distributed Real-Time Systems*, Los Alamitos, CA, USA, Apr. 1996, pp. 229–234.

195. C. Kelling, "Rare event simulation with RESTART in a Petri net modeling environment," in *Proc. of the European Simulation Symposium*, Erlangen, 1995, pp. 370–374.

196. ——, "Simulationsverfahren für zeiterweiterte Petri-netze," Dissertation, Technische Universität Berlin, 1995, advances in Simulation, SCS International.

197. ——, "TimeNET$_{sim}$ – a parallel simulator for stochastic Petri nets," in *Proc. 28th Annual Simulation Symposium*, Phoenix, AZ, USA, 1995, pp. 250–258.

198. ——, "A framework for rare event simulation of stochastic Petri nets using RESTART," in *Proc. of the Winter Simulation Conference*, 1996, pp. 317–324.

199. C. Kelling, R. German, A. Zimmermann, and G. Hommel, "TimeNET – ein werkzeug zur modellierung mit zeiterweiterten Petri-netzen," *Informationstechnik und Technische Informatik (it+ti)*, vol. 37, no. 3, pp. 21–27, June 1995.

200. W. D. Kelton, R. P. Sadowski, and D. T. Sturrock, *Simulation With Arena*. McGraw-Hill, 2003.

201. P. Kemper, "Numerical analysis of superposed GSPNs," *IEEE Transactions on Software Engineering*, vol. 22, no. 9, pp. 615–628, 1996.

202. D. G. Kendall, "Stochastic processes occuring in the theory of queues and their analysis by the method of imbedded markov chains," *Ann. Math. Statist.*, vol. 24, pp. 338–354, 1953.

203. D. Kendelbacher and F. Stein, "EURORADIO - communication base system for ETCS," *Signal und Draht*, vol. 94, no. 6, pp. 6–11, 2002.

204. P. King and R. Pooley, "Using UML to derive stochastic Petri net models," in *Proc. 15th UK Performance Engineering Workshop*, Bristol, UK, July 1999, pp. 45–56.

205. ——, "Derivation of Petri net performance models from UML specifications of communications software," in *Proc. 11th Int. Conf. on Tools and Techniques*

for Computer Performance Evaluation, Schaumburg, Illinois, USA, 2000, pp. 262–276.

206. S. Kirkpatrick, C. D. Gelatt Jr., and M. P. Vecchi, "Optimization by simulated annealing," *Science*, vol. 220, no. 4598, pp. 671–680, 1983.

207. L. Kleinrock, *Queuing Systems*. John Wiley and Sons, 1975.

208. A. Knapp, S. Merz, and C. Rauh, "Model checking timed UML state machines and collaborations," in *Proc. 7th Int. Symp. Formal Techniques in Real-Time and Fault-Tolerant Systems (FTRTFT)*, Lecture Notes in Computer Science, W. Damm and E.-R. Olderoog, Eds., no. 2469. Oldenburg, Germany: Springer Verlag, 2002, pp. 395–414.

209. M. Knoke and G. Hommel, "Dealing with global guards in a distributed simulation of colored Petri nets," in *Proc. 9th IEEE Int. Symposium on Distributed Simulation and Real Time Applications (DS-RT 2005)*, Montreal, Canada, Oct. 2005, pp. 51–58.

210. M. Knoke, F. Kühling, A. Zimmermann, and G. Hommel, "Towards correct distributed simulation of high-level Petri nets with fine-grained partitioning," in *2nd Int. Symp. Parallel and Distributed Processing and Applications (ISPA'04)*, Lecture Notes in Computer Science, J. Cao, Ed., vol. 3358. Hong Kong, China: Springer Verlag, Dec. 2004, pp. 64–74.

211. M. Knoke, F. Kühling, A. Zimmermann, and G. Hommel, "Performance of a distributed simulation of timed colored Petri nets with fine-grained partitioning." in *Design, Analysis, and Simulation of Distributed Systems Symposium, (DASD 2005)*, D. Tutsch, Ed. San Diego, USA: SCS, Apr. 2005, pp. 63–71.

212. M. Knoke, D. Rasinski, A. Zimmermann, and G. Hommel, "Dynamic remapping strategies for an optimistic simulation of colored Petri nets with fine-grained partitioning," Nov. 2005, submitted.

213. M. Knoke and A. Zimmermann, "Distributed simulation of colored stochastic Petri nets with TimeNET 4.0," in *Proc. 3rd Int. Conf. Quantitative Evaluation of Systems (QEST '06)*, Riverside, CA, USA, Sept. 2006, pp. 117–118.

214. R. König and L. Quäck, *Petri-Netze in der Steuerungs- und Digitaltechnik*. Verlag Technik, 1988.

215. S. M. Koriem and L. M. Patnaik, "A generalized stochastic high-level Petri net model for performance analysis," *The Journal of Systems and Software*, vol. 36, no. 3, pp. 247–266, Mar. 1997.

216. S. Kounev and A. Buchmann, "SimQPN—a tool and methodology for analyzing queueing Petri net models by means of simulation," *Performance Evaluation*, 2006, (in press).

217. D. Krahl, "The Extend simulation environment," in *Proc. Winter Simulation Conference*, 2002, pp. 205–213.

218. L. Lamport, "Time, clocks, and the ordering of events in a distributed system," *Communications of the ACM*, vol. 21, no. 7, pp. 558–565, July 1978.

219. A. M. Law and W. D. Kelton, *Simulation Modeling and Analysis*. McGraw-Hill, 1991.

220. E. D. Lazowska, J. Zahorjan, G. S. Graham, and K. C. Sevcik, *Quantitative system performance: computer system analysis using queueing network models*. Prentice-Hall, Inc., 1984.

221. K. Lemmer and E. Schnieder, "Modelling and control of complex logistic systems for manufacturing," in *Advances in Petri Nets 1992*, Lecture Notes in Computer Science, K. Jensen, Ed. Springer Verlag, 1992, vol. 616, pp. 373–378.

222. C. Lin and D. C. Marinescu, "On stochastic high-level Petri nets," in *Proc. 2nd Int. Workshop on Petri Nets and Performance Models*, Madison, Wisconsin, 1987, pp. 34–43.

223. ——, "Stochastic high-level Petri nets and applications," *IEEE Transactions on Computers*, vol. 37, pp. 815–825, 1988.

224. C. Lindemann and G. Shedler, "Numerical analysis of deterministic and stochastic Petri nets with concurrent deterministic transitions," *Performance Evaluation, Special Issue Proc. of PERFORMANCE '96*, pp. 565–582, 1996.

225. C. Lindemann, A. Thümmler, A. Klemm, M. Lohmann, and O. P. Waldhorst, "Performance analysis of time-enhanced UML diagrams based on stochastic processes," in *Proc. 3rd Workshop on Software and Performance (WOSP)*, Rome, Italy, 2002, pp. 25–34.

226. C. Lindemann, "An improved numerical algorithm for calculating steady-state solutions of deterministic and stochastic Petri net models," *Performance Evaluation*, vol. 18, pp. 79–95, 1993.

227. ——, *Stochastic Modeling using DSPNexpress*. Oldenbourg, 1994.

228. ——, "DSPNexpress: A software package for the efficient solution of deterministic and stochastic Petri nets," *Performance Evaluation*, vol. 22, pp. 3–21, 1995.

229. ——, *Performance Modelling with Deterministic and Stochastic Petri Nets*. Wiley, 1998.

230. C. Lindemann and A. Zimmermann, "An adaptive algorithm for the efficient generation of the tangible reachability graph of a stochastic Petri net," Technische Universität Berlin, Technischer Bericht des Fachbereichs Informatik 8, 1994.

231. J. D. C. Little, "A proof of the queuing formula $l = \lambda w$," *Operations Research*, vol. 9, pp. 383–387, 1961.

232. J. P. López-Grao, J. Merseguer, and J. Campos, "Performance Engineering Based on UML and SPNs: A software performance tool," in *Proc. 7th Int. Symposium On Computer and Information Sciences (ISCIS XVII)*. Orlando, Florida, USA: CRC Press, Oct. 2002, pp. 405–409.

233. J. Martínez, P. Muro, and M. Silva, "Modeling, validation and software implementation of production systems using high level Petri nets," in *Proc. Int. Conf. on Robotics and Automation*, Raleigh, North Carolina, 1987, pp. 1180–1185.

234. J. Martínez and M. Silva, "A language for the description of concurrent systems modelled by coloured Petri nets: Application to the control of flexible manufacturing systems," in *Proc. of the 1984 IEEE Workshop on Languages for Automation*, New Orleans, 1984, pp. 72–77.

235. F. Mattern, "Virtual time and global states of distributed systems," in *Proc. Workshop on Parallel and Distributed Algorithms*, M. Cosnard, P. Quinton, M. Raynal, and Y. Robert, Eds. Elsevier Science Publishers, 1989, pp. 215–226.

236. A. Mazzeo, N. Mazzocca, S. Russo, and V. Vittorini, "A systematic approach to the Petri net based specification of concurrent systems," *Real-Time Systems*, vol. 13, pp. 219–236, 1997.

237. C. McGeoch, "Analysing algorithms by simulation: variance reduction techniques and simulation speedups," *ACM Computing Surveys*, vol. 24, no. 2, pp. 195–212, 1992.

238. P. M. Merlin, "A methodology for the design and implementation of communication protocols," *IEEE Transactions on Communication*, vol. 24, pp. 614–621, 1976.

239. J. Merseguer, "On the use of UML State Machines for software performance evaluation," in *Proc. 10th IEEE Real-Time and Embedded Technology and Applications Symposium (RTAS)*, 2004.

240. J. Merseguer, S. Bernardi, J. Campos, and S. Donatelli, "A compositional semantics for UML state machines aimed at performance evaluation," in *Proceedings of the 6th Int. Workshop on Discrete Event Systems (WODES)*. IEEE Computer Society Press, Oct. 2002, pp. 295–302.

241. N. Metropolis, A. W. Rosenbluth, M. N. Rosenbluth, A. H. Teller, and E. Teller, "Equation of state calculations by fast computing machines," *J. Chem. Phys.*, vol. 21, no. 6, pp. 1087–1092, 1953.

242. U. Meyer, K. Kastell, A. Fernandez-Pello, D. Perez, and R. Jakoby, "Performance advantage and use of a location based handover algorithm," in *Proc. IEEE Vehicular Technology Conference (VTC2004)*, Los Angeles, Sept. 2004.

243. M. Meyer zu Hörste and E. Schnieder, "Modelling and simulation of train control systems using Petri nets," in *FM'99 Formal Methods. World Congress on Formal Methods in the Development of Computing Systems.*, Lecture Notes in Computer Science, J. M. Wing, J. Woodcock, and J. Davies, Eds., vol. 1709. Berlin: Springer, 1999, p. 1867.

244. J. Misra, "Distributed discrete-event simulation," *ACM Computing Surveys*, vol. 18, no. 1, pp. 39–65, 1986.

245. I. Mitrani, *Probabilistic Modeling*. Cambridge University Press, 1998.

246. M. K. Molloy, "On the integration of delay and throughput measures in distributed processing models," PhD thesis, UCLA, Los Angeles, CA, 1981.

247. ——, "Discrete time stochastic Petri nets," *IEEE Transactions on Software Engineering*, vol. 11, no. 4, pp. 417–423, Apr. 1985.

248. ——, "Performance analysis using stochastic Petri nets," *IEEE Transactions on Computers*, vol. 31, no. 9, pp. 913–917, 1982.

249. K. E. Moore, J. C. Chiang, and S. D. Hammer, "ALPHA/sim simulation software tutorial," in *Winter Simulation Conference*, 1999, pp. 267–275.

250. MORANE Project Group, *Radio Transmission FFFIS for Euroradio*, Brussels, 1998.

251. S. Natkin, "Les reseaux de Petri stochastiques et leur application a l'evaluation des systémes informatiques," PhD thesis, CNAM, Paris, 1980.

252. D. M. Nicol and W. Mao, "Automated parallelization of timed Petri-net simulations," *Journal of Parallel and Distributed Computing*, vol. 29, no. 1, pp. 60–74, Aug. 1995.

253. W. Obal and W. Sanders, "Importance sampling simulation in UltraSAN," *Simulation*, vol. 62, no. 2, pp. 98–111, 1994.

254. W. D. Obal and W. H. Sanders, "An environment for importance sampling based on stochastic activity networks," in *Proc. 13th Symp. on Reliable Distributed Systems*, Dana Point, CA, October 1994, pp. 64–73.

255. Object Management Group, *"UML profile for schedulability, performance, and time,"* www.uml.org, March 2002.

256. ——, *"Unified Modeling Language Specification v.2.0,"* www.uml.org, Oct. 2004.

257. J. Osburg, "Performance investigation of arbitrary train control techniques," *Signal und Draht*, vol. 94, no. 1+2, pp. 27–30, 2002.

258. C. J. Pérez-Jiménez, "Tecnicas de aproximacion de throughput en redes de Petri estocasticas," Ph.D. dissertation, Departamento de Informatica e Ingenieria de Sistemas, Universidad de Zaragoza, Spain, June 2002.

259. C. J. Pérez-Jiménez, J. Campos, and M. Silva, "State machine reduction for the approximate performance evaluation of manufacturing systems modelled with cooperating sequential processes," in *1996 IEEE International Conference on Robotics and Automation*, Minneapolis, Minnesota, USA, Apr. 1996, pp. 1159–1165.

260. C. J. Pérez-Jiménez and J. Campos, "A response time approximation technique for stochastic general P/T systems," in *Proc. Symp. Industrial and Manufacturing Systems, 2nd IMACS Int. Multiconf. on Computational Engineering in Systems Applications (CESA '98)*, Nabeul-Hammamet, Tunisia, Apr. 1998.

261. C. J. Pérez-Jiménez, J. Campos, and M. Silva, "On approximate performance evaluation of manufacturing systems modelled with weighted T-systems," in *Proc. of the Symp. on Discrete Events and Manufacturing Systems, IMACS-IEEE SMC Multiconf. on Computational Engineering of Systems Applications (CESA 96)*, Lille, France, July 1996, pp. 201–207.

262. J. L. Peterson, *Petri Net theory and the modeling of systems*. Englewood Cliffs, New Jersey: Prentice Hall, 1981.

263. C. Petri, "Kommunikation mit Automaten," Dissertation, Schriften des Institutes für Instrumentelle Mathematik, Bonn, 1962.

264. Petri net home page, http://www.informatik.uni-hamburg.de/TGI/PetriNets.

265. B. Plateau, "On the stochastic structure of parallelism and synchronisation models for distributed algorithms," in *ACM Sigmetrics Conf. on Measurement and Modeling of Computer Systems*. Austin, Texas, USA: ACM, 1985, pp. 147–154.

266. ——, "PEPS: A package for solving complex markov models of parallel systems," in *Proc. 4th Int. Conf. on Modeling Techniques and Tools for Performance Evaluation*, Palma de Mallorca, Spain, 1988, pp. 341–360.

267. B. Plateau and K. Atif, "Stochastic automata network for modeling parallel systems," *IEEE Transactions on Software Engineering*, vol. 10, no. 17, pp. 1093–1108, Oct. 1991.

268. R. Pooley and P. King, "The Unified Modeling Language and Performance Engineering," in *IEE Proceedings - Software*, vol. 146/2, Mar. 1999.

269. A. A. B. Pritsker and J. J. O'Reilly, *Simulation with Visual SLAM and Awesim*. John Wiley and Sons, 1999.

270. J.-M. Proth and X. Xie, *Petri nets: A tool for design and management of manufacturing systems*. John Wiley and Sons, 1996.

271. A. Puliafito, M. Scarpa, and K. S. Trivedi, "Petri nets with k simultaneously enabled generally distributed timed transitions," *Performance Evaluation*, vol. 32, no. 1, 1998.

272. N. Raghavan, "Performance analysis and design of supply chains: a Petri net approach," *Journal of the Operations Research Society*, vol. 51, no. 10, pp. 1158–1169, 2000.

273. P. J. Ramadge and W. M. Wonham, "Supervisory control of a class of discrete-event processes," *SIAM J. Contr. Optimization*, vol. 25, no. 1, pp. 206–230, 1987.

274. ——, "The control of discrete event systems," *Proceedings of the IEEE*, vol. 77, no. 1, pp. 81–98, 1989.

275. C. Ramchandani, "Analysis of asynchonous concurrent systems by timed Petri nets," Ph.D. dissertation, MIT, Cambridge, MA., 1974.

276. L. Recalde, M. Silva, J. Ezpeleta, and E. Teruel, "Petri nets and manufacturing systems: An examples-driven tour," in *Lectures on Concurrency and Petri Nets*, Lecture Notes in Computer Science, J. Desel, W. Reisig, and G. Rozenberg, Eds., vol. 3098. Springer, 2004, pp. 742–788.

277. C. L. Reeves, *Modern Heuristic techniques for Combinatorial Problems*. Wiley, 1993.

278. A. L. Reibman and K. S. Trivedi, "Transient analysis of cumulative measures of Markov model behavior," *Stochastic Models*, vol. 5, no. 4, pp. 683–710, 1989.

279. W. Reisig, *Petri nets*. Springer Verlag Berlin, 1985.

280. R. Righter and J. C. Walrand, "Distributed simulation of discrete event systems," *Proc. of the IEEE*, vol. 77, no. 1, pp. 99–113, 1989.

281. D. Rodriguez, A. Zimmermann, and M. Silva, "Two heuristics for the improvement of a two-phase optimization method for manufacturing systems," in *Proc. Int. Conf. Systems, Man, and Cybernetics (SMC'04)*, The Hague, Netherlands, Oct. 2004, pp. 1686–1692.

282. R. A. Sahner, K. S. Trivedi, and A. Puliafito, *Performance and Reliability Analysis of Computer Systems: An Example-Based Approach Using the SHARPE Software Package*. Kluwer Academic Publishers, 1995.

283. B. Samadi, "Distributed simulation, algorithms and performance analysis," University of California, Tech. Rep., 1985.

284. W. H. Sanders and J. F. Meyer, "A unified approach for specifying measures of performance, dependability, and performability," in *Dependable Computing for Critical Applications*, Dependable Computing and Fault-Tolerant Systems, A. Avizienis and J. Laprie, Eds. Springer Verlag, 1991, vol. 4, pp. 215–237.

285. W. H. Sanders, W. D. Obal, M. A. Qureshi, and F. K. Widjanarko, "The UltraSAN modeling environment," *Performance Evaluation*, vol. 24, pp. 89–115, 1995.

286. E. Schnieder, *Petri-Netze in der Automatisierungstechnik*. Oldenbourg Verlag, 1992.

287. R. Schrenk, "GSM-R: Quality of service tests at customer trial sites," *Signal und Draht*, vol. 92, no. 9, pp. 61–64, 2000.

288. B. Shansagimow and E. Schnieder, "Analyse und Auswahl optimaler Varianten von Automatisierungssystemen mit Hilfe von Petrinetzdarstellungen," *Automatisierungstechnik – at*, vol. 40, no. 6, pp. 228–234, 1992.

289. J. Sifakis, "Use of Petri nets for performance evaluation," in *Measuring, Modelling and Evaluating Computer Systems*, H. Beilner and E. Gelenbe, Eds. Elsevier, 1977, pp. 75–93.

290. M. Silva, *Las Redes de Petri en la Automática y la Informática*. Madrid, Spain: Editorial AC, 1985.

291. M. Silva and E. Teruel, "A systems theory perspective of discrete event dynamic systems: The Petri net paradigm," in *Proc. Symp. on Discrete Events and Manufacturing Systems, Computational Engineering in Systems Applications (IMACS CESA '96 Multiconference)*, Lille, France, 1996, pp. 1–12.

292. ——, "Petri nets for the design and operation of manufacturing systems," *European Journal of Control*, vol. 3, no. 3, pp. 182–199, 1997.

293. M. Silva, E. Teruel, and J. M. Colom, "Linear algebraic and linear programming techniques for the analysis of place/transition net systems," in *Lectures on*

Petri Nets I: Basic Models, Lecture Notes in Computer Science, G. Reisig, W.; Rozenberg, Ed. Springer Verlag, 1998, vol. 1491, pp. 309–373.

294. M. Silva, E. Teruel, R. Valette, and H. Pingaud, "Petri nets and production systems." in *Lectures on Petri Nets II: Applications, Advances in Petri Nets*, Lecture Notes in Computer Science, W. Reisig and G. Rozenberg, Eds., vol. 1492. Springer, 1998, pp. 85–124.

295. M. Silva and R. Valette, "Petri nets and flexible manufacturing," in *Advances in Petri Nets 1989*, Lecture Notes in Computer Science, G. Rozenberg, Ed. Springer Verlag, 1989, vol. 424, pp. 374–417.

296. M. Singhal and F. Mattern, "An optimality proof for asynchronous recovery algorithms in distributed systems," *Information Processing Letters*, vol. 55, no. 3, pp. 117–121, 1995.

297. R. S. Sreenivas and B. H. Krogh, "On condition/event-systems with discrete state realisations," in *Proc. Discrete Event Dynamic Systems 1*, 1991, pp. 209–236.

298. W. Stewart, K. Atif, and B. Plateau, "The numerical solution of stochastic automata network," *European Journal of Operations Research*, vol. 86, pp. 503–525, 1995.

299. W. J. Stewart, *Introduction to the Numerical Solution of Markov Chains*. Princeton University Press, 1994.

300. F. J. W. Symons, "Modeling and analysis of communication protocols using numerical petri nets," PhD thesis, University of Essex, UK, 1978.

301. H. Szu and R. Hartley, "Fast simulated annealing," *Phys. Lett. A*, vol. 122, no. 3-4, pp. 157–162, 1987.

302. M. Telek and A. Horvath, "Transient analysis of age-MRSPNs by the method of supplementary variables," *Performance Evaluation*, vol. 45, no. 4, pp. 205–221, 2001.

303. E. Teruel, G. Franceschinis, and M. D. Pierro, "Well-defined generalized stochastic Petri nets: A net-level method to specify priorities," *IEEE Transactions on Software Engineering*, vol. 29, no. 11, pp. 962–973, Nov. 2003.

304. N. Treves, "A comparative study of different techniques for semi-flows computation in place/transition nets," in *Advances in Petri Nets 1989*, Lecture Notes in Computer Science, G. Rozenberg, Ed. Springer Verlag, 1990, vol. 424, pp. 434–452.

305. K. S. Trivedi, *Probability and Statistics with Reliability, Queuing and Computer Science Applications*, 2nd ed. Wiley, 2002.

306. K. S. Trivedi, B. R. Haverkort, A. Rindos, and V. Mainkar, "Techniques and tools for reliability and performance evaluation: problems and perspectives," in *Proc. 7th Int. Conf. on Computer performance evaluation: modelling techniques and tools*. Vienna, Austria: Springer Verlag, 1994, pp. 1–24.

307. J. Trowitzsch and A. Zimmermann, "Real-time UML state machines: An analysis approach," in *Workshop on Object Oriented Software Design for Real Time and Embedded Computer Systems, Net.ObjectDays 2005*, Erfurt, Germany, Sept. 2005.

308. ——, "Towards quantitative analysis of real-time UML using stochastic Petri nets," in *Proc. 13th Int. Workshop on Parallel and Distributed Real-Time Systems*. Denver, Colorado: IEEE, Apr. 2005.

309. ——, "Using UML state machines and Petri nets for the quantitative investigation of ETCS," in *Proc. Int. Conf. on Performance Evaluation Methodologies and Tools (VALUETOOLS 2006)*, Pisa, Italy, 2006.

310. B. Tuffin and K. S. Trivedi, "Implementation of importance splitting techniques in stochastic Petri net package," in *Computer Performance Evaluation, Modelling Techniques and Tools — 11th Int. Conf., TOOLS 2000*, Lecture Notes in Computer Science, C. U. S. Boudewijn R. Haverkort, Henrik C. Bohnenkamp, Ed., vol. 1786. Schaumburg, IL, USA: Springer Verlag, 2000, pp. 216–pp.

311. W. M. P. van der Aalst, K. M. van Hee, and H. A. Reijers, "Analysis of discrete-time stochastic Petri nets," *Statistica Neerlandica*, vol. 54, no. 2, pp. 237–255, 2000.

312. W. van der Aalst and A. Waltmans, "Modelling logistic systems with EXSPECT," in *Dynamic Modelling of Information Systems*, H. Sol and K. v. Hee, Eds. Amsterdam: Elsevier Science Publishers, 1991, pp. 269–288.

313. P. J. M. van Laarhoven and E. H. L. Aarts, *Simulated Annealing: Theory and Applications*. Boston: D. Reidel Publishing Company, 1987.

314. P. J. M. van Laarhoven, E. H. L. Aarts, and J. K. Lenstra, "Job shop scheduling by simulated annealing," *Operations Research*, vol. 40, no. 1, pp. 113–125, 1992.

315. J. Vautherin, "Parallel specification with coloured Petri nets and algebraic data types," in *Proc. 7th European Workshop on Application and Theory of Petri Nets*, Oxford, UK, July 1986, pp. 5–23.

316. J. L. Villaroel, J. Martínez, and M. Silva, "GRAMAN: A graphic system for manufacturing system design," in *IMACS Symposium on System Modelling and Simulation*, S. Tzafestas, Ed. Elsevier Science Publ., 1989, pp. 311–316.

317. M. Villén-Altamirano and J. Villén-Altamirano, "RESTART: A straightforward method for fast simulation of rare events," in *Proc. Winter Simulation Conference*, 1994, pp. 282–289.

318. ——, "Analysis of RESTART simulation: Theoretical basis and sensitivity study," *European Transactions on Telecommunications*, vol. 13, no. 4, pp. 373–385, 2002.

319. ——, "Optimality and robustness of RESTART simulation," in *Proc. 4th Workshop on Rare Event Simulation and Related Combinatorial Optimisation Problems*, Madrid, Spain, Apr. 2002.

320. ——, "On the efficiency of RESTART for multidimensional systems," *ACM Transactions on Modeling and Computer Simulation*, vol. 16, no. 3, pp. 251–279, July 2006.

321. M. Villén-Altamirano, J. Villén-Altamirano, J. Gamo, and F. Fernández-Cuesta, "Enhancement of the accelerated simulation method RESTART by considering multiple thresholds." in *Proc. 14th Int. Teletraffic Congress*. Elsevier Science Publishers B. V., 1994, pp. 797–810.

322. N. Viswanadham and J. Narahari, Y.and Johnson, "Deadlock prevention and deadlock avoidance in flexible manufacturing sytems," *IEEE Transactions on Robotics and Automation*, vol. 6, pp. 713–723, 1990.

323. N. Viswanadham and R. Narahari, Ram, "Performability of automated manufacturing systems," in *Advances in Manufacturing and Automation Systems*. Academic Press, 1991.

324. A. M. Warda and M. Sczittnick, "Hierarchical modeling and distributed simulation with HIT," in *Workshop on Parallel and Distributed Simulation*, 1997, pp. 148–155.

325. W. Whitt, "The queueing network analyzer," *The Bell System Technical Journal*, vol. 62, no. 9, pp. 2779–2815, 1983.

326. F. Wieland, "The threshold of event simultaneity," *Trans. of the Society for Computer Simulation International*, vol. 16, no. 1, pp. 1–9, 1999.

327. F. Wolff, "Entwurf und Implementierung eines Konzepts zur Parametrisierung von objektbasierten farbigen Petrinetzen," Master's thesis, Technische Universität Berlin, 2004.

328. K. Wolter, "Second order fluid stochastic Petri nets: an extension of GSPNs for approximate and continuous modelling," in *Proc. World Congress on Systems Simulation*, Singapore, Sept. 1997, pp. 328–332.

329. K. Wolter and A. Zisowsky, "On Markov reward modelling with FSPNs." *Performance Evaluation*, vol. 44, no. 1-4, pp. 165–186, 2001.

330. Y. Wu, J. Zeng, and G. Sun, "Distributed simulation algorithms of generalized differential Petri nets," in *Proceedings of the 2002 Int. Conf. on Machine Learning and Cybernetics (ICMLC02)*. Beijing, China: IEEE Computer Society, Nov. 2002, pp. 1013–1017.

331. S.-T. Yee, J. Tew, A. Zimmermann, M. Knoke, and A. Huck, "New methodology for developing supply chain models in support of OTD," General Motors Research and Development Center, Warren, Research Report MSR-121, 2002.

332. B. P. Zeigler, H. Praehofer, and T. G. Kim, *Theory of Modeling and Simulation*, 2nd ed. London: Academic Press, 2000.

333. Y. Zeng, W. T. Cai, and S. J. Turner, "Causal order based Time Warp: A trade-off of optimism," in *Proceedings of the 35th Winter Simulation Conference (WSC'03)*, D. M. Ferrin and D. J. Morrice, Eds. New Orleans, LA, USA: ACM, Dec. 2003.

334. Y. Zeng, W. Cai, and S. J. Turner, "Batch based cancellation: A rollback optimal cancellation scheme in time warp simulations." in *Proc. 18th Workshop on Parallel and Distributed Simulation (PADS'04)*. Kufstein, Austria: IEEE Computer Society, 2004, pp. 78–86.

335. A. Zenie, "Colored stochastic Petri nets," in *Proc. 1st Int. Workshop on Petri Nets and Performance Models*, 1985, pp. 262–271.

336. M. Zhou and K. Venkatesh, *Modeling, Simulation, and Control of Flexible Manufacturing Systems*, Intelligent Control and Intelligent Automation. World Scientific, 1999.

337. R. Zijal, "Analysis of discrete time deterministic and stochastic Petri nets," Dissertation, Technische Universität Berlin, Oct. 1997.

338. R. Zijal, G. Ciardo, and G. Hommel, "Discrete deterministic and stochastic Petri nets," in *9. ITG/GI-Fachtagung: Messung, Modellierung und Bewertung von Rechen- und Kommunikationssystemen (MMB'97)*. Freiberg, Germany: VDE-Verlag, 1997, pp. 103–117.

339. A. Zimmermann, S. Bode, and G. Hommel, "Performance and dependability evaluation of manufacturing systems using Petri nets," in *1st Workshop on Manufacturing Systems and Petri Nets, 17th Int. Conf. on Application and Theory of Petri Nets*, Osaka, Japan, 1996, pp. 235–250.

340. A. Zimmermann, K. Dalkowski, and G. Hommel, "A case study in modeling and performance evaluation of manufacturing systems using colored Petri nets," in *Proc. of the 8th European Simulation Symposium*, Genoa, Italy, 1996, pp. 282–286.

341. A. Zimmermann and J. Freiheit, "TimeNET$_{ms}$ — an integrated modeling and performance evaluation tool for manufacturing systems," in *IEEE Int. Conf. on Systems, Man, and Cybernetics*, San Diego, USA, 1998, pp. 535–540.

342. ——, "Tool support for model-based online control of manufacturing systems," in *Proc. 4th World Multiconference on Systemics, Cybernetics and Informatics (SCI 2000)*, vol. VIII, Orlando, USA, 2000, pp. 695–700.

343. A. Zimmermann, J. Freiheit, and G. Hommel, "Discrete time stochastic Petri nets for modeling and evaluation of real-time systems," in *Proc. Int. Workshop on Parallel and Distributed Real-Time Systems (WPDRTS01*, San Francisco, 2001, pp. 282–286, invited paper.

344. A. Zimmermann, J. Freiheit, and A. Huck, "A Petri net based design engine for manufacturing systems," *Int. Journal of Production Research, special issue on Modeling, Specification and Analysis of Manufacturing Systems*, vol. 39, no. 2, pp. 225–253, 2001.

345. A. Zimmermann, R. German, J. Freiheit, and G. Hommel, "Timenet 3.0 tool description," in *Int. Conf. on Petri Nets and Performance Models (PNPM 99), Tool descriptions*. Zaragoza, Spain: University of Zaragoza, 1999.

346. A. Zimmermann and G. Hommel, "A train control system case study in model-based real time system design," in *Proc. 11th Int. Workshop on Parallel and Distributed Real-Time Systems (WPDRTS03)*, Nice, France, 2003.

347. ——, "Towards modeling and evaluation of ETCS real-time communication and operation," *Journal of Systems and Software*, vol. 77, pp. 47–54, 2005.

348. A. Zimmermann, M. Knoke, and G. Hommel, "Complete event ordering for time-warp simulation of stochastic discrete event systems," in *Proc. 2006 Spring Simulation Multiconference (DASD)*, Huntsville, USA, 2006, pp. 459–466.

349. A. Zimmermann, M. Knoke, A. Huck, and G. Hommel, "Towards version 4.0 of TimeNET," in *13th GI/ITG Conference on Measurement, Modeling, and Evaluation of Computer and Communication Systems (MMB 2006)*, March 2006, pp. 477–480.

350. A. Zimmermann, A. Kühnel, and G. Hommel, "A modelling and analysis method for manufacturing systems based on Petri nets," in *Computational Engineering in Systems Applications (CESA '98)*, Nabeul-Hammamet, Tunisia, 1998, pp. 276–281.

351. A. Zimmermann, D. Rodriguez, and M. Silva, "Modelling and optimisation of manufacturing systems: Petri nets and simulated annealing," in *Proc. European Control Conference (ECC'99)*, Karlsruhe, 1999.

352. ——, "A two phase optimisation strategy for DEDS: Application to a manufacturing system," in *Discrete Event Systems - Analysis and Control (Proc. Int. Workshop on Discrete Event Systems)*, R. Boel and G. Stremersch, Eds. Kluwer Academic Publishers, 2000, pp. 291–298.

353. ——, "A two phase optimisation method for Petri net models of manufacturing systems," *Journal of Intelligent Manufacturing*, vol. 12, no. 5/6, pp. 409–420, Oct. 2001, special issue "Global Optimization Meta-Heuristics for Industrial Systems Design and Management".

354. A. Zimmermann, "Modellierung und Bewertung von Fertigungssystemen mit Petri-Netzen," Dissertation, Technische Universität Berlin, Sept. 1997.

355. ——, "Modeling of manufacturing systems and production routes using coloured Petri nets," *Int. Journal of Robotics and Automation*, vol. 13, no. 3, pp. 96–100, 1998.

356. ——, "Colored Petri net modeling, evaluation, and control of a manufacturing cell," in *Seminar on Advanced Robotics and its Applications*, Shanghai, China, Oct. 2000, pp. 49–54.

357. ——, "Applied restart estimation of general reward measures," in *Proc. 6th Int. Workshop on Rare Event Simulation (RESIM 2006)*, Bamberg, Germany, Oct. 2006, pp. 196–204.

358. A. Zimmermann, "Extended reward measures in the simulation of embedded systems with rare events," in *Proc. 7th Int. Workshop Embedded Systems - Modelling, Technology, Applications*. Berlin, Germany: Springer Verlag, June 2006, pp. 43–52.

359. A. Zimmermann, J. Freiheit, R. German, and G. Hommel, "Petri net modelling and performability evaluation with TimeNET 3.0," in *11th Int. Conf. on Modelling Techniques and Tools for Computer Performance Evaluation*, Lecture Notes in Computer Science, vol. 1786, Schaumburg, Illinois, USA, 2000, pp. 188–202.

360. A. Zimmermann, J. Freiheit, and G. Hommel, "Fertigungssysteme Modellieren, Bewerten und Steuern mit TimeNET$_{ms}$," in *6. Fachtagung Entwicklung und Betrieb komplexer Automatisierungssysteme (EKA'99)*, E. Schnieder, Ed., Braunschweig, Germany, 1999, pp. 31–48.

361. A. Zimmermann and G. Hommel, "Modelling and evaluation of manufacturing systems using dedicated Petri nets," *Int. Journal of Advanced Manufacturing Technology*, vol. 15, pp. 132–137, 1999.

362. A. Zimmermann and M. Knoke, *TimeNET 4.0 User Manual*, Technische Universität Berlin, 2007, http://pdv.cs.tu-berlin.de/~timenet.

363. A. Zimmermann, M. Knoke, S.-T. Yee, and J. D. Tew, "Model-based performance engineering of General Motors' vehicle supply chain," in *IEEE Int. Conf. on Systems, Man and Cybernetics (SMC 2007)*, Montreal, Canada, Oct. 2007, accepted for publication.

Index